# Pests, Pesticides, Pollinators and Sustainable Farming

# Pests, Pesticides, Pollinators and Sustainable Farming

Guest Editors

**Antonios E. Tsagkarakis**
**Georgios Papadoulis**
**Argyro Kalaitzaki**

MDPI

Basel • Beijing • Wuhan • Barcelona • Belgrade • Novi Sad • Cluj • Manchester

*Guest Editors*
Antonios E. Tsagkarakis
Department of Crop Science
Agricultural University
of Athens
Athens
Greece

Georgios Papadoulis
Department of Crop Science
Agricultural University
of Athens
Athens
Greece

Argyro Kalaitzaki
Institute for Olive Tree,
Subtropical Crops &
Viticulture
Hellenic Agricultural
Organization "DEMETER"
Chania
Greece

*Editorial Office*
MDPI AG
Grosspeteranlage 5
4052 Basel, Switzerland

This is a reprint of the Special Issue, published open access by the journal *Agronomy* (ISSN 2073-4395), freely accessible at: https://www.mdpi.com/journal/agronomy/special_issues/Z7N7N3L3I3.

For citation purposes, cite each article independently as indicated on the article page online and as indicated below:

| Lastname, A.A.; Lastname, B.B. Article Title. *Journal Name* **Year**, *Volume Number*, Page Range. |
|---|

**ISBN 978-3-7258-6558-1 (Hbk)**
**ISBN 978-3-7258-6559-8 (PDF)**
**https://doi.org/10.3390/books978-3-7258-6559-8**

Cover image courtesy of Antonios Tsagkarakis
Aphis spiraecola with ants, predators and parasitoids.

# Contents

# About the Editors

**Antonios E. Tsagkarakis**

Antonios E. Tsagkarakis is an Assistant Professor of Agricultural and Productive Entomology in the Department of Crop Science at the Agricultural University of Athens (AUA), Greece. He holds a B.Sc., M.Sc., and Ph.D. in Entomology from AUA, where his research focuses on productive insects, precision insect control, and sustainable pest management. He completed a postdoctoral fellowship at the University of Florida (USA), contributing to research on citrus pest management and the effects of insecticides on pathogen transmission. At AUA, he teaches Apiculture and Sericulture, Pathology of Productive Insects, and Agricultural Entomology, while supervising graduate and undergraduate students. He leads and participates in Horizon Europe and other international projects on apiculture, sericulture innovation, and agricultural entomology, and has authored numerous scientific publications. His research advances understanding of pest ecology, pollinator health, and biodiversity, supporting environmentally sound pest control practices. He has specialized training in electrical penetration graph (EPG) analysis and collaborates extensively with European and U.S. institutions. He also serves on the Evaluation and Accreditation Council of the Hellenic Authority of Higher Education, the Council of the International Bee Research Association (IBRA), and other scientific societies.

**Georgios Papadoulis**

Georgios Papadoulis is Professor Emeritus of Agricultural Zoology and Entomology at the Agricultural University of Athens (AUA), where he has served as Rector, Head of Department, and Director of the Laboratory of Agricultural Zoology and Entomology. He has made extensive contributions to integrated pest management, biological control, acarology, pest ecology, and entomofaunal biodiversity in Mediterranean agroecosystems. He has authored numerous peer-reviewed publications and has been widely recognized for his leadership and research in agricultural entomology. His work has advanced understanding of pest–natural enemy interactions, predatory mites, invasive species monitoring, and faunistic surveys of economically important arthropods. He has taught courses in Integrated Pest Management, Agricultural Zoology, and Entomology, mentoring generations of agronomists and researchers. He has co-supervised graduate projects and fostered collaborations focused on biodiversity assessment and sustainable crop protection. Through his research and academic service, Prof. Papadoulis has linked classical zoological and taxonomic studies with applied agricultural practices, promoting environmentally sound and pollinator-friendly pest management. His scholarly achievements and mentorship have established him as a leading figure in agricultural entomology and sustainable farming systems.

**Argyro Kalaitzaki**

Argyro Kalaitzaki is a Collaborating Researcher in Entomology at the Institute of Olive Tree, Subtropical Plants and Viticulture (ELGO DIMITRA) in Chania, Greece. Her research focuses on sustainable crop protection, combining fundamental and applied studies in insect pest ecology, biological control, and integrated pest management (IPM). She holds a Ph.D. in Entomology and has built a strong record of scientific publications and research collaborations addressing the population dynamics of pests in Mediterranean crops such as olive, citrus, avocado, and vegetables. Her work explores the ecology of parasitoids and beneficial organisms to enhance natural pest suppression and supports the design of environmentally compatible IPM strategies. Dr. Kalaitzaki also investigates

Auchenorrhyncha species, potential vectors of *Xylella fastidiosa*, and applies geoinformatics and sensor technologies to improve pest monitoring and optimize fruit fly (Tephritidae) control. She has participated in several national and international projects, including the PRIMA-FreeClimb initiative on climate-resilient fruit crops and regional preparedness actions for *Xylella fastidiosa*. Her publications reflect expertise in pest dynamics, biological control agents, and precision pest management, contributing to sustainable agriculture that supports both productivity and environmental integrity.

*agronomy*

MDPI

*Editorial*

# Editorial for the Special Issue "Pests, Pesticides, Pollinators and Sustainable Farming"

**Antonios Tsagkarakis [1,*], Georgios Papadoulis [2] and Argyro Kalaitzaki [3]**

1 Laboratory of Sericulture and Apiculture, Agricultural University of Athens, Iera Odos 75, 118 55 Athens, Greece
2 Laboratory of Agricultural Zoology and Entomology, Agricultural University of Athens, Iera Odos 75, 118 55 Athens, Greece; gpapadoulis@aua.gr
3 Institute of Olive Tree, Subtropical Crops and Viticulture, Hellenic Agricultural Organization 'DIMITRA', 73 100 Chania, Greece; kalaitzaki@elgo.gr
* Correspondence: atsagarakis@aua.gr; Tel.: +30-2105294578

## 1. Introduction

The global agricultural sector continues to face the pressing challenge of ensuring food security while maintaining ecological integrity. Modern crop production systems must address the dual crises of pest pressure and pollinator decline, both of which are influenced by climate change, habitat loss, and extensive synthetic pesticide use. Conventional pest management, while historically successful in securing yields, often results in ecological disruption, pest resistance, and negative effects on beneficial organisms [1–3]. Thus, sustainable farming systems must embrace integrative approaches that combine effective pest suppression with pollinator conservation and ecosystem resilience [4].

This Special Issue of Agronomy, entitled "Pests, Pesticides, Pollinators and Sustainable Farming," brings together thirteen papers—twelve original research articles and one review—that collectively reflect the growing scientific and technological advances supporting the transition toward environmentally sound agricultural practices. The contributions span field ecology, chemical ecology, biotechnology, and precision agriculture, all converging on the shared objective of reconciling pest management with biodiversity conservation.

## 2. Highlights of This Special Issue

The importance of pollinator services and landscape diversification is reflected in studies demonstrating that sowing flowering plant mixtures in field margins enhances pollinator and natural enemy populations, leading to improved ecosystem functionality and potential yield gains [5,6]. Similar findings across cropping systems highlight how habitat diversification strengthens pollination services and biological control [7,8].

Efforts to develop alternative pest management tools based on natural products and microorganisms are gaining momentum globally. For example, essential oils and microbial biocontrol agents have shown promise in suppressing major pests while improving plant vigor [9,10]. These approaches align with the global transition toward bio-based pest management, minimizing chemical inputs and ecological disturbance [11,12].

Several contributions to this Special Issue also address the use of innovative technologies and ecological knowledge for pest monitoring and detection. Recent developments in unmanned aerial systems (UAS), hyperspectral imaging, and machine learning offer powerful tools for early pest detection and precision management [13,14]. Such tools can greatly enhance the efficiency of integrated pest management (IPM) programs and reduce pesticide reliance.

Ecological risks associated with pesticide exposure remain an urgent issue. Longitudinal studies have demonstrated that chronic pesticide exposure contributes to pollinator health decline and increased pathogen prevalence [15,16]. These findings reinforce the call for policies that integrate pesticide risk assessment with pollinator protection frameworks.

Finally, global perspectives on pest control are increasingly emphasizing preventive and ecologically based approaches. Locust management, for instance, has evolved from chemical-intensive eradication to early warning systems and habitat-based prevention strategies [17–19]. This shift exemplifies how sustainable pest control and ecosystem stewardship can coexist within resilient agricultural systems.

## 3. Conclusions

Taken together, the papers in this Special Issue illustrate viable pathways toward resilient, biodiversity-friendly farming systems through field experimentation, technological innovation, and ecological understanding. They reaffirm that pest control and pollinator protection are complementary objectives within sustainable agriculture. By integrating ecological principles, innovative technologies, and international collaboration, the agricultural sector can achieve both productivity and environmental stewardship.

We extend our sincere gratitude to all authors for their valuable contributions, to the reviewers for their time and expertise in ensuring scientific rigor, and to the Agronomy editorial team for their continuous support and professionalism. The collective efforts represented in this Special Issue advance our understanding of how to achieve productive, resilient, and ecologically sustainable farming systems for the future.

**Author Contributions:** Conceptualization, A.T.; methodology, A.T., G.P. and A.K.; writing—original draft preparation, A.T., G.P. and A.K.; writing—review and editing, A.T., G.P. and A.K.; visualization, A.T.; supervision, A.T. All authors have read and agreed to the published version of the manuscript.

**Conflicts of Interest:** The authors declare no conflicts of interest.

**List of Contributions:**

1. Kati, V.; Stathakis, T.; Economou, L.; Mylonas, P.; Barda, M.; Angelioudakis, T.; Bratidou Parlapani, A.; Tsamis, I.; Karamouna, F. Processing Tomato Crop Benefits from Flowering Plants in Field Margins That Support Pollinators and Natural Enemies. *Agronomy* **2025**, *15*, 1558. https://doi.org/10.3390/agronomy15071558.
2. Sarma, A.K.; Neog, B.; Deka, M.K.; Carabet, A.; Stef, R. Variable Transect Method Outperformed in Sampling Hymenopteran Flower Visitors in *Brassica campestris* L. var. *toria* Ecosystem. *Agronomy* **2025**, *15*, 1281. https://doi.org/10.3390/agronomy15061281.
3. Nikolova, M.; Lyubenova, A.; Yankova-Tsvetkova, E.; Georgiev, B.; Gavrilov, G.; Gavrilova, A. *Satureja kitaibelii* Essential Oil and Extracts: Bioactive Compounds and Pesticide Properties. *Agronomy* **2025**, *15*, 357. https://doi.org/10.3390/agronomy15020357.
4. Papantzikos, V.; Mantzoukas, S.; Koutsompina, A.; Karali, E.M.; Eliopoulos, P.A.; Servis, D.; Bitivanos, S.; Patakioutas, G. Use of *Beauveria bassiana* and *Bacillus amyloliquefaciens* Strains as *Gossypium hirsutum* Seed Coatings: Evaluation of the Bioinsecticidal and Biostimulant Effects in Semi-Field Conditions. *Agronomy* **2024**, *14*, 2335. https://doi.org/10.3390/agronomy14102335.
5. Munawar, A.; Zhang, H.; Zhang, J.; Zhang, X.; Shi, X.-X.; Chen, X.; Li, Z.; He, X.; Zhong, J.; Zhu, Z.; et al. Entomopathogenic Fungus Treatment Affects Trophic Interactions by Altering Volatile Emissions in Tomato. *Agronomy* **2025**, *15*, 1161. https://doi.org/10.3390/agronomy15051161.
6. Panopoulou, C.; Antonopoulos, A.; Arapostathi, E.; Stamouli, M.; Katsileros, A.; Tsagkarakis, A. Using Multispectral Data from UAS in Machine Learning to Detect Infestation by *Xylotrechus*

*chinensis* (Chevrolat) (Coleoptera: Cerambycidae) in Mulberries. *Agronomy* **2024**, *14*, 2061. https://doi.org/10.3390/agronomy14092061.

7. Mo, L.; Xie, R.; Ye, F.; Wang, G.; Wu, P.; Yi, X. Enhanced Tomato Pest Detection via Leaf Imagery with a New Loss Function. *Agronomy* **2024**, *14*, 1197. https://doi.org/10.3390/agronomy14061197.
8. Thanou, Z.; Stamouli, M.; Magklara, A.; Theodorou, D.; Stamatakou, G.; Konidis, G.; Koufopoulou, P.; Lyberopoulos, C.; Tribonia, S.; Vetsos, P.; et al. Faunistic Study of Auchenorrhyncha in Olive Orchards in Greece, Including First Records of Species. *Agronomy* **2024**, *14*, 2792. https://doi.org/10.3390/agronomy14122792.
9. Holý, K.; Kovaříková, K. Spring Abundance, Migration Patterns and Damaging Period of *Aleyrodes proletella* in the Czech Republic. *Agronomy* **2024**, *14*, 1477. https://doi.org/10.3390/agronomy14071477.
10. Wen, J.; Shan, Z.; Zou, Y.; Lin, X.; Cui, Z.; Yan, R.; Cao, F. Developing an Effective Push–Pull System for Managing Outbreaks of the Invasive Pest *Bactrocera dorsalis* (Diptera: Tephritidae) in *Nephelium lappaceum* Orchards. *Agronomy* **2024**, *14*, 890. https://doi.org/10.3390/agronomy14050890.
11. Sun, Y.; Chen, D.; Chen, X.; Wu, X. Stress Response of Citrus Leaves under Mechanical Damage and Huanglongbing Disease Infection Using Plasmonic $TiO_2$ Nanotube Substrate-Based Imprinting Mass Spectrometry Imaging. *Agronomy* **2024**, *14*, 1797. https://doi.org/10.3390/agronomy14081797.
12. Kegley, S.E.; Radford, R.; Brown, T.J.; Anderson, J.; Cox, D.; Ellis, S.; Marcy, G.W. Longitudinal Analysis of Honey Bee Colony Health as a Function of Pesticide Exposure. *Agronomy* **2024**, *14*, 2505. https://doi.org/10.3390/agronomy14112505.
13. Panopoulou, C.; Tsagkarakis, A. From Surveillance to Sustainable Control: A Global Review of Strategies for Locust Management. *Agronomy* **2025**, *15*, 2268. https://doi.org/10.3390/agronomy15102268.

## References


1. Goulson, D.; Nicholls, E.; Botías, C.; Rotheray, E.L. Bee declines driven by combined stress from parasites, pesticides, and lack of flowers. *Science* **2015**, *347*, 1255957. [CrossRef] [PubMed]
2. Pimentel, D. Environmental and Economic Costs of the Application of Pesticides Primarily in the United States. *Environ. Dev. Sustain.* **2005**, *7*, 229–252. [CrossRef]
3. Carvalho, F.P. Pesticides, environment, and food safety. *Food Energy Secur.* **2017**, *6*, 48–60. [CrossRef]
4. Pretty, J.; Bharucha, Z.P. Integrated Pest Management for Sustainable Intensification of Agriculture in Asia and Africa. *Insects* **2015**, *6*, 152–182. [CrossRef] [PubMed]
5. Kremen, C.; Miles, A. Ecosystem services in biologically diversified versus conventional farming systems: Benefits, externalities, and trade-offs. *Ecol. Soc.* **2012**, *17*, 40. [CrossRef]
6. Garibaldi, L.A.; Steffan-Dewenter, I.; Winfree, R.; Aizen, M.A.; Bommarco, R.; Cunningham, S.A.; Kremen, C.; Carvalheiro, L.G.; Harder, L.D.; Afik, O.; et al. Wild pollinators enhance fruit set of crops regardless of honey bee abundance. *Science* **2013**, *339*, 1608–1611. [CrossRef] [PubMed]
7. Kennedy, C.M.; Lonsdorf, E.; Neel, M.C.; Williams, N.M.; Ricketts, T.H.; Winfree, R.; Bommarco, R.; Brittain, C.; Burley, A.L.; Cariveau, D.; et al. A Global Quantitative Synthesis of Local and Landscape Effects on Wild Bee Pollinators in Agroecosystems. *Ecol. Lett.* **2013**, *16*, 584–599. [CrossRef] [PubMed]
8. Kleijn, D.; Winfree, R.; Bartomeus, I.; Carvalheiro, L.G.; Henry, M.; Isaacs, R.; Klein, A.M.; Kremen, C.; M'Gonigle, L.K.; Rader, R.; et al. Delivery of Crop Pollination Services Is an Insufficient Argument for Wild Pollinator Conservation. *Nat. Commun.* **2015**, *6*, 7414. [CrossRef] [PubMed]
9. Isman, M.B. Botanical insecticides, deterrents, and repellents in modern agriculture and increasingly regulated world. *Annu. Rev. Entomol.* **2020**, *65*, 233–249. [CrossRef]
10. Chandler, D.; Bailey, A.S.; Tatchell, G.M.; Davidson, G.; Greaves, J.; Grant, W.P. The development, regulation, and use of biopesticides for integrated pest management. *Philos. Trans. R. Soc. B* **2011**, *366*, 1987–1998. [CrossRef] [PubMed]
11. Raveau, R.; Fontaine, J.; Lounès-Hadj Sahraoui, A. Essential Oils as Potential Alternative Biocontrol Products against Plant Pathogens and Weeds: A Review. *Foods* **2020**, *9*, 365. [CrossRef] [PubMed]
12. Glare, T.; Caradus, J.; Gelernter, W.; Jackson, T.; Keyhani, N.; Köhl, J.; Marrone, P.; Morin, L.; Stewart, A. Have Biopesticides Come of Age? *Trends Biotechnol.* **2012**, *30*, 250–258. [CrossRef] [PubMed]

13. Vijayakumar, S.; Shanmugapriya, P.; Saravanane, P.; Ramesh, T.; Murugaiyan, V.; Ilakkiya, S. Precision Weed Control Using Unmanned Aerial Vehicles and Robots: Assessing Feasibility, Bottlenecks, and Recommendations for Scaling. *NDT* **2025**, *3*, 10. [CrossRef]
14. Liakos, K.G.; Busato, P.; Moshou, D.; Pearson, S.; Bochtis, D. Machine Learning in Agriculture: A Review. *Sensors* **2018**, *18*, 2674. [CrossRef] [PubMed]
15. Woodcock, B.A.; Bullock, J.M.; Shore, R.F.; Heard, M.S.; Pereira, M.G.; Redhead, J.; Ridding, L.; Dean, H.; Sleep, D.; Henrys, P.; et al. Country-Specific Effects of Neonicotinoid Pesticides on Honey Bees and Wild Bees. *Science* **2017**, *356*, 1393–1395. [CrossRef] [PubMed]
16. Sánchez-Bayo, F.; Goka, K. Pesticide Residues and Bees—A Risk Assessment. *PLoS ONE* **2014**, *9*, e94482. [CrossRef] [PubMed]
17. Lecoq, M. Desert Locust Management: From Ecology to Anthropology. *J. Orthoptera Res.* **2005**, *14*, 179–186. [CrossRef]
18. Lecoq, M. Integrated Pest Management for Locusts and Grasshoppers: Are Alternatives to Chemical Pesticides Credible? *J. Orthoptera Res.* **2010**, *19*, 131–132. [CrossRef]
19. Panopoulou, C.; Tsagkarakis, A. From Surveillance to Sustainable Control: A Global Review of Strategies for Locust Management. *Agronomy* **2025**, *15*, 2268. [CrossRef]

agronomy

MDPI

*Article*

# Processing Tomato Crop Benefits from Flowering Plants in Field Margins That Support Pollinators and Natural Enemies

**Vaya Kati [1,*,†], Theodoros Stathakis [2,*], Leonidas Economou [2], Philippos Mylonas [2], Myrto Barda [2], Theodoros Angelioudakis [3], Athanasia Bratidou Parlapani [1], Ilias Tsamis [4] and Filitsa Karamaouna [2]**

1 Laboratory of Weed Science, Scientific Directorate of Pesticides' Control and Phyto-Pharmacy, Benaki Phytopathological Institute, 8 Stefanou Delta Str., GR-145 61 Kifissia, Attica, Greece; abratidou@minagric.gr
2 Laboratory of Efficacy Control of Pesticides, Scientific Directorate of Pesticides' Control and Phytophamacy, Benaki Phytopathological Institute, 8 Stefanou Delta Str., GR-145 61 Kifissia, Attica, Greece; l.economou@bpi.gr (L.E.); ph.mylonas@bpi.gr (P.M.); myrto.barda@yahoo.gr (M.B.); f.karamaouna@bpi.gr (F.K.)
3 Laboratory of Agricultural Zoology and Entomology, Faculty of Crop Science, Agricultural University of Athens, 75 Iera Odos Str., GR-118 55 Athens, Greece; aggelt22@gmail.com
4 D. Nomikos S.A., 32 Kifissias Ave., GR-151 25 Maroussi, Attica, Greece; ilias.tsamis@dnomikos.gr
* Correspondence: vayakati@agro.auth.gr (V.K.); theostathakis1@gmail.com (T.S.); Tel.: +30-2310-998626 (V.K.); +30-6970868720 (T.S.)
† Current address: Laboratory of Agronomy, Faculty of Agriculture, Aristotle University of Thessaloniki, GR-541 24 Thessaloniki, Greece.

**Abstract:** In a two-year experiment, we examined whether increasing plant diversity in the margins of processing tomato fields could attract pollinators and natural enemies of pests compared to weed flora, and questioned the effect on crop yield. Two plant mixtures sown in winter (WM) and spring (SM) were compared with weed vegetation along a tomato crop (CT) and an adjacent irrigation channel (CC). Flower cover was higher in the sown mixtures than the weedy margins, and brought in more visits of pollinating bees (including potential tomato pollinators) than the latter. Flowering species were mainly *Eruca vesicaria* (WM, SM), *Coriandrum sativum* and *Lathyrus sativus* (WM), *Fagopyron esculentum* and *Phacelia tanacetifolia* (SM), and *Ammi majus*, *Rapistrum rugosum* (CC, CT). Parasitoids (Eulophidae, Braconidae, Scelionidae) were more abundant in the sown and CC margins compared to the CT margin, while the abundance of predators (Aeolothripidae, *Orius* sp., Thomisidae) was similar among all types of margins. Fruit weight was higher in the field with the sown margins, while pest incidence in the crop was not affected by the margin type. Our findings provide new insights into the contribution of managed and existing field margins in attracting beneficial arthropods, and their implications on yield.

**Keywords:** field margin management; weed flora; sown flowering mixtures; Hymenoptera pollinators; beneficial arthropods; crop yield

## 1. Introduction

Increasing intensification of agriculture to meet the growing demand in food production could dramatically affect farmland biodiversity [1]. Moreover, additional stressors resulting from the ongoing climatic change are reported to negatively affect plant diversity and alter the onset of flowering [2], challenging the ecosystem services of pollination [3–5] and biological control [6,7]. Both services are valued at billions of euros worldwide [8,9]. Furthermore, pollination is directly affecting food security since it is vital for 75% of global food crops [10].

Sustainable management strategies in agricultural landscapes are emerging as the way forward to halt the functional biodiversity loss associated with agricultural intensification [11,12]. In this respect, the European Union (EU) has launched mandatory or voluntary agro-environmental schemes specifically designed to promote ecological balance and sustainability in agroecosystems as part of the new Common Agricultural Policy (CAP) program and European Green Deal [13]. Such agro-ecological measures include, among other, universally accepted field management practices to enhance the ecosystem services of pollination and biological control based on the improvement of plant diversity in degraded field margins with sown mixtures of selected flowering plants [14]. Extensive research over the past few decades has demonstrated the positive relationship between plant diversity along field margins, and beneficial arthropod abundance and species richness in various crops [15–24]. However, a meta-analysis on the impact of field margins on biological pest control identified knowledge gaps regarding their effect on crop yield [25]. Indeed, an earlier quantitative synthesis of data from 35 studies by Albrecht et al. [26] indicated that flower strips along field margins had variable effectiveness on crop yield. What is more evident is a clear advantage regarding crop productivity on smaller farms (<2 ha) gained through higher flower visitor density, while larger fields required higher flower visitor richness for a measurable benefit [27]. It is thus obvious that plant selection and synthesis in mixtures are crucial for the desired impact, depending on crop species, field size and landscape complexity, as some species may also support herbivorous pests, potentially harming the crops.

Field grown tomatoes (*Solanum lycopersicum* L.) are primarily self-pollinating but can benefit from insect-mediated pollination [10]. Although our knowledge on insect pollinators of tomato remains limited, species like *Bombus* and *Anthophora* are known to improve tomato fruit set, especially under high temperatures that reduce pollen viability [28]. *Bombus pascuorum* (Scopoli) and *B. terrestris* (L.), *Megachile willughbiella* (Kirby), *Hylaeus gibbus* Saunders, and buzzing *Lasioglossum* species are reported to benefit the pollination of tomato in Central Europe due to their 'buzzing' vibration effect on the flowers [29]. A recent study in Greece showed that selected native non-crop flowering plants sown along the margins of a processing tomato field could increase pollinator and beneficial arthropod abundance compared to locally occurring weeds, but did not elucidate the effect on crop pollination and yield [19]. Following on that study and considering the scarcity of seeds from non-crop flowering plants in Greece, here, we studied the effect of field margin management with mixtures of selected flowering cultivated plant species that have available seeds in the Greek market. Our specific objectives were to evaluate the effectiveness of these plant species sown in mixtures along the field margins of processing tomato crops in attracting pollinators and natural enemies compared to weed flora, and whether such species could be a suitable alternative to non-crop flowering plants. Secondly, in addition to the previous study [19], here, we aimed to quantitatively assess the effect of this field margin management practice on crop yield. The study aligns with the EU's sustainable agriculture policies, aiming to protect and enhance biodiversity within agroecosystems and is conducted as part of the Operation Pollinator biodiversity project in Greece, which has been implemented in various crops since 2010 [18].

## 2. Materials and Methods

### 2.1. Selection of Plant Species and Mixture Composition

Two flowering plant mixtures were composed for sowing in winter (WM) and spring (SM) (Table 1). The species selection and their proportion in the mixtures was based on previous experience obtained during implementation in processing tomato and other crops in various areas of Greece [18–20,30]. Briefly, the general criteria included the selection of annual species known to exist in the area, which do not pose a threat as potential noxious weeds, or act as specialized hosts for tomato pests, are attractive to pollinators and natural enemies and belong to a diverse range of families. Seed availability defined the final species selection and mixture composition.

**Table 1.** Plant species selected for the mixtures sown in winter (WM) or spring (SM) along the margin of a processing tomato field, and the corresponding percentage.

| | | (%) | | g/100 m$^2$ | |
|---|---|---|---|---|---|
| **Family** | **Species** | **WM** | **SM** | **WM** | **SM** |
| Apiaceae | *Anethum graveolens* L. | 26 | | 6 | |
| | *Coriandrum sativum* L. | 18 | | 32 | |
| Brassicaceae | *Eruca vesicaria* (L.) Cav. | 23 | 26 | 9 | 14 |
| Fabaceae | *Pisum sativum* L. | 7 | | 368 | |
| | *Lathyrus sativus* L. | 12 | 12 | 301 | 361 |
| Polygonaceae | *Fagopyrum esculentum* Moench | | 33 | | 214 |
| Boraginaceae | *Phacelia tanacetifolia* Benth. | | 29 | | 31 |
| Poaceae | *Triticum aestivum* L. | 14 | | 150 | |
| | TOTAL | 100 | 100 | 866 | 620 |

### 2.2. Seed Rate Calculation

The calculation of the seed rate per species was based on the following parameters: target plant number/m$^2$, percentage of each species in the mixture, thousand grain weight per species, seed germination capacity based on growth chamber germination assays, and estimated plant survival rate in the field. Based on the above, the seed weight in grams for each species in a mixture (Ws) is the output of the equation

$$Ws = Ps \times (1/ESR) \times (1/Pg) \times Tp \times A \times (TGW/1000) \quad (1)$$

where Ps = % of each species (Table 1), Tp = total target number of plants/m$^2$ (set to be 80 plants/m$^2$), A = area to be sown per field (84 m$^2$/per plant mixture), TGW = thousand grain weight/species (g), ESR = estimated survival rate of germinated seeds (set to be 65% for all dicots and 80% for wheat), and Pg = seed germination percentage based on petri-dish assay results per species.

### 2.3. Experimentation Site

Two-year field experiments (2021–2022) were established in the main processing tomato area in Larissa, (Thessaly, Greece) at a different location each year (Figure 1A). The two plant mixtures (WM and SM) were sown along one of the field margins, as shown in Figure 1B. The plot size was 14 m$^2$ (7 m long × 2 m wide), with six plots per plant mixture (total sown area per mixture: 84 m$^2$). Two sets of control plots consisting of weed flora were assigned along the margin next to the crop (CT) or along an adjacent irrigation channel (CC), to cater for the different arable weed communities present around the field (Figure 1B). Each control site had six plots (total area per control site: 84 m$^2$).

Field preparation of the sown margins included shallow soil cultivation to control weeds, and hand raking to smooth the surface. Sowing was performed by hand. A bulking material (corn meal at 200 cm$^3$/m$^2$) was used to augment the seed quantity and enable its even distribution on soil surface. After sowing, the seeds were covered by hand raking, and the soil surface was rolled to ensure good seed/soil contact, using a manually operated seed roller. Sowing of the WM was performed in early February of each year while the SM was sown in early April. The H1015 Heinz jointless processing tomato hybrid (2nd early, 105–114 days to maturity) was transplanted mid-March of each year.

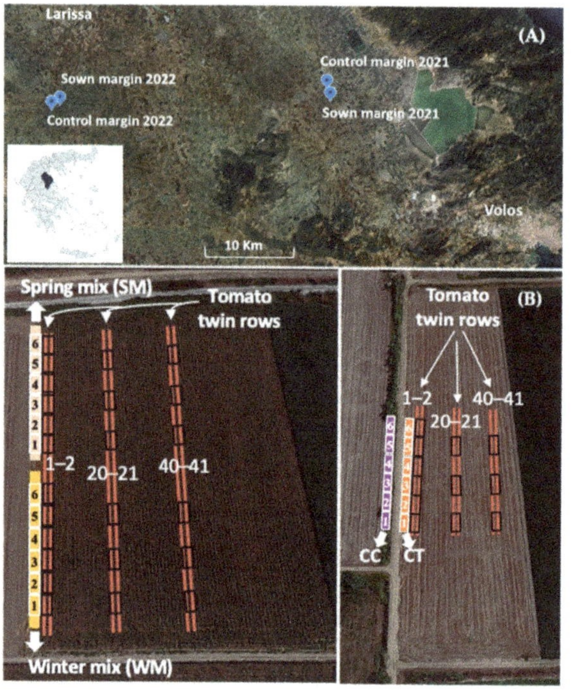


**Figure 1.** (**A**) Experimental sites in the county of Larissa, Greece, and (**B**) layout of the sown plant mixtures and the control weed flora, with designated plots for measurements.

## *2.4. Arthropod Measurements*

Measurements of flowering and attracted beneficial arthropods were performed during the main flowering period of the sown plants, from mid-May to late-June in 2021, and from early-June to mid-July in 2022.

The total plant cover and flower cover (total and per species) were visually estimated and expressed as percentage cover/plot in all plots of the sown mixtures (WM, SM) and the two control sites (CC, CT). Plant species were identified in situ or, when necessary, in the lab using the botanical identification key Flora Europaea [31].

Hymenoptera pollinator (*Apis mellifera* L., wild bees) visits on the flowers of the sown margins and the control plots were recorded with visual observation of landings for 4′/plot between 10:30 and 14:30 h. The observations and corresponding counts refer to the foraging visits of pollinators and not their absolute numbers which could not be accurate due to their high abundance and their mobility from flower to flower during observation time. All the observations throughout the sampling period were made by the same observer to eliminate potential bias between different observers. Measurements were conducted in three of the six plots (1st, 3rd, and 5th) of each treatment (WM and SM), as well as for the control sites (CT and CC).

Pollinator visits were also recorded on the crop flowers with visual observation as before, at three sites based on the distance from the sown field margin, starting from the first two twin crop rows next to the margin and moving infield to the 20th–21st and 40th–41st twin rows. Each sampling site had three replications of approximately 14 $m^2$ (2 twin rows, 7 m long).

Wild bee specimens that required identification after the visual observation measurements were captured with a sweeping net and were identified later in the lab. The identification of pollinators was based on identification keys [32,33].

Beneficial arthropods (parasitoids, predators) were recorded with suction sampling (16″/plot) using a modified leaf blower (Echo ES-2400, 24 $cm^3$, Kioritz Corporation, Tokyo, Japan), adapted as described in [34]. Measurements were conducted in the other three plots (2nd, 4th, and 6th) of the sown margins and the control sites, and in the three corresponding plots of the first two twin crop rows.

Additionally, the herbivore insects collected via suction sampling were counted. Only individuals belonging to groups that could be potentially harmful for tomato crop were taken into consideration, i.e., aphids (Aphidoidea), leafhoppers (Auchenorrhyncha), whiteflies (Aleyrodidae), stink bugs (Pentatomidae), and thrips (Thysanoptera).

The collected arthropod samples were kept in the freezer (−18 °C) and sorted according to family, genus, and species (where possible) under a stereomicroscope.

### 2.5. Crop Yield Parameters

The effect of the sown flower margin intervention on the quality characteristics of tomato fruit was assessed based on the following parameters: fruit weight (digital laboratory scale), BRIX [digital refractometer Maselli UR 24 (Maselli Misure S.P.A., Parma, Italy)], pH (pH-meter Metrohm 713), and color, including brightness (L*), redness or greenness (a*), yellowness or blueness (b*) and the ratio a/b [colorimeter Hunterlab DP-9000 D25A (Hunter associates laboratory, Reston, VA, USA)]. All analyses were performed based on in-house protocols at the premises of D. Nomikos S.A., Agricultural Department, Panagia, 35010, Domokos, Greece. Samples of 20 fruits/plot were collected from the fields with the sown margins and the control field with the weed flora, and were transported immediately for analysis. The samples were collected from twin crop rows aligned with the 1st, 3rd, and 5th sown or weed plots, at three field sites (i.e., 1st–2nd next to the margin, the 20th–21st row and 40th–41st row).

### 2.6. Statistical Analysis

Generalized linear mixed models (GLMM) were conducted to examine the effect of field margin management (treatment), plant cover, flower cover, and number of plant species in bloom, as fixed factors, on the abundance of the main arthropod functional groups (honeybees, wild bees, parasitoids, predators and insect pests) in the field margins. Experimental year and sampling date were treated as random factors. The models used a negative binomial distribution with a log link function. Fisher's LSD method was used for pairwise comparisons. Kruskal–Wallis ANOVA was used for comparison of plant and flower cover among field margin treatments, and for examination of the effect of field margin management (treatment) on the abundance of parasitoids, predators, and insect pests in the processing tomato crop. A one-way ANOVA ($\alpha = 0.05$) was performed to examine the effect of field margin management (treatment) on crop yield parameters. The means were separated using Tukey's HSD test. Statistical analysis was carried out using SPSS version 21.0 for Windows software (IBM Corp; Armonk, NY, USA).

# 3. Results

## 3.1. Plant Cover

Overall, the WM margin provided complete plant cover that differed significantly compared to all other treatments ($\chi^2 = 88.565$, $df = 3$, $p < 0.001$) (Figure 2). In general, all plant species in the mixtures sown in winter (WM) or spring (SM) along the margin of the selected processing tomato field, emerged and provided a dense plant cover both years (98–100% of the plot area). Exceptions were *P. tanacetifolia* in 2021 and *F. esculentum* in 2022 in the SM, with very low corresponding densities. This was attributed to the low germination ability of *P. tanacetifolia* in 2021 and to a late frost that reduced the survival of *F. esculentum* in 2022. However, the remaining sown species resulted in a high plant cover, which was mainly due to *E. vesicaria* and secondly to *L. sativus*.

The two control sites with weed flora, either next to the irrigation channel (CC) or next to the crop (CT), had comparable plant cover in 2021. However, in 2022 from mid- and late June because of an accidental glyphosate application for weed control on that site by field workers. The plant cover in that margin was partially recovered only around mid-July (Supplementary Figure S1).

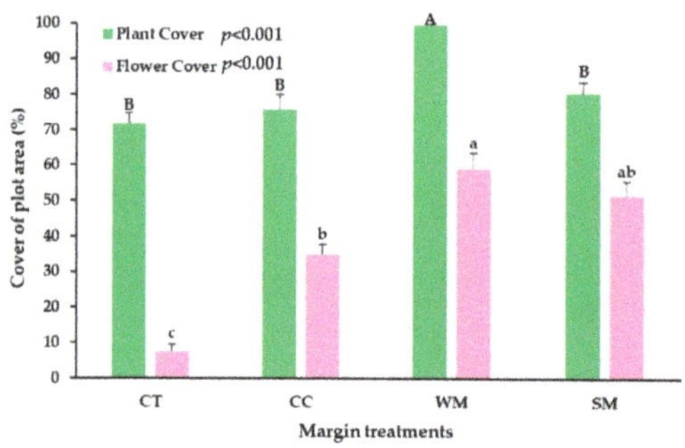


**Figure 2.** Mean total flower cover % of plot area of the processing tomato field margin with weed flora or sown with plant mixtures. Treatments: CT—control weed margin next to tomato crop; CC—control weed margin along a neighboring irrigation channel; WM—winter sown mixture; SM—spring sown mixture. Identical letters above the error bar indicate no statistically significant differences among treatments.

## 3.2. Flower Cover

The flower cover was the highest for the sown mixtures (WM, SM), while from the control margins the CC flower cover was comparable to the SM margin and the CT was significantly lower from all other treatments ($\chi^2 = 56.088$, $df = 3$, $p < 0.001$) (Figure 2). In 2021, the sown mixtures combined provided continuous floral resources from mid-May to mid-June. The mean flower cover of the WM mixture was higher (68–100% of the plot area) compared to the other sites, until late-May. In early-June, flower cover was similar in the sown mixtures (WM and SM) and the CC weed margin site, which maintained an intermediate flower cover (28–61% of plot area) throughout the experiment. In mid-June the highest flower cover was provided by the SM (92% of the plot area). The CT weed margin had the lowest flower cover throughout the season (>10% of the plot area). In 2022, the combined flowering of the two sown mixtures lasted from June to mid-July. Early in the season (early to mid-June) the flower cover of the sown mixtures was comparable to that of the weed flora (18–32%). The flower cover of the sown mixtures was higher compared to the weed flora later in the season, with a peak in late-June (63% and 33% of the plot area for the sown mixtures and weed flora, respectively) (Supplementary Figure S2).

The main flowering species in each sown mixture varied over time during both years (Figure 3). In 2021, the order of flowering species in the WM was *E. vesicaria*, *L. sativus*, *P. sativum*, *C. sativum*, *A. graveolens*, and in the SM, it was *L. sativus*, *F. esculentum*, *P. tanacetifolia*, and *E. vesicaria*. The main dicotyledonous weed species that emerged and reached flowering together with the sown species was *Rapistrum rugosum* (L.) All. in the WM and *Heliotropium europaeum* L. in the SM. Other weeds that emerged included *Solanum nigrum* L., *Chenopodium album* L., and *Sorghum halepense* (L.) Pers. Overall, weeds emerged in low numbers and did not affect the establishment and flowering capacity of the sown mixtures. That year, the CC margin was dominated by the Brassicaceae *R. rugosum* mid-May, and later by Apiaceae species. In the CT margin, flower cover was provided by a few *R. rugosum* and *Sisymbrium* sp. plants (Figure 3 and Supplementary Figure S3).

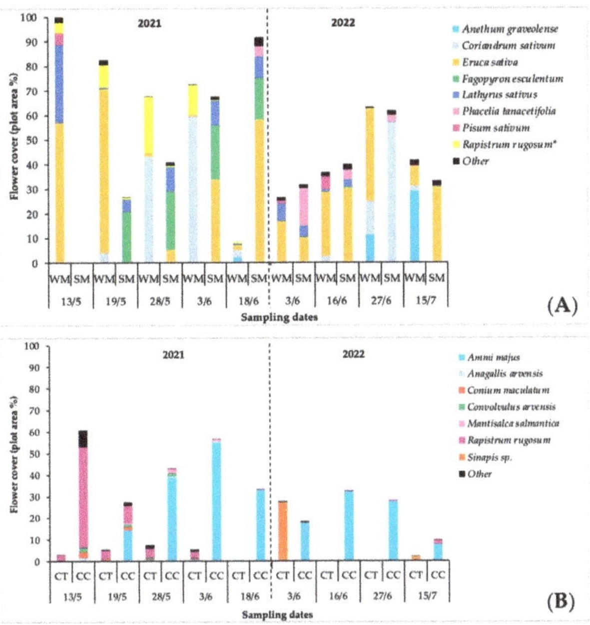


**Figure 3.** Mean flower cover per species (% of plot area) in the tomato field margin during spring 2021 and 2022 of (**A**) sown mixtures in winter (WM) or spring (SM), and of (**B**) weed species in CT (field margin with weed flora along the control tomato crop) and CC sites (weed flora along the irrigation channel next to the control field). Main weed species that emerged in the sown margins are indicated with an asterisk (*). Other flowering species include weeds with a low abundance and a small contribution to the overall flower cover (<2%). In 2021, these included WM: *Papaver rhoeas* L., SM: *S. nigrum*, *Chrozophora tinctoria* (L.) A.Juss., *Euphorbia* sp., *H. europaeum*, *C. arvensis*, *R. rugosum*, CT: *Calendula* sp., *Cirsium* sp., *Mantisalca salmantica* (L.) Briq. and Cavill., *Sonchus* sp., *P. rhoeas*, *E. elaterium*, *F. officinalis*, *Consolida regalis* Gray, CC: *Euphorbia* sp., *Geranium* sp., *Malva* sp., *Sonchus* sp., and in 2022 WM: *C. arvensis*, *E. elaterium*, *S. nigrum*, *F. officinalis*, *P. oleraceae*, *S. arvensis*, *R. rugosum*, SM: *C. arvensis*, *E. elaterium*, *S. nigrum*, *S. arvensis*, CT: *C. arvensis*, *Euphorbia humifusa* Willd., *Matricaria chamomilla* L., *Malva* sp., *P. oleracea*, *Tribulus terrestris* L., and in CC: *Lactuca serriola* L., *M. chamomilla*, *Malva* sp.

In 2022, *E. vesicaria* was more dominant in the SM due to the low emergence of *P. tanacetifolia* and the absence of *Fagopyrum esculentum* that was lost after the late frost. The order of flowering among the sown species in the WM was the same as the previous year:

*E. vesicaria* > *L. sativus* > *P. sativum* > *C. sativum* > *A. graveolens*, while in the SM, the order was *L. sativus* ≥ *P. tanacetifolia* ≥ *E. vesicaria*. Several flowering weeds emerged in low numbers in the sown mixtures and had a very low contribution to the flower cover. These were, in WM: *Convolvulus arvensis* L., *Ecballium elaterium* (L.) A.Rich., *S. nigrum*, *Fumaria officinalis* L., *Portulaca oleracea* L., *Sinapis arvensis* L., *R. rugosum*, and in SM: *C. arvensis*, *E. elaterium*, *S. nigrum*, *S. arvensis*. As mentioned in the plant cover section, in 2022, flower cover in the CT control site was recorded only in early June and mid-July, with *S. arvensis* and *S. alba* L. being the main flowering species. The flower cover in the CC site was almost solely provided by *Ammi majus* L. throughout the season (Figure 3 and Supplementary Figures S4 and S5).

### 3.3. Hymenoptera Pollinators

The total abundance of honey bees was significantly higher in the sown margins (WM and SM) compared to the weed flora along the irrigation channel (CC) and the field margin of the tomato crop (CT) ($F_{3,91}$ = 4.190, $p$ = 0.008). The flower cover percentage had a significant effect on honey bee abundance ($F_{1,91}$ = 17.709, $p$ = 0.000), while the number of plant species in bloom did not have any significant effect ($F_{1,91}$ = 1.817, $p$ = 0.181). On the contrary, the abundance of wild bees did not differ statistically among treatments ($F_{3,91}$ = 2.106, $p$ = 0.105) (Figure 4 and Supplementary Table S1). On a date-to-date examination, it is obvious that honey bees had the highest abundance in late June (27 June 2022) in SM, when the available floral resources of *E. vesicaria* reached the highest flower cover percentage (Supplementary Figure S6). Sporadic visits of wild bees (1–2 individuals/plot/4 min of observation time) were recorded on tomato flowers regardless of the row distance from the field margins throughout the experiment. These wild bees belonged to the genera *Andrena*, *Lasioglossum* and *Nomiapis*. The wild bee genera recorded on flowers of the sown mixtures, weed flora in the control sites, and tomato flowers are shown in Table 2 (Supplementary Figures S7 and S8).

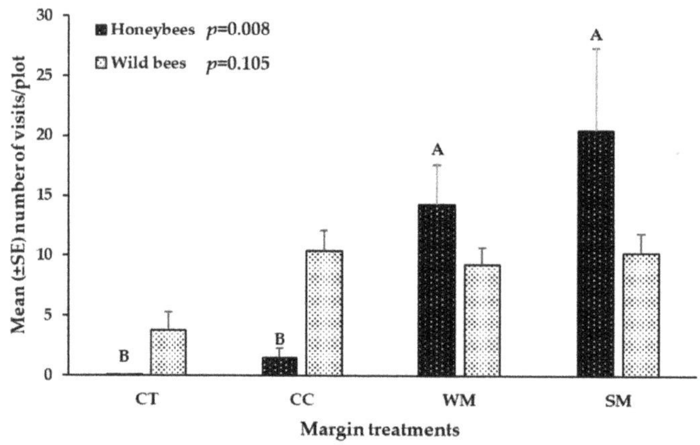


**Figure 4.** Total abundance of Hymenoptera pollinators (mean ± SE) over 2021 and 2022 in sown flower mixtures and weed flora of processing tomato field margins. Treatments: CT—control weed margin next to tomato crop; CC—control weed margin along a neighboring irrigation channel; WM—winter sown mixture; SM—spring sown mixture. Identical letters above the error bar indicate no statistically significant differences among treatments.

**Table 2.** Wild hymenoptera pollinators and associated plant species of sown mixtures and weed flora in processing tomato field margins: CT—control weed margin next to tomato crop; CC—control weed margin along an irrigation channel; WM—winter sown mixture; SM—spring sown mixture, Larisa plain, Greece, 2021 and 2022.

| Plant Species | Bee Genus | Treatment |
|---|---|---|
| *Ammi majus* | *Andrena, Hylaeus, Lasioglossum, Nomiapis* | CC |
| *Anethum graveolens* | *Andrena, Hylaeus* | WM |
| *Centauria* sp. | *Halictus* | CC |
| *Coriandrum sativum* | *Andrena, Ceratina, Nomiapis* | WM |
| *Ecballium elaterium* | *Lasioglossum, Ceratina* | SM |
| *Eruca vesicaria* | *Amegilla, Andrena, Ceratina, Eucera, Halictus, Lasioglossum* | SM |
| *Fagopyrum esculentum* | *Andrena, Lasioglossum* | SM |
| *Lathyrus sativus* | *Eucera, Megachile, Osmia* | SM |
| *Phacelia tanacetifolia* | *Ceratina* | SM |
| *Rapistrum rugosum* | *Andrena, Eucera, Lasioglossum, Nomada* | CT |
| *Solanum lycopersicum* | *Andrena, Lasioglossum, Nomiapis* | WM |

### 3.4. Natural Enemies

The total abundance of parasitoid wasps was significantly higher in the sown margins (WM and SM) and weed flora along the irrigation channel (CC) compared to the weedy margin next to the tomato crop (CT) ($F_{3,89}$ = 4.214, $p$ = 0.008). The plant cover, flower cover percentage and the number of plant species in bloom did not have a significant effect on parasitoid abundance ($F_{1,89}$ = 0.942, $p$ = 0.334, $F_{1,89}$ = 0.050, $p$ = 0.823 and $F_{1,89}$ = 0.073, $p$ = 0.787, respectively). The abundance of predatory arthropods did not differ statistically among treatments ($F_{3,89}$ = 0.930, $p$ = 0.430); however, the SM margin harbored higher numbers of predators. The flower cover percentage and number of plant species in bloom had a positive effect on the abundance of predatory arthropods ($F_{1,89}$ = 4.323, $p$ = 0.040 and $F_{1,89}$ = 5.224, $p$ = 0.025) (Figure 5A and Supplementary Table S2).

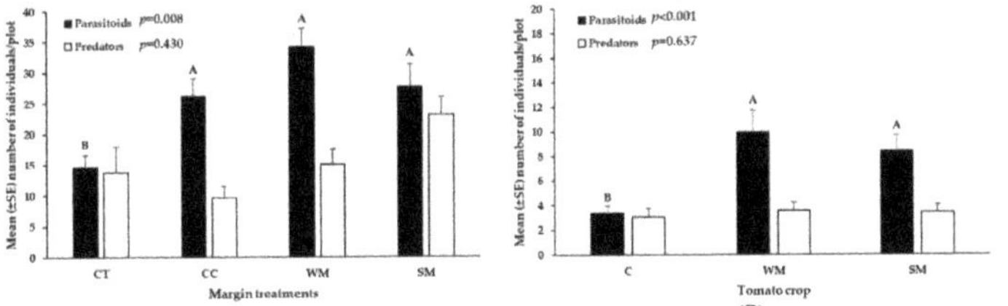


**Figure 5.** Total abundance (mean ± SE) of Hymenoptera parasitoids and predatory arthropods (**A**) in field margins and (**B**) in tomato crop with different field margin management: CT—control weed margin next to tomato crop; CC—control weed margin along the irrigation channel; WM—winter sown mixture; SM—spring sown mixture; C—control tomato field. Identical letters above the error bar indicate no statistically significant differences among treatments.

The total abundance of natural enemies in the tomato crop followed a similar pattern to that of the field margins. Higher numbers of parasitoid Hymenoptera were observed in the tomato fields with sown margins compared to the control fields ($\chi^2$ = 15.566, $df$ = 2, $p$ < 0.001), while the abundance of predators did not differ significantly among treated tomato fields ($\chi^2$ = 0.902, $df$ – 2, $p$ = 0.637) (Figure 5B).

Looking at their temporal abundance fluctuation of parasitoid Hymenoptera, the CC control margin was superior to the WM sown margin at the beginning of May 2021 (54 ± 5.3 and 39 ± 17.1, respectively). The WM sown margin had higher numbers of parasitoid wasps compared to both control margins during the rest of the sampling period of both experimental years. The SM margin reached peak levels during early June 2021 (36.7 ± 11) and mid-June 2022 (54 ± 10.4). The highest numbers of predators were observed in SM sown margin from the beginning until mid-June on both sampling periods (Supplementary Figure S9).

In the tomato crop, parasitoids in the fields with sown margins had increased numbers from mid to late May, and mid-June during the first sampling period (2021), while predators had generally low populations. In 2022, predatory arthropods were more abundant reaching peak levels during mid-July in the tomato fields with sown margins (10.7 ± 0.3 for WM and 10 ± 1.2 for SM). At the same time, the highest numbers of parasitoid wasps were observed (8.7 ± 1.3 for WM and 14.3 ± 2.8 for SM) (Supplementary Figure S10).

Parasitic wasps in the samples belonged to 7 different superfamilies (Ceraphronoidea, Chalcidoidea, Chrysidoidea, Cynipoidea, Diaprioidea, Ichneumonoidea, Platygastroidea) and 24 different families. The samples from the WM belonged to 20 families, while the ones from the SM mixture belonged to 19 families and from the CT and CC sites to 20 and 19 families, respectively. The WM and SM sown margins had similar community structure dominated by Eulophidae (35–36%), Braconidae (17–19%), and Scelionidae (19%), while Encyrtidae held 6.5–8%. The weed flora along the irrigation channel (CC) attracted mostly eulophids (44.7%), while braconids, scelionids and encyrtids held subequal proportions (11–12%). The weed margin next to the crop (CT) harbored mostly Eulophidae (37%), Mymaridae (14.6%) and Braconidae (12.6%). Within the tomato crop, the parasitoid families with the highest proportion were Braconidae (35%) for fields with WM margin, Scelionidae (23%) for fields with SM margin and Mymaridae (19.7%) for the control fields (Figure 6A).

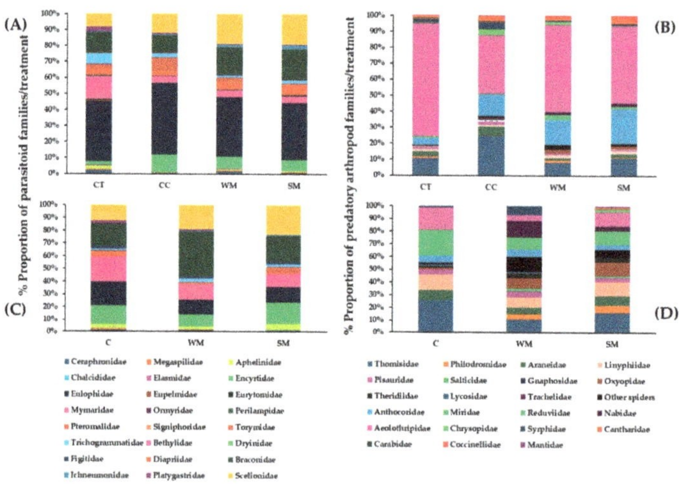


**Figure 6.** Community composition of natural enemies. In field margins: (**A**) Hymenoptera parasitoids; (**B**) predatory arthropods. In tomato crop with different field margin management: (**C**) Hymenoptera parasitoids; (**D**) predatory arthropods. CT—weedy margin next to tomato crop; CC—weedy margin along the irrigation channel; WM—winter sown mixture; SM—spring sown mixture; C—control tomato field.

The communities of predatory arthropods hosted by the WM and SM sown margins were also similar; they consisted mostly of predatory thrips (Aeolothripidae) (48.5–54.7%), pirate bags (Anthocoridae) (15–21.5%), and crab spiders (Thomisidae) (8–11%). The proportions of these groups in CC sites were 36.5%, 13.5%, and 24.5%, respectively. Predatory thrips were superdominant (71%) in CT margins (Figure 6B). In tomato fields with the sown margins, several predatory groups were detected such as predatory Heteroptera (Anthocoridae, Miridae, Nabidae), spiders (Linyphiidae, Oxyopidae, Thomisidae), and predatory thrips, without any obvious dominance. The control fields hosted mostly crab spiders (25.7%), predatory mirids (20%), and Aeolothripidae (17.6%) (Figure 6C,D).

### 3.5. Insect Pests

Regarding the abundance of insect pests, the sown margins and the weedy margin next to the crop harbored significantly more insect pests than the margin along the irrigation channel ($F_{3,89} = 3.950$, $p = 0.011$), while there were no statistical differences between the tomato fields ($\chi^2 = 5.129$, $df = 2$, $p = 0.077$) (Figure 7).

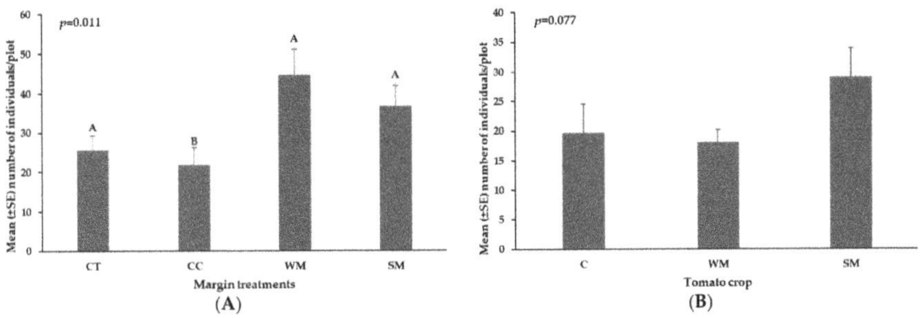


**Figure 7.** Total abundance (mean ± SE) of insect pests (**A**) in field margins and (**B**) in tomato crop with different field margin management: CT—weedy margin next to tomato crop; CC—weedy margin along the irrigation channel; WM—winter mix; SM—spring mix; C—control tomato field. Identical letters above the error bar indicate no statistically significant differences among treatments.

In the SM sown margins, the most abundant group was aphids (44.6%), in the WM margins and the CC sites, thrips (49% and 51.8%, respectively), and in the CT sites, leafhoppers (51%). Within tomato crops, the control fields (C) and those with the SM margin hosted mainly aphids (51% and 52%, respectively), while the fields with WM margin hosted mostly thrips (33.8%) (Supplementary Figure S11).

### 3.6. Crop Yield

The yield parameters (means) of processing tomato (weight, BRIX, pH, L, color AB) are shown in Table 3. In both experimental years, the average fruit weight from fields with sown margins was similar to the expected values (75–80 g/fruit) and significantly higher compared to the control fields with weed vegetation ($F_{2,24} = 93.179$, $p < 0.0001$ for 2021, $F_{2,24} = 54.136$, $p < 0.0001$ for 2022). The total soluble solids (TSS) of fruits harvested from WM, SM and control fields were similar to the expected value (5.2 °Bx) in 2021; however, in 2022, the fruits from the control field had significantly higher levels of TSS (6.1 °Bx) compared to those with the sown margins ($F_{2,24} = 21.116$, $p < 0.0001$). The pH values ranged between 4.4 and 4.5 without significant differences. Regarding the fruit color parameters among treatments, a significantly higher $L$ value was observed in the control field in 2021 ($F_{2,24} = 10.377$, $p = 0.011$) and significantly lower a/b value in 2022 ($F_{2,24} = 8.714$, $p = 0.001$) (Table 3 and Supplementary Table S3).

**Table 3.** Yield parameters (mean ± SE) of processing tomato hybrid H1015 fruits (weight, BRIX, pH, *L*, a/b) harvested from fields with different margin management: WM—winter sown mixture; SM—spring sown mixture; C—control tomato field. Identical letters indicate no statistically significant differences among treatments.

| | Treatment | | | | | |
|---|---|---|---|---|---|---|
| | 2021 | | | 2022 | | |
| Yield Parameter | WM | SM | C | WM | SM | C |
| Weight (g/fruit) | 74.9 ± 0.7 a | 76.8 ± 0.9 a | 63.4 ± 0.6 b | 79.3 ± 1.2 A | 76.2 ± 2.2 A | 57.3 ± 1.3 B |
| BRIX | 5.2 ± 0.04 a | 5.2 ± 0.02 a | 5.2 ± 0.07 a | 5.2 ± 0.1 B | 5.3 ± 0.1 B | 6.1 ± 0.1 A |
| pH | 4.4 ± 0.03 a | 4.5 ± 0.03 a | 4.5 ± 0.04 a | 4.5 ± 0.02 A | 4.5 ± 0.02 A | 4.5 ± 0.03 A |
| *L* | 27.8 ± 0.1 b | 28.0 ± 0.3 b | 29.6 ± 0.4 a | 28.3 ± 0.3 A | 28.1 ± 0.3 A | 28.1 ± 0.6 A |
| a/b | 2.5 ± 0.01 a | 2.5 ± 0.04 a | 2.4 ± 0.05 a | 2.5 ± 0.04 A | 2.5 ± 0.03 A | 2.3 ± 0.03 B |

## 4. Discussion

The establishment of flower strips inside the crop or at the field margins is among the agro-ecological practices outlined in the EU's recent common agricultural policy [35] to maintain functional biodiversity in agroecosystems [36]. Indigenous wild flora species are generally preferred for the flower strips, to avoid genetic erosion from imported species and to exploit their adaptation to local conditions, ensuring a successful establishment (19). However, as indicated by our results, a feasible and effective alternative for the implementation of this agro-ecological practice could be the use of cultivated flowering plants, such as the Apiaceae (*C. sativum* and *A. graveolens*), Brassicaceae (*E. vesicaria*) and Fabaceae (*L. sativus, P. sativum*) species tested here, with easily available seeds in the market. Nevertheless, field margins with weed flora adjacent to the experimental tomato crop were also assessed to examine the ability of natural flora to safeguard beneficial arthropods for agriculture in comparison to the sown flowering mixtures, as it has been conducted in previous studies in tomato and other crops [18,19,37]. Assessing field margins for the dual purpose of supporting both pollinators and natural enemies is in line with the Integrated Pest and Pollinator Management (IPPM) framework proposed recently by Egan et al. [38] to enhance IPM compatibility with crop pollination management and enable their coordination towards a unified economic decision making and implementation. However, diversified plant resources provided by crop or wild species should be characterized on their capacity to support functionally important arthropods, such as pollinators and natural enemies, while also considering potential trade-offs such as enhancing crop pests or causing weed related problems [18,39] before employing them at the field scale as an IPPM tool [40].

Our two-year field margin assessment indicated that the sown flowering mixtures attracted more honey bee visits compared to the weed flora, as also reported by other studies [20,41]. Furthermore, the flower cover percentage was positively correlated with the number of attracted bee visits, but not the number of flowering plant species, as also recently reported by Liira and Jürjendal [42]. The latter study provided evidence that areas with low flower diversity are likely to attract higher number of bees, and that agricultural policies for the conservation of pollinators should not focus on floral biodiversity, as it is commonly supported [43,44]. However, there is undisputable evidence that diversity in floral traits has a positive effect on bee species diversity because it caters for their variable nectar and pollen needs, while also ensuring that accessibility requirements are met, which depend on flower and bee anatomy [45].

Most of the bees recorded on the margins of the field were honey bees (*A. mellifera*). However, honey bees are unable to perform buzz pollination, i.e., to vibrate the tomato flower in the frequency needed to release the pollen from the anthers to the stigma [46]. Nevertheless, wild bee genera (e.g., *Amegilla*) recorded in lower numbers of visits in our study, include species which use floral buzzing while foraging [47], and are thus considered potential pollinators of tomato flowers [48]. In fact, in our study, only wild bees were recorded to visit tomato flowers (*Andrena*, *Lasioglossum* and *Nomiapis*), although in small numbers. However, upscaling the practice of flowering field margins in tomato crops could possibly enhance foraging on tomato flowers by these pollinators and subsequently contribute to the improvement of the fruit set percentage and fruit characteristics [49]. Similar wild bee genera were attracted on the sown flowering strip of similar experimentation in tomato (the WM included also *Glebionis coronaria* (L.) Cass. ex Spach), in the southern part of continental Greece (area of Orchomenos), i.e., mainly included *Andrena* spp., *Colletes* sp., *Halictus* sp., *Hylaeus* spp., *Lasioglossum* spp., *Pseudapis* sp. (identified later as *Nomiapis* sp.; Barda, personal communication), and *Sphecodes* sp. [19], contributing to tomato pollination with *Hylaeus* spp. and *Lasioglossum* spp. [29]. Further experimentation should focus on the long-term effects of the sown flower mixtures to insect pollinators.

*Eruca vesicaria*, the species which held the higher percentage of the flower cover and longest flowering duration in the mixtures, attracted honey bees and wild bees (*Amegilla, Andrena, Ceratina, Eucera, Halictus, Lasioglossum*). Indeed, *Eruca vesicaria* has been proven to attract a wide variety of pollinators [50–52]. *Coriandrum sativum* and *A. graveolens* also provided a long-lasting flowering throughout the season, which overlapped with that of the tomato crop and lasted until the fruit set, providing an attractive habitat for potential pollinators of the crop. *Coriandrum sativum* attracted both honey bees and wild bees, including genera recorded on tomato flowers (*Andrena, Nomiapis*), confirming its favorable profile as a tool to promote the ecosystem services of pollination as well as biological pest control [19,53–55]. Records of wild bee attraction by *C. sativum* include the families Andrenidae, Halictidae and Colletidae (Algeria) [53], and *Megachile* spp. (Pakistan) [56]. Likewise, the second Apiaceae, *A. graveolens*, brought in honey bees along with wild bee genera (*Andrena, Hylaeus*) known to pollinate tomato [29]. *Lathyrus sativus* attracted early in the season bees of *Eucera*, as also stated at the similar study by Kati et al. [19], but also of *Megachile*, which are among the most efficient pollinators for tomato [29], and *Osmia*.

In the SM mixture, the flowering of *F. esculentum*, evident only in the first year, attracted both honeybees and wild bees of the genera *Andrena* and *Lasioglossum*. In other studies, reported foraging of wild bees on *F. esculentum* flowers include *Bombus* spp. [57,58], *Xylocopa* spp., *Halictus* spp. in Greece [19], as well as other Halictidae species in Florida [51]. The capacity of *F. esculentum* to upgrade the floral resources of field margins for the benefit of pollinators and beneficial arthropods is supported by [59,60]. The poor establishment of *P. tanacetifolia* in the SM mixture did not allow for the full evaluation of its potential for use in flowering mixtures in the study area. Its low flower cover was foraged by both honey bees and wild bees as also reported elsewhere [18,61]. However, Petanidou [61] expressed concern on the suitability of this species for use as a flower resource for pollinating bees in Greece.

Furthermore, the contribution of wild plants already present in the field margins to attracting pollinators should not be overlooked, as long as they are not among species that are likely to infest the crop as weeds. For example, *A. majus* is a wild plant that is not threatening as a weed in the area of our study. This species is known to be a nectar-bearing plant [62] and has been reported to attract species of *Andrena* [63]. In our study, it supported wild bee genera (*Andrena, Hylaeus, Lasioglossum, Nomiapis*), some of which were also recorded on tomato flowers and include species that can buzz-pollinate [29].

Both sown flowering mixtures and the undisturbed natural vegetation along the irrigation channel outcompeted the weed margin along the tomato field (CT) as reservoirs of parasitoid wasps. The positive effect of flower-rich strips on the abundance of parasitoid Hymenoptera has been demonstrated by several studies [19,60,64–66]. However, the fact that the total flower cover of the plot area and the number of plant species in bloom did not affect the population density of parasitoids, indicate a complex interplay among different variables that can affect parasitoids' abundance. Each species has indeed a different potential as an insectary plant regarding the abundance and diversity of parasitoids [67,68], while on a given plant individual, the parasitoids' abundance may be affected by host abundance and the plant's flowering state [69].

The dominant parasitoid family in all field margins was Eulophidae. This family includes parasitoid wasps against tomato crop pests, such as leafminers *Liriomyza* spp. (Diptera: Agromyzidae) and the tomato leafminer *Tuta absoluta* Meyrick (Lepidoptera: Gelechiidae) [70,71]. Braconidae and Scelionidae were also abundant in the field margins. Braconids are known for their contribution to controlling aphid populations and several species are promising biological control agents of *T. absoluta* and the cotton bollworm *Helicoverpa armigera* Hübner (Lepidopetra: Noctuidae) [72–74]. Scelionids are mainly egg parasitoids, with a wide range of hosts, which have been proven effective in the biological control of stink bugs (Hemiptera: Pentatomidae) such as *Nezara viridula* (Linnaeus) and *Halyomorpha halys* (Stål) [75,76].

The benefits from the sown flowering plants to parasitoids have been demonstrated by several studies. The main flowering species of both sown mixtures, *E. vesicaria*, has been reported to host a high diversity of potential natural enemies (parasitoids and predators) of aphids and thrips, mainly due to its early and long-lasting blooming [68]. Flowers of *F. esculentum* (component of SM) positively affected the longevity of three *T. absoluta* parasitoids, *Necremnus artynes* Walker, *N. tutae* (Ribes & Bernardo) (Eulophidae) and *Bracon nigricans* Szépligeti (Braconidae) [77,78]. Nectar of *A. graveolens*, *C. sativum* (components of WM) and *F. esculentum* improved the survival of the stink bug parasitoid *Trissolcus japonicus* (Ashmead) (Scelionidae) [79].

A higher abundance of parasitoid Hymenoptera was observed also within the tomato fields with sown margins compared to the control fields. Samples included mostly braconid wasps (mainly Aphidiinae) associated with aphids' abundance; scelionids and eulophids were also present. Fairyflies (Mymaridae) had increased proportions, probably due to the presence of leafhoppers, as they are common egg-parasitoids of Auchenorrhyncha (Hemiptera) [80].

The SM sown margin surpassed numerically the other field margins in predatory arthropods, with the most abundant groups in all cases being Aeolothripidae, Anthocoridae and Thomisidae. All the Aeolothripidae individuals were assigned to the species *Aeolothrips intermedius* Bagnallis which is considered the primary native predator of *Thrips tabaci* and other species of Thysanoptera in Europe [81] and has been reported as a very effective predator in the biological control of thrips [82,83]. Minute pirate bugs (Anthocoridae), especially *Orius* spp., are generalist predators that have proved successful predators of the western flower thrips, *Frankliniella occidentalis* (Pergande) and other thrips [84], but they can also prey on other key agricultural pests, such as whiteflies, aphids and spider mites [85–87]. Both Aeolothripidae and Anthocoridae are omnivorous predators that also feed on different plant resources such as pollen and nectar [88,89] and the increasing plant diversity allows them to optimize their fitness by exploiting various plant-based resources such as nutrition and oviposition sites [81,90–92]. Of the two Thomisidae spider species, *Thomisus onustus* Walckenaer and *Runcinia grammica* (C. L. Koch) found in the samples,

*R. grammica*, which was most abundant, is a polyphagous predator feeding on a wide range of arthropods (Diptera, Hymenoptera, Lepidoptera) [93]. Crab spiders (Thomisidae) are ambush predators exploiting flowers' ability to attract insect visitors by sitting on inflorescences and then attacking and consuming insect prey [94,95]. These ecological strategies (omnivory and ambush hunting) explain the positive effect of flower cover and the number of plant species in bloom on the abundance of predators in our study.

Zoophytophagous mirids, *Deraeocoris ruber* (L.) and *Macrolophus pygmaeus* (Rambur) were present in the field margins and the tomato crop. *Macrolophus pygmaeus* is a key predator for biological control of *T. absoluta*, whiteflies, aphids and other pests on tomato crop [96]. Coccinellid species (*Coccinella septempunctata* L., *Hippodamia variegata* Goeze, *Propylea quatuordecimpunetata* L., *Scymnus auritus* (Thunberg) and *S. suturalis* Thunberg) were sampled only from the field margins and they are mainly aphidophagous and coccidophagous [97,98]. Apart from Thomisidae, several spider species of other families were collected both from the field margins and the tomato crop, i.e., *Argiope bruennichi* (Scopoli), *Hypsosinga pygmaea* (Sundevall), *H. sanguinea* (C. L. Koch), *Mangora acalypha* (Walckenaer), *Neoscona adianta* (Walckenaer) (Araneidae), *Aphantaulax cincta* (L. Koch) (Gnaphosidae), *Agyneta pseudorurestris* Wunderlich, *Microlinyphia pusilla* (Sundevall) (Linyphiidae), *Oxyopes heterophthalmus* (Latreille), *O. lineatus* Latreille (Oxyopidae), *Thanatus atratus* Simon, *Tibellus oblongus* (Walckenaer) (Philodromidae), *Phylloneta impressa* (L. Koch), *Theridion cinereum* Thorell (Theridiidae), and *Trachelas minor* O. Pickard-Cambridge (Trachelidae). The web-building families Araneidae, Linyphiidae, and Theridiidae are widely abundant in agricultural landscapes and trap mostly dipteran and homopteran pests on their webs, while the active hunters, Oxyopidae and Philodromidae, are wandering spiders that prey on a wide variety of insect pests [99].

Overall, the good establishment of *E. vesicaria* and *C. sativum* in combination with their ability to attract pollinators and other beneficial insects makes them superior candidate plants for agro-ecological practices that utilize annual plants for the conservation of functional groups in agroecosystems in the studied region. Potential tomato pests such as aphids, thrips and leafhoppers were detected in suction samples of all field margins, which were less abundant in the control site near the irrigation channel (CC), probably due to the composition of the flower species. Although no direct monitoring of tomato pests was conducted on the crop, the total pest abundance in the suction samples was comparable across the field margins. Moreover, no outbreak or damage by these or other tomato pests was reported by the farmers in both cultivation periods, supporting on one hand, the lack of an adverse effect of the sown mixtures on the crop and, on the other, the potential of this plant synthesis to function as a banker plant mixture building up reservoirs of natural enemies throughout the crop season [30,100].

Our results provide evidence that attracting pollinators in tomato fields by increasing the floral resources along the crop margin can have a positive impact on fruit weight. Similar findings were reported in previous research which demonstrated that insect pollination, particularly by native and buzz-pollinating bees, leads to increased fruit set, larger fruit size, and improved physicochemical properties compared to self-pollination or wind pollination in open-field tomato crops [46,101,102]. For example, Bashir et al. [102] reported that open-pollinated tomatoes, which received visits from pollinators and the simultaneous effect of wind, had significantly larger fruit sizes and higher seed weights compared to those subjected only to wind or self-pollination. More specifically, open-pollinated fruits had a mean weight (g/fruit) of 109.76, outperforming wind-pollinated (72.97) and self-pollinated (49.03) fruits. On the other hand, studies associating BRIX levels with various pollination methods in tomato are scarce. The existing evidence suggests that insect pollination can enhance tomato fruit quality, as indicated by higher BRIX values that imply changes in sugar composition, and therefore fruit sweetness [103]. In our study, BRIX levels were similar for the fruits harvested from all fields in 2021, while fruits from the control field had higher values in 2022. Fruits in that field had also the smallest size (57.3 g/fruit) compared to those from the field with the sown margins (76.2–79.3 g/fruit, for SM and WM respectively). Previous studies highlighted the complexity between fruit size and BRIX relationship, and reported that there is often an inverse correlation, particularly in small-fruited tomatoes [104]. Fruit quality parameters associated with the fruit color brightness (L*) and the hue index (a*/b*) were not consistent over the years, but ranged between the expected values for the tomato hybrid of our study. Ultimately, fruit quality is likely to be affected by the interaction of genetic, environmental, and agronomic factors, which requires further investigation.

## 5. Conclusions

Our results suggest that selected flowering plants can replace undesirable weeds commonly found in disturbed field margins, to serve as habitats for pollinating insects and beneficial arthropods with a potential benefit for functional biodiversity and the tomato fruit quality attributes. Introducing selected plants from the families of Brassicaceae, Fabaceae, and Apiaceae in field margins can benefit the processing tomato by attracting wild bee species (e.g., from the genera *Amegilla, Andrena, Hylaeus, Lasioglossum, Nomiapis)* that can support the pollination of tomato flowers, and parasitoids (Eulophidae, Braconidae, Scelionidae) and predatory arthropods (Aeolothripidae, Anthocoridae, Thomisidae) that could contribute to the biological control of tomato pests. Future research should elaborate further on the possible effect of insect pollination on tomato crop yield and quantify the potential of the sown margins to support biological control of tomato pests.

**Supplementary Materials:** The following supporting information can be downloaded at: https://www.mdpi.com/article/10.3390/agronomy15071558/s1, Figure S1: Mean plant cover (% of plot area) of the processing tomato field margin sown with plant mixtures and the weed vegetation; Figure S2: Mean flower cover (% of plot area) of the processing tomato field margin sown with plant mixtures and the weed vegetation; Figure S3: Flowering margins in the studied processing tomato fields for different sampling dates in late spring (May–June) 2021; Figure S4: Flowering margins in the studied processing tomato fields for different sampling dates in early summer (June–July) 2022; Figure S5: Close-up images of the flowering margins in the studied processing tomato fields on different sampling dates in early summer (June–July) 2022; Figure S6: Temporal distribution of Hymenoptera pollinators in the processing tomato field margin sown with plant mixtures, and the weed vegetation; Figure S7: Pollinators of tomato flower and various flowering species, either sown in winter (WM) or spring (SM) along the margin of a processing tomato field, or present in the weed vegetation of the margin in the control tomato field (CT) and in the weed vegetation along a nearby irrigation channel (CC) during May–June 2021; Figure S8: Pollinators of tomato flower and various flowering species, either sown in winter (WM) or spring (SM) along the margin of a processing tomato field, or present in the weed vegetation of the margin in the control tomato field (CT) and in the weed vegetation along a nearby irrigation channel (CC) during June-mid July 2022; Figure S9: Temporal distribution of natural enemies in the processing tomato field margin sown with plant mixtures, and the weed vegetation; Figure S10: Temporal distribution of natural enemies in the processing tomato crop with different field margin managements; Figure S11: Community composition of insect pests (A) in field margins; (B) in tomato crop with different field margin management. Table S1: Summary of the statistics for GLMM testing the effect of field margin management (treatment), flower cover, and number of plant species in bloom, on the abundance of Hymenoptera pollinators (honeybees, wild bees) in field margins; Table S2: Summary of the statistics for GLMM testing the effect of field margin management (treatment), plant cover, flower cover, and number of plant species in bloom, on the abundance of natural enemies (parasitoids, predators) and insect pests in field margins; Table S3: Summary of the statistics for a one-way ANOVA testing the effect of field margin management (treatment) on the crop yield parameters of processing tomato fruits.

**Author Contributions:** Conceptualization, V.K. and F.K.; methodology, V.K., F.K., L.E., P.M., T.S., T.A. and I.T.; formal analysis, T.S. and L.E.; investigation, V.K., F.K., L.E., P.M., M.B., T.S., I.T. and A.B.P.; resources, I.T., V.K. and F.K.; data curation, L.E., T.S., T.A. and P.M.; writing—original draft preparation, V.K., T.S. and F.K.; writing—review and editing, T.S., F.K. and V.K.; visualization, V.K., F.K., T.S., T.A. and M.B.; supervision, V.K. and F.K.; project administration, M.B., A.B.P., V.K. and F.K.; funding acquisition, F.K. and V.K. All authors have read and agreed to the published version of the manuscript.

**Funding:** This research was funded by NOMIKOS S.A., Processing Tomato Company, Greece (private grant).

**Data Availability Statement:** The data presented in this study are available on request from the corresponding authors.

**Acknowledgments:** This paper is dedicated to the memory of Fotis Andrinopoulos (1979–2025) as a token of our sincere appreciation, respect, and gratitude for his continuous support and collaboration during the biodiversity project Operation Pollinator of Syngenta, a predecessor of the current study. His enthusiasm, creativity, and luminous personality will be greatly missed.

**Conflicts of Interest:** Author I.T. was employed by the company Nomikos S.A., Processing Tomato Company, Greece. The remaining authors declare that the research was conducted in the absence of any commercial or financial relationships that could be construed as a potential conflict of interest.

## References

1. Jeanneret, P.; Lüscher, G.; Schneider, M.K.; Pointereau, P.; Arndorfer, M.; Bailey, D.; Balázs, K.; Báldi, A.; Choisis, J.-P.; Dennis, P.; et al. An increase in food production in Europe could dramatically affect farmland biodiversity. *Commun. Earth Environ.* **2021**, *2*, 183. [CrossRef]
2. Rafferty, N.E.; Ives, A.R. Effects of experimental shifts in flowering phenology on plant–pollinator interactions. *Ecol. Lett.* **2011**, *14*, 69–74. [CrossRef]
3. Memmott, J.; Carvell, C.; Pywell, R.F.; Craze, P.G. The potential impact of global warming on the efficacy of field margins sown for the conservation of bumble-bees. *Philos. Trans. R. Soc. B Biol. Sci.* **2010**, *365*, 2071–2079. [CrossRef] [PubMed]
4. Goulson, D.; Nicholls, E.; Botías, C.; Rotheray, E.L. Bee declines driven by combined stress from parasites, pesticides, and lack of flowers. *Science* **2015**, *347*, 1255957. [CrossRef]
5. Elias, M.A.S.; Borges, F.J.A.; Bergamini, L.L.; Franceschinelli, E.V.; Sujii, E.R. Climate change threatens pollination services in tomato crops in Brazil. *Agric. Ecosyst. Environ.* **2017**, *239*, 257–264. [CrossRef]
6. Thomson, L.J.; Macfadyen, S.; Hoffmann, A.A. Predicting the effects of climate change on natural enemies of agricultural pests. *Biol. Control* **2010**, *52*, 296–306. [CrossRef]
7. Guzmán, C.; Aguilar-Fenollosa, E.; Sahún, R.M.; Boyero, J.R.; Vela, J.M.; Wong, E.; Jaques, J.A.; Montserrat, M. Temperature-specific competition in predatory mites: Implications for biological pest control in a changing climate. *Agric. Ecosyst. Environ.* **2016**, *216*, 89–97. [CrossRef]
8. Naylor, R.; Ehrlich, P.R. Natural pest control services and agriculture. In *Nature's Services: Societal Dependence on Natural Ecosystems*; Daily, G.C., Ed.; Island Press: Washington, DC, USA, 1997; pp. 151–174.
9. Gallai, N.; Salles, J.-M.; Settele, J.; Vaissière, B.E. Economic valuation of the vulnerability of world agriculture confronted with pollinator decline. *Ecol. Econ.* **2009**, *68*, 810–821. [CrossRef]
10. Klein, A.-M.; Vaissière, B.E.; Cane, J.H.; Steffan-Dewenter, I.; Cunningham, S.A.; Kremen, C.; Tscharntke, T. Importance of pollinators in changing landscapes for world crops. *Proc. R. Soc. B Biol. Sci.* **2006**, *274*, 303–313. [CrossRef]
11. Emmerson, M.; Morales, M.B.; Oñate, J.J.; Batáry, P.; Berendse, F.; Liira, J.; Aavik, T.; Guerrero, I.; Bommarco, R.; Eggers, S.; et al. Chapter Two—How agricultural intensification affects biodiversity and ecosystem services. In *Advances in Ecological Research*; Large-Scale Ecology: Model Systems to Global Perspectives; Dumbrell, A.J., Kordas, R.L., Woodward, G., Eds.; Academic Press: Cambridge, MA, USA, 2016; Volume 5, pp. 43–97.

12. Dudley, N.; Alexander, S. Agriculture and Biodiversity: A Review. *Biodiversity* **2017**, *18*, 45–49. [CrossRef]
13. Peeters, A.; Lefebvre, O.; Balogh, L.; Barberi, P.; Batello, C.; Bellon, S.; Gaifami, T.; Gkisakis, V.; Lana, M.; Migliorini, P.; et al. A green deal for implementing agroecological systems—Reforming the common agricultural policy of the European Union. *J. Sustain. Agric. Syst.* **2020**, *70*, 83–93. [CrossRef]
14. Geiger, F.; Bengtsson, J.; Berendse, F.; Weisser, W.W.; Emmerson, M.; Morales, M.B.; Ceryngier, P.; Liira, J.; Tscharntke, T.; Winqvist, C.; et al. Persistent negative effects of pesticides on biodiversity and biological control potential on European farmland. *Basic Appl. Ecol.* **2010**, *11*, 97–105. [CrossRef]
15. Bianchi, F.J.J.A.; Wäckers, F.L. Effects of flower attractiveness and nectar availability in field margins on biological control by parasitoids. *Biol. Control* **2008**, *46*, 400–408. [CrossRef]
16. Tschumi, M.; Albrecht, M.; Collatz, J.; Dubsky, V.; Entling, M.H.; Najar-Rodriguez, A.J.; Jacot, K. Tailored flower strips promote natural enemy biodiversity and pest control in potato crops. *J. Appl. Ecol.* **2016**, *53*, 1169–1176. [CrossRef]
17. Campbell, A.J.; Wilby, A.; Sutton, P.; Wäckers, F. Getting more power from your flowers: Multi-functional flower strips enhance pollinators and pest control agents in apple orchards. *Insects* **2017**, *8*, 101. [CrossRef]
18. Karamaouna, F.; Kati, V.; Volakakis, N.; Varikou, K.; Garantonakis, N.; Economou, L.; Birouraki, A.; Markellou, E.; Liberopoulou, S.; Edwards, M. Ground cover management with mixtures of flowering plants to enhance insect pollinators and natural enemies of pests in olive groves. *Agric. Ecosyst. Environ.* **2019**, *274*, 76–89. [CrossRef]
19. Kati, V.; Karamaouna, F.; Economou, L.; Mylona, P.V.; Samara, M.; Mitroiu, M.-D.; Barda, M.; Edwards, M.; Liberopoulou, S. Sown wildflowers enhance habitats of pollinators and beneficial arthropods in a tomato field margin. *Plants* **2021**, *10*, 1003. [CrossRef]
20. Barda, M.; Karamaouna, F.; Kati, V.; Perdikis, D. Do patches of flowering plants enhance insect pollinators in apple orchards? *Insects* **2023**, *14*, 208. [CrossRef]
21. Scheper, J.; Badenhausser, I.; Kantelhardt, J.; Kirchweger, S.; Bartomeus, I.; Bretagnolle, V.; Clough, Y.; Gross, N.; Raemakers, I.; Vilà, M.; et al. Biodiversity and pollination benefits trade off against profit in an intensive farming system. *Proc. Natl. Acad. Sci. USA* **2023**, *120*, e2212124120. [CrossRef]
22. Sanchez, J.A.; de Pedro, L.; López-Gallego, E.; Pérez-Marcos, M.; Ramírez-Soria, M.J.; Perera-Fernández, L.G.; Atenza, J.F. How plant composition in margins influences the assemblage of pests and predators and its effect on biocontrol in melon fields. *Sci. Rep.* **2024**, *14*, 13094. [CrossRef]
23. Jachowicz, N.; Sigsgaard, L. Highly diverse flower strips promote natural enemies more in annual field crops: A review and meta-analysis. *Agric. Ecosyst. Environ.* **2025**, *381*, 109412. [CrossRef]
24. Pérez-Méndez, N.; Alcaraz, C.; Catala-Forner, M. Ecological restoration of field margins enhances biodiversity and multiple ecosystem services in rice agroecosystems. *Agric. Ecosyst. Environ.* **2025**, *382*, 109484. [CrossRef]
25. Crowther, L.I.; Wilson, K.; Wilby, A. The impact of field margins on biological pest control: A meta-analysis. *BioControl* **2023**, *68*, 387–396. [CrossRef]
26. Albrecht, M.; Kleijn, D.; Williams, N.M.; Tschumi, M.; Blaauw, B.R.; Bommarco, R.; Campbell, A.J.; Dainese, M.; Drummond, F.A.; Entling, M.H.; et al. The effectiveness of flower strips and hedgerows on pest control, pollination services and crop yield: A quantitative synthesis. *Ecol. Lett.* **2020**, *23*, 1488–1498. [CrossRef]
27. Garibaldi, L.A.; Carvalheiro, L.G.; Vaissière, B.E.; Gemmill-Herren, B.; Hipólito, J.; Freitas, B.M.; Ngo, H.T.; Azzu, N.; Sáez, A.; Åström, J.; et al. Mutually beneficial pollinator diversity and crop yield outcomes in small and large farms. *Science* **2016**, *351*, 388–391. [CrossRef] [PubMed]
28. Greenleaf, S.S.; Kremen, C. Wild bee species increase tomato production and respond differently to surrounding land use in Northern California. *Biol. Conserv.* **2006**, *133*, 81–87. [CrossRef]
29. Teppner, H. Pollinators of tomato, *Solanum lycopersicum* (Solanaceae), in Central Europe. *Phyton Ann. Rei Bot.* **2005**, *45*, 217–235.
30. Barda, M.S.; Karamaouna, F.; Kati, V.; Stathakis, T.I.; Economou, L.P.; Perdikis, D.C. Flowering plant patches to support the conservation of natural enemies of pests in apple orchards. *Agric. Ecosyst. Environ.* **2025**, *381*, 109405. [CrossRef]
31. Tutin, T.G.; Heywood, V.H.; Burges, N.A.; Valentine, D.H.; Walters, S.M.; Webb, D.A. *Flora Europaea*; Cambridge University Press: Cambridge, UK, 1964; Volumes 1–5.
32. Michener, C.D. *The Bees of the World*; Johns Hopkins University Press: Baltimore, MA, USA, 2007.
33. Collins, A.G. Key to the Genera of British Bees. 2012. Available online: https://bwars.com/sites/default/files/diary_downloads/Britain's_Bees_Chapter_4_Keys_to_Genera.pdf (accessed on 30 May 2025).
34. Stewart, A.J.A.; Wright, A.F. A New inexpensive suction apparatus for sampling arthropods in grassland. *Ecol. Entomol.* **1995**, *20*, 98–102. [CrossRef]
35. CAP 2023-27. Agriculture and Rural Development. Available online: https://agriculture.ec.europa.eu/common-agricultural-policy/cap-overview/cap-2023-27_en (accessed on 25 February 2025).

36. Cole, L.J.; Kleijn, D.; Dicks, L.V.; Stout, J.C.; Potts, S.G.; Albrecht, M.; Balzan, M.V.; Bartomeus, I.; Bebeli, P.J.; Bevk, D.; et al. A Critical analysis of the potential for EU Common Agricultural Policy measures to support wild pollinators on farmland. *J. Appl. Ecol.* **2020**, *57*, 681–694. [CrossRef]
37. Karamaouna, F.; Kati, V.; Economou, L.; Troyanos, G.; Samara, M.; Liberopoulou, S.; Barda, M.; Mitroiu, M.-D.; Edwards, M. Selected flowering plants as a habitat for pollinators and natural enemies in field margins of a watermelon crop—Implications for crop yield. *Int. J. Pest Manag.* **2024**, *70*, 920–936. [CrossRef]
38. Egan, P.A.; Dicks, L.V.; Hokkanen, H.M.T.; Stenberg, J.A. Delivering Integrated Pest and Pollinator Management (IPPM). *Trends Plant Sci.* **2020**, *25*, 577–589. [CrossRef] [PubMed]
39. Lundin, O.; Ward, K.L.; Williams, N.M. Identifying native plants for coordinated habitat management of arthropod pollinators, herbivores and natural enemies. *J. Appl. Ecol.* **2019**, *56*, 665–676. [CrossRef]
40. Lundin, O.; Rundlöf, M.; Jonsson, M.; Bommarco, R.; Williams, N.M. Integrated pest and pollinator management–expanding the concept. *Front. Ecol. Environ.* **2021**, *19*, 283–291. [CrossRef]
41. Decourtye, A.; Mader, E.; Desneux, N. Landscape enhancement of floral resources for honey bees in agro-ecosystems. *Apidologie* **2010**, *41*, 264–277. [CrossRef]
42. Liira, J.; Jürjendal, I. Are bees attracted by flower richness? implications for ecosystem service-based policy. *Ecol. Indic.* **2023**, *154*, 110927. [CrossRef]
43. Haaland, C.; Naisbit, R.E.; Bersier, L.-F. Sown wildflower strips for insect conservation: A review. *Insect Conserv. Divers.* **2011**, *4*, 60–80. [CrossRef]
44. Buhk, C.; Oppermann, R.; Schanowski, A.; Bleil, R.; Lüdemann, J.; Maus, C. Flower strip networks offer promising long term effects on pollinator species richness in intensively cultivated agricultural areas. *BMC Ecol.* **2018**, *18*, 55. [CrossRef]
45. Rowe, L.; Gibson, D.; Bahlai, C.A.; Gibbs, J.; Landis, D.A.; Isaacs, R. Flower traits associated with the visitation patterns of bees. *Oecologia* **2020**, *193*, 511–522. [CrossRef]
46. Cooley, H.; Vallejo-Marín, M. Buzz-Pollinated Crops: A global review and meta-analysis of the effects of supplemental bee pollination in tomato. *J. Econ. Entomol.* **2021**, *114*, 505–519. [CrossRef]
47. Cardinal, S.; Buchmann, S.L.; Russell, A.L. The evolution of floral sonication, a pollen foraging behavior used by bees (Anthophila). *Evolution* **2018**, *72*, 590–600. [CrossRef] [PubMed]
48. Amala, U.; Shivalingaswamy, T.M. Role of Native buzz pollinator bees in enhancing fruit and seed set in tomatoes under open field conditions. *J. Entomol. Zool. Stud.* **2017**, *5*, 1742–1744.
49. Toni, H.C.; Djossa, B.A.; Ayenan, M.A.T.; Teka, O. Tomato (*Solanum lycopersicum*) pollinators and their effect on fruit set and quality. *J. Hortic. Sci. Biotechnol.* **2021**, *96*, 1–13. [CrossRef]
50. El-Berry, A.A.; Moustafa, M.A.; Abdel-Gawaad, A.A.; El-Bialey, S. Pollinators other than honey bees visiting certain vegetable plants in Egypt. *Z. Angew. Entomol.* **1974**, *77*, 106–110. [CrossRef]
51. Shakeel, M.; Ali, H.; Ahmad, S.; Said, F.; Khan, K.A.; Bashir, M.A.; Anjum, S.I.; Islam, W.; Ghramh, H.A.; Ansari, M.J.; et al. Insect pollinators diversity and abundance in *Eruca sativa* Mill. (arugula) and *Brassica rapa* L. (field mustard) crops. *Saudi J. Biol. Sci.* **2019**, *26*, 1704–1709. [CrossRef]
52. Sihag, R.C. Pollination ecology of rocket (*Eruca vesicaria* (L.) Cav. ssp. sativa (Mill.) Thell) in the semi-arid environments of Northwest India: Native bees are the major pollinators. *Ecologies* **2023**, *4*, 580–594. [CrossRef]
53. Bendifallah, L.; Louadi, K.; Doumandji, S. Bee fauna potential visitors of coriander flowers *Coriandrum sativum* L. (Apiaceae) in the Mitidja area (Algeria). *J. Apic. Sci.* **2013**, *57*, 59–70. [CrossRef]
54. Amy, C.; Noël, G.; Hatt, S.; Uyttenbroeck, R.; Van de Meutter, F.; Genoud, D.; Francis, F. Flower strips in wheat intercropping system: Effect on pollinator abundance and diversity in Belgium. *Insects* **2018**, *9*, 114. [CrossRef]
55. Mena, G.T.; Gospodarek, J. White mustard, sweet alyssum, and coriander as insectary plants in agricultural systems: Impacts on ecosystem services and yield of crops. *Agriculture* **2024**, *14*, 550. [CrossRef]
56. Usman, M. Incidence of different insect visitors and their relative abundance associated with coriander (*Coriandrum sativum*) in district Charsadda. *Pure Appl. Biol.* **2018**, *7*, 539–546. [CrossRef]
57. Sasaki, H.; Wagatsuma, T. Bumblebees (Apidae: Hymenoptera) are the main pollinators of common buckwheat, *Fagopyrum esculentum*, in Hokkaido, Japan. *Appl. Entomol. Zool.* **2007**, *42*, 659–661. [CrossRef]
58. Campbell, J.W.; Irvin, A.; Irvin, H.; Stanley-Stahr, C.; Ellis, J.D. Insect visitors to flowering buckwheat, *Fagopyrum esculentum* (Polygonales: Polygonaceae), in North-Central Florida. *Fla. Entomol.* **2016**, *99*, 264–268. [CrossRef]
59. Denys, C.; Tscharntke, T. Plant-insect communities and predator-prey ratios in field margin strips, adjacent crop fields, and fallows. *Oecologia* **2002**, *130*, 315–324. [CrossRef]
60. Balzan, M.V.; Bocci, G.; Moonen, A.-C. Augmenting flower trait diversity in wildflower strips to optimise the conservation of arthropod functional groups for multiple agroecosystem services. *J. Insect Conserv.* **2014**, *18*, 713–728. [CrossRef]

61. Petanidou, T. Introducing plants for bee-keeping at any cost?—Assessment of *Phacelia tanacetifolia* as nectar source plant under xeric Mediterranean conditions. *Plant Syst. Evol.* **2016**, *238*, 155–168. [CrossRef]
62. Sowmya, K.S. Effect of number of *Apis cerana* visits on quantitative parameters in nectar plant, *Ammi majus* L. *Insect Environ.* **2012**, *18*, 15.
63. Mavromoustakis, G.A. The bees (Hymenoptera, Apoidea) of Cyprus. *Ann. Mag. Nat. Hist.* **1957**, *10*, 321–337. [CrossRef]
64. Balzan, M.V.; Moonen, A.-C. Field margin vegetation enhances biological control and crop damage suppression from multiple pests in organic tomato fields. *Entomol. Exp. Appl.* **2014**, *150*, 45–65. [CrossRef]
65. Pellissier, M.E.; Jabbour, R. Herbivore and parasitoid insects respond differently to annual and perennial floral strips in an alfalfa ecosystem. *Biol. Control* **2018**, *123*, 28–35. [CrossRef]
66. Rizzo, M.C.; Massa, B. Ecology of the Eulophid parasitoid community living on hosts of spontaneous flora linked to citrus grove (Hymenoptera: Chalcidoidea: Eulophidae). In *Parasitic Wasps: Evolution, Systematics, Biodiversity and Biological Control*; Melika, G., Thuroczy, C., Eds.; Agroinform: Budapest, Hungary, 2002; pp. 351–361.
67. Denis, C.; Riudavets, J.; Gabarra, R.; Molina, P.; Arnó, J. Selection of insectary plants for the conservation of biological control agents of aphids and thrips in fruit orchards. *Bull. Entomol. Res.* **2021**, *111*, 517–527. [CrossRef]
68. Stathakis, T.; Economou, L.; Barda, M.; Angelioudakis, T.; Kati, V.; Karamaouna, F. Potential of hedgerows with aromatic plants as reservoirs of natural enemies of pests in orange orchards. *Insects* **2023**, *14*, 391. [CrossRef]
69. Kishinevsky, M.; Keasar, T.; Bar-Massada, A. Parasitoid abundance on plants: Effects of host abundance, plant species, and plant flowering state. *Arthropod Plant Interact.* **2017**, *11*, 155–161. [CrossRef]
70. Yefremova, Z. Order Hymenoptera, family Eulophidae. *Arthr. Fauna UAE* **2008**, *1*, 345–360.
71. Gebiola, M.; Bernardo, U.; Ribes, A.; Gibson, G.A.P. An integrative study of *Necremnus* Thomson (Hymenoptera: Eulophidae) associated with invasive pests in Europe and North America: Taxonomic and ecological implications. *Zool. J. Linn. Soc.* **2015**, *173*, 352–423. [CrossRef] [PubMed]
72. Izquierdo, J.J.; Solans, P.; Vitalle, J. Parasitoides y depradadores de *Helicoverpa armigera* (Hübner) en cultivos de tomate para consume en fresco. *Bol. San. Veg. Plagas* **1994**, *20*, 521–530.
73. Denis, C.; Riudavets, J.; Alomar, O.; Agustí, N.; Gonzalez-Valero, H.; Cubí, M.; Matas, M.; Rodríguez, D.; van Achterberg, K.; Arnó, J. Naturalized *Dolichogenidea gelechiidivoris* complement the resident parasitoid complex of *Tuta Absoluta* in North-Eastern Spain. *J. Appl. Entomol.* **2022**, *146*, 461–464. [CrossRef]
74. Ivezić, A.; Popović, T.; Trudić, B.; Krndija, J.; Barošević, T.; Sarajlić, A.; Stojačić, I.; Kuzmanović, B. Biological control agents in greenhouse tomato production (*Solanum lycopersicum* L.): Possibilities, challenges and policy insights for Western Balkan region. *Horticulturae* **2025**, *11*, 155. [CrossRef]
75. Yang, Z.-Q.; Yao, Y.-X.; Qiu, L.-F.; Li, Z.-X. A new species of *Trissolcus* (Hymenoptera: Scelionidae) parasitizing eggs of *Halyomorpha halys* (Heteroptera: Pentatomidae) in China with comments on its biology. *Ann. Entomol. Soc. Am.* **2009**, *102*, 39–47. [CrossRef]
76. Gard, B.; Bout, A.; Pierre, P. Release strategies of *Trissolcus basalis* (Scelionidae) in protected crops against *Nezara viridula* (Pentatomidae): Less is more. *Crop Prot.* **2022**, *161*, 106069. [CrossRef]
77. Balzan, M.V.; Wäckers, F.L. Flowers to selectively enhance the fitness of a host-feeding parasitoid: Adult feeding by *Tuta absoluta* and its parasitoid *Necremnus artynes*. *Biol. Control* **2013**, *67*, 21–31. [CrossRef]
78. Arnó, J.; Oveja, M.F.; Gabarra, R. Selection of flowering plants to enhance the biological control of *Tuta absoluta* using parasitoids. *Biol. Control* **2018**, *122*, 41–50. [CrossRef]
79. McIntosh, H.R.; Skillman, V.P.; Galindo, G.; Lee, J.C. Floral resources for *Trissolcus japonicus*, a parasitoid of *Halyomorpha halys*. *Insects* **2020**, *11*, 413. [CrossRef] [PubMed]
80. Huber, J.T. Systematics, biology, and hosts of the Mymaridae and Mymarommatidae (Insecta: Hymenoptera): 1758–1984. *Entomography* **1986**, *4*, 185–243.
81. Trdan, S.; Andjus, L.; Raspudić, E.; Kač, M. Distribution of *Aeolothrips intermedius* Bagnall (Thysanoptera: Aeolothripidae) and its potential prey Thysanoptera species on different cultivated host plants. *J. Pest. Sci.* **2005**, *78*, 217–226. [CrossRef]
82. Riley, D.; Sparks, A.; Srinivasan, R.; Kennedy, G.; Fonsah, G.; Scott, J.; Olson, S. Chapter 3—Thrips: Biology, ecology, and management. In *Sustainable Management of Arthropod Pests of Tomato*; Wakil, W., Brust, G.E., Perring, T.M., Eds.; Academic Press: San Diego, CA, USA, 2018; pp. 49–71. ISBN 978-0-12-802441-6.
83. Abenaim, L.; Bedini, S.; Greco, A.; Giannotti, P.; Conti, B. Predation capacity of the banded thrips *Aeolothrips intermedius* for the biological control of the onion thrips *Thrips tabaci*. *Insects* **2022**, *13*, 702. [CrossRef]
84. Mouratidis, A.; De Lima, A.P.; Dicke, M.; Messelink, G.J. Predator-prey interactions and life history of *Orius laevigatus* and *O. majusculus* feeding on flower and leaf-inhabiting thrips. *Biol. Control* **2022**, *172*, 104954. [CrossRef]
85. Venzon, M.; Janssen, A.; Sabelis, M.W. Prey Preference and reproductive success of the generalist predator *Orius laevigatus*. *Oikos* **2002**, *97*, 116–124. [CrossRef]

86. Harwood, J.D.; Desneux, N.; Yoo, H.J.S.; Rowley, D.L.; Greenstone, M.H.; Obrycki, J.J.; O'neil, R.J. Tracking the role of alternative prey in soybean aphid predation by *Orius insidiosus*: A molecular approach. *Mol. Ecol.* **2007**, *16*, 4390–4400. [CrossRef]
87. Arnó, J.; Roig, J.; Riudavets, J. Evaluation of *Orius majusculus* and *O. laevigatus* as predators of *Bemisa tabaci* and estimation of their prey preference. *Biol. Control* **2008**, *44*, 1–6. [CrossRef]
88. Patrzich, R.; Klumpp, M. Vergleich Der phytophagen und räuberischen thripse (Thysanoptera) auf unterschiedlich bewirtschafteten weizenfeldern in Hessen/Comparison of phytophagous and predatory thrips (Thysanoptera) in winter wheat fields in Hesse with different systems of management. *J. Plant Dis. Prot.* **1991**, *98*, 464–470.
89. Coll, M.; Guershon, M. Omnivory in terrestrial arthropods: Mixing plant and prey diets. *Annu. Rev. Entomol.* **2002**, *47*, 267–297. [CrossRef]
90. Gruss, I.; Twardowski, J.P.; Cierpisz, M. The effects of locality and host plant on the body size of *Aeolothrips intermedius* (Thysanoptera: Aeolothripidae) in the Southwest of Poland. *Insects* **2019**, *10*, 266. [CrossRef]
91. Pumariño, L.; Alomar, O.; Lundgren, J.G. Effects of floral and extrafloral resource diversity on the fitness of an omnivorous bug, *Orius insidiosus*. *Entomol. Exp. Appl.* **2012**, *145*, 181–190. [CrossRef]
92. Hinds, J.; Barbercheck, M.E. Diversified floral provisioning enhances performance of the generalist predator, *Orius insidiosus* (Hemiptera: Anthocoridae). *Biol. Control* **2020**, *149*, 104313. [CrossRef]
93. Huseynov, E.F. oglu Natural prey of the crab spider *Runcinia Grammica* (Araneae: Thomisidae) on *Eryngium* plants. *Bull. Br. Arachnol. Soc.* **2007**, *14*, 93–96. [CrossRef]
94. Morse, D.H. Prey capture by the crab spider *Misumena vatia* (Clerck) (Thomisidae) on three common native flowers. *Am. Midl. Nat.* **1981**, *105*, 358–367. [CrossRef]
95. Welti, E.A.R.; Putnam, S.; Joern, A. *Crab spiders* (Thomisidae) attract insect flower-visitors without UV signalling. *Ecol. Entomol.* **2016**, *41*, 611–617. [CrossRef]
96. De Backer, L.; Caparros Megido, R.; Haubruge, É.; Verheggen, F. *Macrolophus pygmaeus* (Rambur) as an efficient predator of the tomato leafminer *Tuta absoluta* (Meyrick) in Europe. A Review. *Biotechnol. Agron. Soc. Environ.* **2014**, *18*, 536–543.
97. Obrycki, J.J.; Kring, T.J. Predaceous Coccinellidae in biological control. *Annu. Rev. Entomol.* **1998**, *43*, 295–321. [CrossRef]
98. Magro, A.; Hemptinne, J.L. The pool of Coccinellids (Coleoptera: Coccinellidae) to control Coccids (Homoptera: Coccoidea) in Portuguese citrus groves. *Bol. San. Veg. Plagas* **1999**, *25*, 311–320.
99. Bogya, S. Spiders (Araneae) as Polyphagous Natural Enemies in Orchards. Ph.D. Thesis, Wangenigen University & Research, Wangenigen, The Netherlands, 1999.
100. Altieri, M.A.; Nicholls, C.I. Vegetational designs to enhance biological control of insect pests in agroecosystems. In *Natural Enemies of Insect Pests in Neotropical Agroecosystems: Biological Control and Functional Biodiversity*; Souza, B., Vázquez, L.L., Marucci, R.C., Eds.; Springer International Publishing: Cham, Switzerland, 2019; pp. 3–13. ISBN 978-3-030-24733-1.
101. Silva-Neto, C.M.; Bergamini, L.L.; Elias, M.A.S.; Moreira, G.L.; Morais, J.M.; Bergamini, B.A.R.; Franceschinelli, E.V. High species richness of native pollinators in Brazilian tomato crops. *Braz. J. Biol.* **2016**, *77*, 506–513. [CrossRef] [PubMed]
102. Bashir, M.A.; Alvi, A.M.; Khan, K.A.; Rehmani, M.I.A.; Ansari, M.J.; Atta, S.; Ghramh, H.A.; Batool, T.; Tariq, M. Role of pollination in yield and physicochemical properties of tomatoes (*Lycopersicon esculentum*). *Saudi J. Biol. Sci.* **2018**, *25*, 1291–1297. [CrossRef] [PubMed]
103. Zhang, H.; Han, C.; Breeze, T.D.; Li, M.; Mashilingi, S.K.; Hua, J.; Zhang, W.; Zhang, X.; Zhang, S.; An, J. Bumblebee pollination enhances yield and flavor of tomato in Gobi Desert greenhouses. *Agriculture* **2022**, *12*, 795. [CrossRef]
104. Roohanitaziani, R.; de Maagd, R.A.; Lammers, M.; Molthoff, J.; Meijer-Dekens, F.; van Kaauwen, M.P.W.; Finkers, R.; Tikunov, Y.; Visser, R.G.F.; Bovy, A.G. Exploration of a resequenced tomato core collection for phenotypic and genotypic variation in plant growth and fruit quality traits. *Genes* **2020**, *11*, 1278. [CrossRef]

agronomy

Article

# Variable Transect Method Outperformed in Sampling Hymenopteran Flower Visitors in *Brassica campestris* L. var. *toria* Ecosystem

**Arup Kumar Sarma [1,*], Borsha Neog [2], Mukul Kumar Deka [3], Alin Carabet [4] and Ramona Stef [4,*]**

1 AAU-Zonal Research Station, Shillongani, Assam Agricultural University, Nagaon 782002, Assam, India
2 Department of Statistics, Assam Agricultural University, Jorhat 785013, Assam, India; borsha.neog@aau.ac.in
3 Department of Entomology, Assam Agricultural University, Jorhat 785013, Assam, India; mukul.deka@aau.ac.in
4 Faculty of Agriculture, University of Life Sciences "King Mihai I" from Timisoara, Calea Aradului 119, 300645 Timisoara, Romania; alincarabet@usvt.ro
* Correspondence: arup.sarma@aau.ac.in (A.K.S.); ramonastef@usvt.ro (R.S.)

**Abstract:** *Brassica campestris* L. var. *toria*, a major oilseed crop cultivated in India, is primarily an entomophilic species. Hymenopteran flower-visiting species provide important ecological services like pollination or pest control in *Brassica* crops. In this context, a study was conducted during 2015–2017 in three localities in Assam, a state in northeast India that falls under two global biodiversity hotspots—Indo–Burma and Himalayan—to bring data on the diversity of hymenopteran flower visitors of toria crops by using multiple sampling techniques and to compare the efficiency of these techniques. Altogether, nine sampling treatments were used. To assess the sampling effectiveness of the different treatments, the data from the two cropping periods of toria in each locality were analysed cumulatively and comparatively. Variable transect outperformed the other sampling methods with the highest number of hymenopteran flower visitor species recorded in toria crops at 54, representing 84.4% of the total number of species, and was followed by standard transect (34 species, 53.1%), elevated yellow trap (22 species, 34.4%), and observation plot (21 species, 32.8%). However, the importance of multiple sampling methods in this diversity study was noticed; one method alone could not sample all the species recorded. The cluster of traps and netting with transect walks was proven to be complementary and considered useful for future research studies in the upstream basin of the Burhidihing River of Assam, India.

**Keywords:** Assam; *Brassica campestris* L.; hymenopteran; diversity; pollinator; sampling method

## 1. Introduction

Thirty-five percent of the global production of crops, including at least 800 cultivated plants, depends on animal pollination [1,2]. Pollinators play a crucial role in flowering plant reproduction and in the production of most fruits and vegetables; without the assistance of these pollinators, most plants could not reproduce [3–5]. The importance of pollinators has also been reported from an economic point of view, so that in 2009, the economic benefit was EUR 153 billion and represented 9.5% of the world's agricultural economic production [6,7]. In 2016, a report by IPBES [8] estimated that pollinators annually determine an added value of USD 235–577 billion to global food production.

Previous statements show that animal-mediated pollination represents one of the most important biotic interactions in terrestrial ecosystems, being essential both for their functioning and for the conservation of biodiversity [7].

In terrestrial ecosystems, plant pollination can also be accomplished by abiotic factors, such as wind and water [9]. The most widespread types of pollination are anemophilic and entomophilic. Anemophilic pollination predominates in dense plant communities such as grasslands and forests [10], while insect pollination is dominant in areas of high biodiversity [11].

*Brassica campestris* L. var. *toria* is an important oilseed crop that requires insect pollinators to ensure its reproductive success [12]. The pollination of *Brassica campestris* var. *toria* is mainly entomophilic, being an attractive crop for insects and providing abundant pollen and nectar [13,14], although there are indications of a secondary contribution of anemophilous pollination under specific climatic conditions [15]. *Brassica campestris* L. is a xenogamous species, almost self-sterile, and produces more seeds when cross-pollination occurs [16,17]. The yield of this crop is strongly influenced by the abundance and diversity of floral visitors [17]. According to Brittain et al. [18], pollinating insects determine an annual production increase of 580 million tons in oilseed plants, and this statement is also endorsed by subsequent studies by Bandenes-Perez et al. and Woodcock et al. [17,19]. The quality and quantity of *Brassica campestris* L. var. *toria* production is influenced by the diversity and abundance of floral visitors, which in turn depend on the structure and composition of the local pollinator community [20–24]. In the success of pollination of agricultural crops, in addition to the diversity and abundance of floral visitors, it is important to know the functional differences between species in facilitating pollination [19,25]. Studies conducted to date support that pollination with significant economic impact is carried out by a relatively small number of species [19,26–28]. Therefore, the stable and efficient pollination of agricultural crops in different climatic and ecological contexts is ensured by the functional diversity of pollinators [18,29].

In the ecosystem of *Brassica campestris* var. toria, the main entomophilic species belongs to the orders Hymenoptera and Diptera due to their efficiency in pollen transport [30]. An important role in the pollination of agroecosystems is played by wild pollinators, which provide essential pollination services and represent a natural biological reserve, offering protection and compensation in the case of decline in pollinator communities due to pests and diseases [31].

The species-level identification of pollinators is extremely important, especially of hymenopterans, as they show differences in feeding behaviour and floral preferences [32]. Understanding the specific relationships between plants and pollinators is essential in developing effective conservation strategies and sustainable agricultural management [33].

Studies regarding the pollinators in the *Brassica campestris* var. *toria* agroecosystem using different sampling methods have been conducted by Sarma et al., Taba et al., Westphal et al., Potts et al., and Sarma et al. [14,15,34–40]. The sampling methods used by researchers for pollinating insects of *Brassica campestris* var. *toria* included direct observation, manual capture by sweep nets, coloured traps, Malaise traps, etc. The data present in the literature indicate that for a comprehensive understanding of ecological relationships involved in the reproduction process of *Brassica campestris* var. *toria*, it is mandatory to combine multiple sampling methods [38].

Methods for sampling pollinator ecological assemblages seek to be efficient, repeatable, and representative; there is a concern that common methods have their limits in terms of revealing species function and so have less value for comparative studies [39].

The sampling methods frequently used in ecosystems and agroecosystems are coloured pan traps, Malaise traps, and direct observations (transects).

Malaise traps are recommended for biodiversity studies due to their ability to capture a wide range of flying insects, rare and hard to detect species, over a long period of time [40,41]. The use of these traps offers the advantage of continuous sampling without researcher interventions, generating a lot of data about biodiversity. However, these traps also have several disadvantages, such as high costs, and the setup and capture of non-pollinator insects require additional effort for sorting and specimen identification, leading to further increased costs [39–41].

The method of direct observation (transect) provides data on pollinator behaviour, floral preferences, and interactions between species [42]. Walking transects can have different sizes, from 100 m × 1 m to 250 m × 4 m, and are completed in 20 to 50 min. These are considered variable when the position and direction of the transects are randomly selected during sampling or are fixed/standard when the same point is visited repeatedly during monitoring [43]. However, the term "variable" is interpreted differently by Westphal et al. [34] and Nielsen et al. [44]: both use the term for a second transect performed in a 1 ha area, where the researcher can move freely among flower patches for about 30 min. This has the significant advantage of allowing for a detailed assessment of insects' behaviour in their natural habitat, providing data on flower visit frequency and the specific preferences of pollinators [39]. The disadvantage of this method is determined by a long observation time and the involvement required from researchers, which can introduce subjectivity in observations and affect the results [40].

The most used sampling method is represented by the use of coloured pan traps filled with soapy water [45–47], whether accompanied or not by collecting flowers with a net [48]. When comparing the two methods, even though they were found to capture similar richness and abundance, the assemblages differed, particularly in relation to aspects of pollination function [39]. The frequent use of coloured traps relies on low costs, ease of use, and reproducibility, and can cover large areas simultaneously [49]. One disadvantage of pan traps is that they do not offer behavioural data with respect to insect–plant interactions [50]. The selection of the sampling method represents an important and necessary component in determining relevant species in an area [44]; otherwise, the collected specimens do not represent the groups of insects present in an ecosystem or the rare collection of a group of insects, which is due to the lack of an efficient collecting technique for this group [51]. The use of a variety of pan colours has been advocated by Gollan et al. [52] for sampling overall bee biodiversity, but specific colours may be more effective when targeting certain groups or species [50]. A literature review that compares six sampling methods (Malaise traps, pan traps, bait, sweep nets, timed observations, and aspirators) for bee species populations found little consensus regarding which method would be most reliable for sampling multiple species in tropical forests and agroecosystems [53]. The complementary use of several techniques allows for the minimising of individual limits and ensures a proper picture of a pollinator community [45]. Moreover, a key area for future pollination research is to more deeply understand the roles played by the full complement of pollinators [54,55].

Published research highlights the existence of pressures on pollinator communities, so their effective identification is becoming increasingly important. Pollinator diversity and abundance are affected by habitat degradation, excessive pesticide use, climate change, and land use change [56–58]. Studies by Landanverde-Gonzalez et al. [7] show that in traditional, low-intensity milpa agriculture in tropical Mexico, pan traps and transect walk methods of sampling were found to be similarly efficient for pollinator diversity analyses, although each collected different components of the bee community, emphasising the fact that bee species richness and abundance were negatively corelated with the area of chili crops due to agricultural intensification while being positively related to the amount of forest cover. Millard et al. [59] highlight that the selection of sampling methods is essential in biodiversity studies and recommend the use of standardised and complementary methods in ecosystems affected by anthropogenic activities.

Although there are many studies on pollinator diversity and decline, there is a lack of research that comparatively evaluates the efficiency of sampling methods in agroecosystems, especially in India. Therefore, this present study aimed to compare the effectiveness of nine methods used for species monitoring of hymenopteran floral visitors of *Brassica campestris* L. var. *toria* crops under the particular conditions of the Assam state area in India in order to select the methods with the greatest potential to be used in Hymenoptera biodiversity-related studies.

## 2. Materials and Methods

This study was conducted in October 2015–March 2016 and October 2016–March 2017 in the state of Assam in India in an area that is within two global biodiversity hotspots: Indo–Burma and the Himalayas [38].

### 2.1. *Study Area*

Dibrugarh District is in the Oriental region near the border of the Palearctic region. This study was conducted in farmers' fields in three villages, viz., Jajimukh (GPS: N 27°17′12.3″; E 94°50′09.0″, recorded using GPS Map Camera mobile app for Android), Lejai Panimirigaon (GPS: N 27°17′06.9″; E 94°47′08.5″), and Kutuha-Bhagamur (GPS: N 27°16′49.6″; E 94°49′58.0″) in Dibrugarh District of the Upper Brahmaputra Valley Zone (UBVZ) of Assam, India. The localities where the study was carried out are located near the Burhidihing River, which is prone to flooding during the monsoon season.

The study area is characterised by a humid subtropical climate and is influenced by southwest monsoons, with an annual mean precipitation of 2518.3 mm [60]. Climate conditions vary accordingly to the three seasons: summer (March–May—characterised by a mean temperature of 26–38 °C, with reduced precipitations), monsoon (June–September—mean temperature from 25 °C to 32 °C, with abundant precipitation—mean of 1731.6 mm/season), and winter (October–February—mean temperature 6–25 °C, with low level of precipitations (42.2 mm mean/season)) [60].

The *Brassica campestris* var. *toria* crop ranks first in both area and production among oilseed crops grown in Assam, and is grown on hundreds of hectares near the Burhidihing River. For this present study, all sampling sites were selected in *Brassica campestris* var. *toria* crop fields positioned on both banks of the Burhidihing River. The distance between the experimental fields did not exceed 3 km. No managed honeybee colonies were available in the selected villages. No insecticide was applied to the crops at any stage.

### 2.2. *The Agroecosystem of Brassica campestris var. toria*

In 2015, two sampling sites were selected in each locality, one with *Brassica campestris* var. *toria* crops sown in mid-October and one with *Brassica campestris* var. *toria* sown at the beginning of December. In 2016, all sites were set up in the same areas selected the previous year.

#### 2.2.1. Normal Cropping Period for *Brassica campestris* var. *toria*

The crop variety that was sown in mid-October was TS 36, produced and recommended by the Agricultural University of Assam, Jorhat, India [61]. The TS 36 variety has a vegetation period of 90–95 days and is suitable for late sowing from 15 October to 15 November. This variety can be cultivated after rice, normally achieving yields up to 1200 kg/ha, and is tolerant to drought stress [62]. This cropping system of *Brassica campestris* var. *toria* is common in hundreds of hectares in the uplands near the Burhidihing River. The peak flowering period was between mid-November and mid-January.

#### 2.2.2. Late Cropping Period for *Brassica campestris* var. *toria*

The crop variety that was sown in the first week of December was TS-67 [62]. It needs 90–95 days to reach maturity, medium height, and a small number of primary brunches, and exerts a high production capability up to 1000–1200 kg/ha when sown in an optimal period and up to 700–800 kg/ha when sown up to the first week of December [61]. This variety is well known for its capacity to attract pollinator insects due to the high quantity of nectar and pollen production [38].

The TS-67 variety was sown after the rice harvest in the first week of December. The peak flowering period was from mid-January to mid-March.

### 2.3. *Sampling Methodology*

#### 2.3.1. Period of Observation and Field Data Collection

Pollinator data collection in the *Brassica campestris* var. *toria* ecosystem was performed eight times at one-week intervals at each site during the crop flowering period.

#### 2.3.2. Methods and Sampling Design

The methodology and sampling design were adapted from those described by Belavadi and Ganeshaiah [63], Nielsen et al. [44], and Westphal et al. [34] as follows:

##### Ground Bowl Traps (GBTs)

At each study site, 30 coloured bee bowls in clusters of three colours (yellow, white, and blue) at ground level at 15 m intervals were used by following the methodology of Belavadi and Ganeshaiah [63].

##### Elevated Bowl Traps (EBTs)

In each of the six experimental sites, 30 bowl traps were set up in 10 clusters separated by 15 m. Each cluster contained three bowl traps (white, yellow, and blue). These colours accounted for the different colour preferences among bee species, increasing the efficiency of the trap clusters in attracting bees [45,47]. Each trap was filled with 200 mL of water and a few drops of liquid detergent. The pan traps were mounted on a pole at the vegetation height and left active for 48 h during each of the eight rounds of sampling conducted in a cropping period.

For both EBT and GBT sampling, the same bowl type with a 16 cm inner diameter × 7 cm inner depth was used.

##### Observation Plots (OP)

In each study site, ten rectangular (1 m × 2 m) observation plots were randomly established and marked. Direct observations at each plot lasted 6 min. Each observed hymenopteran species was collected with a sweeping net for further identification. Plot observations were randomised at each visit to ensure data accuracy.

##### Standardised Transect Walks (ST)

A permanently marked 250 m-long and 4 m-wide corridor (transect) was divided into 10 plots, each 25 m long, and used for the standardised transect walk. The sub-plots were marked permanently with a bamboo stick before crop flowering. Each sub-plot was surveyed for 5 min, during which all hymenopteran species visiting the toria flowers were collected by netting and preserved for further identification.

The methods described in Sections 2.3.1 and 2.3.2 were performed at each experimental site on a 1 ha plot delimited inside the greater crop surface.

Variable Transect Walks (VT)

For this method, an adjacent 1 ha plot was selected, located in the same crop for each experimental site. These plots were surveyed by slowly walking around on variable line transects for 30 min, while all specimens of flower visitors observed were collected by netting.

Sampling in the case of OP, ST, and VT methods was performed randomly, at different intervals of the day (from 10 a.m. to 4 p.m.) with favourable weather conditions (temperature above 15 °C, no rain, no wind, and cloud cover of less than 50%).

Altogether, the nine sampling methods, considered here as treatments (T1 to T9), were T1—YTG (yellow trap in the ground); T2—BTG (blue trap in the ground); T3—WTG (white trap in the ground); T4—YTE (yellow trap elevated to canopy); T5—BTE (blue trap elevated to canopy); T6—WTE (white trap elevated to canopy); T-7—OP (observation plot); T8—ST (standardised transect walk) and T9—VT (variable transect walk).

The samples consisted of hymenopteran specimens collected at every site, and treatments at various sampling moments were stored in ethanol, transferred to a laboratory, and further processed for taxonomic segregation.

### 2.4. Statistical Analysis of Data

The efficiency of each sampling method was evaluated based on the number of species of flower visitors sampled. To assess the sampling effectiveness of the different methods, the data from the two cropping periods were analysed cumulatively and comparatively using SPSS software (version 20). To compare the mean differences between treatments, Tukey's HSD test was used. ANOVA was performed to assess differences in the efficiency of the seven non-transect sampling methods, while a *t*-test was performed for the two transect methods. For the comparison of non-transect methods (YTG, BTG, WTG, YTE, BTE, WTE, and OP), the Kruskal–Wallis test was applied as a non-parametric method used when there are more than two independent groups.

## 3. Results

### Efficiency of All Sampling Methods

Overall, the species of hymenopteran floral visitors in the *Brassica campestris* var. *toria* agroecosystem collected during the study period were classified into nine families: Apidae (fifteen species), Halictidae (nine species), Vespidae (eighteen species), Megachilidae (eight species), Crabronidae (five species), Andrenidae (two species), Sphecidae (four species), Scoliidae (two species) and Colletidae (one species), totalling 64 species (Table 1).

The families Vespidae (eighteen species), Apidae (fifteen species), and Halictidae (nine species) recorded the highest species richness. Colletidae, Scoliidae, and Sphecidae families were present in the toria crop in a small number of species (Table 1).

The results regarding the use of sampling methods for hymenopteran flower visitors in toria crops varied in terms of the number of species caught (Table 1). Treatment 9 (VT) was found to be the most efficient method of sampling, with the highest number of species recorded, 54,—84.4% of the total number of species caught. It can be noted that VT was the only method by which all the pollinator species belonging to Andrenidae, Scoliidae, and Colletidae were collected. In terms of the number of species recorded per sampling method, treatment 9 (VT) was followed by treatment 8 (ST) (with 34 species), treatment 4 (YTE) (22 species), and treatment 7 (OP) (21 species). Notably, these four methods also had higher trends of catching Vespidae and Apidae—the dominant families of Hymenoptera in the toria crop ecosystem. For other families, viz., Megachilidae, Halictidae, and Andrenidae, the extent of catches recorded by two non-transect methods, treatment 4 (YTE) and treatment 7 (OP), was also relatively high.

**Table 1.** The number of species of hymenopteran flower visitors sampled using different methods in toria ecosystem of Assam, India (2015–2017).

| Sl. No. | Family | Total No. of Species Sampled | No. of Species Sampled by Different Sampling Techniques | | | | | | | | |
|---|---|---|---|---|---|---|---|---|---|---|---|
| | | | YTG (T1) | BTG (T2) | WTG (T3) | YTE (T4) | BTE (T5) | WTE (T6) | OP (T7) | ST (T8) | VT (T9) |
| 1 | Apidae | 15 | 5 (33.3) | 3 (20.0) | 4 (26.7) | 8 (53.3) | 3 (20.0) | 4 (26.7) | 7 (46.7) | 10 (66.7) | 13 (86.7) |
| 2 | Andrenidae | 02 | 1 (50.0) | 1 (50.0) | 1 (50.0) | 1 (50.0) | - | - | 1 (50.0) | 1 (50.0) | 2 (100.0) |
| 3 | Megachilidae | 08 | 1 (12.5) | - | - | 4 (50.0) | - | 1 (12.5) | 3 (37.5) | 5 (62.5) | 7 (87.5) |
| 4 | Vespidae | 18 | - | - | - | 3 (18.8) | - | - | 3 (18.8) | 8 (44.4) | 16 (88.9) |
| 5 | Scoliidae | 02 | - | - | - | - | - | - | - | 1 (50.0) | 2 (100.0) |
| 6 | Halictidae | 09 | 4 (44.4) | 3 (33.3) | 3 (33.3) | 5 (55.6) | 2 (22.2) | 2 (22.2) | 5 (55.6) | 5 (55.6) | 7 (77.8) |
| 7 | Crabronidae | 05 | 1 (20.0) | - | 1 (20.0) | 1 (20.0) | 1 (20.0) | - | 1 (20.0) | 3 (60.0) | 4 (80.0) |
| 8 | Colletidae | 01 | - | - | - | - | - | - | - | - | 1 (100.0) |
| 9 | Sphecidae | 04 | - | - | - | - | - | - | 1 (25.0) | - | 3 (75.0) |
| Total | | 64 | 12 | 07 | 09 | 22 | 06 | 07 | 21 | 34 | 54 |
| % species sampled | | | 18.8 | 10.9 | 14.1 | 34.4 | 9.4 | 10.9 | 32.8 | 53.1 | 84.4 |
| % deviation from mean (=19.1) | | | −37.2 | −63.4 | −52.9 | +15.2 | −68.6 | −63.4 | +9.9 | +78.0 | +182.7 |

* Value within parentheses is the percentage of species of the corresponding family as sampled by the method under test.

The proportionate comparisons among different hymenopteran families across the sampling methods can be perceived from Figure 1 (radar chart), while the data dispersion and variation among sampling methods are highlighted in Figure 2 (box plot).

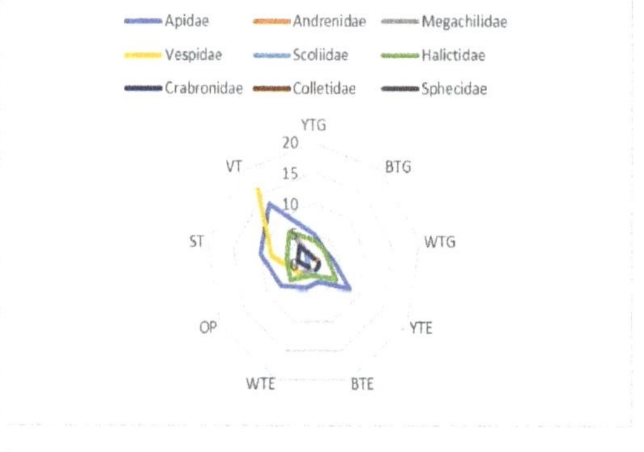


**Figure 1.** Radar chart showing the proportionate comparisons among different hymenopteran families across nine sampling methods.

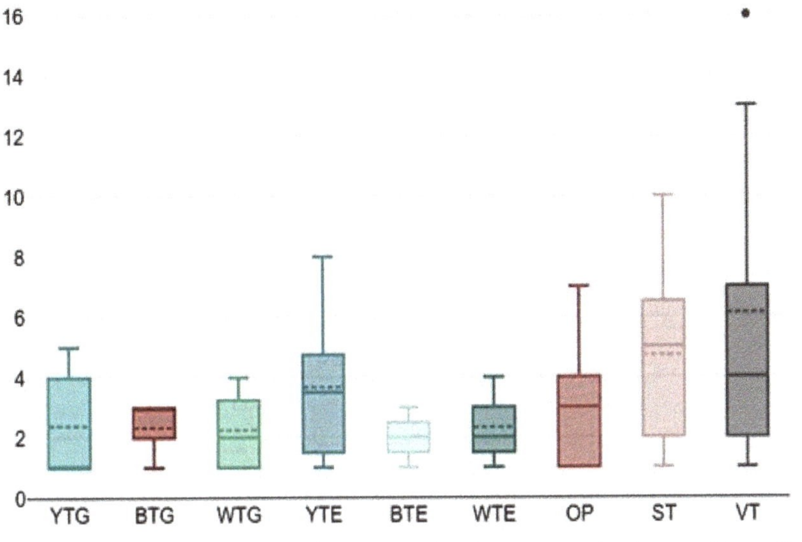


**Figure 2.** Boxplot showing the data dispersion and variation among the nine sampling methods.

The radar chart provides a comparative visualisation of the species richness of the nine Hymenoptera families across data from different sampling treatments. This graphical representation highlights clear differences in hymenopteran family distribution and suggests variation in detection efficiency among sampling methods.

Among all families, Apidae exhibits the highest representation across most treatments, particularly in VT and ST, and the elevated values recorded in these treatments indicate that these methods are particularly effective in capturing the dominant pollinator group.

Vespidae also shows a pronounced peak in the VT transect, suggesting that the method might be more sensitive to detecting fast-moving or aggressive taxa. Similarly, Halictidae (a family often composed of smaller, ground-nesting bees) was detected in a moderate number of species in VT and ST but much less so in other sampling treatments, further supporting the higher efficacy of these two methods for detecting a wider array of taxa.

In contrast, families such as Andrenidae, Megachilidae, Colletidae, and Crabronidae appear in lower species richness across all treatments, potentially indicating their low natural abundance in the study area. This underrepresentation may also reflect habitat preferences or nesting behaviours that are less compatible with the sampling techniques used.

Analysing the boxplot from Figure 2, showing central tendencies of dispersion but also extreme values, it can be observed that the hymenopteran species caught in toria crops exhibits a numerical variation for each method. The use of transect methods showed the highest variation in the number of species collected, suggesting a higher capacity for collecting a wide range of entomofauna diversity. The use of treatment 9 (VT) has recorded very significant differences; out of a total of sixteen species, the species richness reached a mean of six species, suggesting a high capturing potential but with highly variable values from one plot to another.

Assessing treatment 5 (BTE) and treatment 2 (BTG) with respect to floral visitors captured in the toria agroecosystem, the data highlighted the stability of the results regarding the number of species caught but showed lower results in terms of species richness. Treatment 4 (YTE) and treatment 7 (OP) offered balanced results regarding hymenopteran species diversity.

A comparative analysis (Kruskal–Wallis test) of the nine treatments highlighted that there are significant differences regarding sampling methods used to capture hymenopteran species (Table 2). The number of collected species in treatment 9 (VT) was very significantly higher than the number of species collected by some passive (non-transect) methods like treatment 2 (BTG) (p-adj. = 0.031), treatment 5 (blue elevated traps—BTE) (p-adj. = 0.025), and treatment 6 (white elevated traps—WTE) (p-adj. = 0.032). These significant differences between treatment 9 (VT) (64.83) and treatment 2 (BTG) (28.78) and treatment 6 (WTE) (29.56) are also highlighted in Figure 3.

**Table 2.** Pairwise comparison of sampling methods used to collect hymenopterans in *Brassica campestris* var. *toria* agroecosystem (Kruskal–Wallis test).

| Pair 1–Pair 2 | Test Statistic | Std. Error | Std. Test Statistic | Sig. | Adj. Sig. |
|---|---|---|---|---|---|
| BTG–VT | −35.333 | 10.617 | −3.328 | 0.001 | 0.031 |
| BTE–VT | −36.056 | 10.617 | −3.396 | 0.001 | 0.025 |
| WTE–VT | −35.278 | 10.617 | −3.323 | 0.001 | 0.032 |
| WTG–VT | −31.778 | 10.617 | −2.993 | 0.003 | 0.099 |
| YTG–VT | −27.667 | 10.617 | −2.606 | 0.009 | 0.330 |
| BTE–ST | −24.278 | 10.617 | −2.287 | 0.022 | 0.800 |
| WTE–ST | −23.500 | 10.617 | −2.213 | 0.027 | 0.967 |
| BTG–ST | −23.556 | 10.617 | −2.219 | 0.027 | 0.954 |
| WTG–ST | −20.000 | 10.617 | −1.884 | 0.060 | 1.000 |
| YTE–VT | −19.167 | 10.617 | −1.805 | 0.071 | 1.000 |
| BTE–OP | −18.611 | 10.617 | −1.753 | 0.080 | 1.000 |
| BTG–OP | −17.889 | 10.617 | −1.685 | 0.092 | 1.000 |
| WTE–OP | −17.833 | 10.617 | −1.680 | 0.093 | 1.000 |
| OP–VT | −17.444 | 10.617 | −1.643 | 0.100 | 1.000 |
| BTE–YTE | 16.889 | 10.617 | 1.591 | 0.112 | 1.000 |
| BTG–YTE | −16.167 | 10.617 | −1.523 | 0.128 | 1.000 |
| WTE–YTE | 16.111 | 10.617 | 1.517 | 0.129 | 1.000 |
| YTG–ST | −15.889 | 10.617 | −1.497 | 0.135 | 1.000 |
| WTG–OP | −14.333 | 10.617 | −1.350 | 0.177 | 1.000 |
| WTG–YTE | −12.611 | 10.617 | −1.188 | 0.235 | 1.000 |
| ST–VT | −11.778 | 10.617 | −1.109 | 0.267 | 1.000 |
| YTG–OP | −10.222 | 10.617 | −0.963 | 0.336 | 1.000 |
| YTG–YTE | −8.500 | 10.617 | −0.801 | 0.423 | 1.000 |
| BTE–YTG | 8.389 | 10.617 | 0.790 | 0.429 | 1.000 |
| BTG–YTG | 7.667 | 10.617 | 0.722 | 0.470 | 1.000 |
| WTE–YTG | 7.611 | 10.617 | 0.717 | 0.473 | 1.000 |
| YTE–ST | −7.389 | 10.617 | −0.696 | 0.486 | 1.000 |
| OP–ST | −5.667 | 10.617 | −0.534 | 0.594 | 1.000 |
| BTE–WTG | 4.278 | 10.617 | 0.403 | 0.687 | 1.000 |
| WTG–YTG | 4.111 | 10.617 | 0.387 | 0.699 | 1.000 |
| BTG–WTG | −3.556 | 10.617 | −0.335 | 0.738 | 1.000 |
| WTE–WTG | 3.500 | 10.617 | 0.330 | 0.742 | 1.000 |
| YTE–OP | −1.722 | 10.617 | −0.162 | 0.871 | 1.000 |
| BTE–WTE | −0.778 | 10.617 | −0.073 | 0.942 | 1.000 |
| BTE–BTG | 0.722 | 10.617 | 0.068 | 0.946 | 1.000 |
| BTG–WTE | −0.056 | 10.617 | −0.005 | 0.996 | 1.000 |

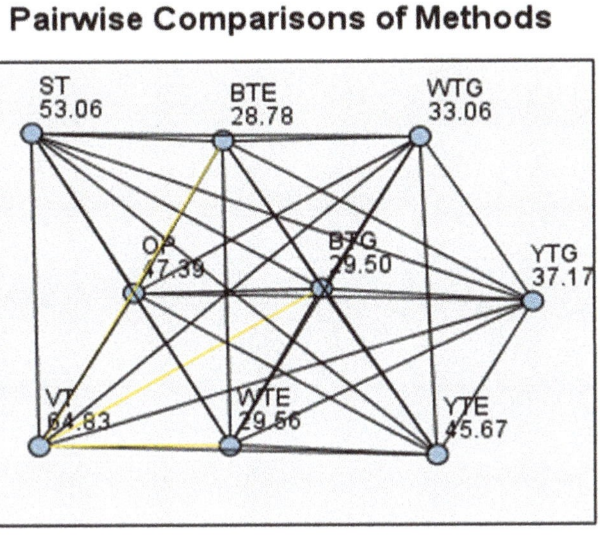


**Figure 3.** Ranking representation of used sampling methods of hymenopteran species in *Brassica campestris* var. *toria* crops.

In a comparison of the VT method with the active sampling method (ST, standardised transect; OP, observation plots), no significant differences were recorded, suggesting that these methods provide similar results over diversity. a. Efficiency of non-transect sampling methods The efficiency of non-transect methods (treatments 1 to 7) showed highly significant differences in each cropping period (Table 3 (a)), as well as at their pooled mean level (Table 3 (b)).

Table 3 (a) shows that the number of collected species, using the non-transect sampling method, is more sensitive in the late-drilled toria crop, with the only exception being recorded when the YTE method was used (8.19—normally drilled toria crop; 8.16—late-drilled toria crop).

The mean number of the collected species in normally drilled toria plots varied between 2.36 in treatment 6 (WTE) and 8.35 in treatment 7 (OP). The results were almost similar in the case of plots with toria drilled in a later period, and the mean number of recorded species varied between 2.64 (treatment 5, BTE) and 8.85 (treatment 7, OP). Non-transect sampling methods YTG (4.63 species), BTG (4.23 species), and WTG (4.14 species) showed a medium efficiency in collecting the crop visitor species (normal or late crops) in *Brassica campestris* var. *toria*.

The mean sampling record per site was higher in Kutuha village as compared to the other two locations (Table 3 (b)). Treatment 7 (OP) outperformed the other non-transect methods at all three locations; however, the results in the case of treatment 4 (YTE) were statistically similar to those registered in treatment 7 (OP) when comparing the mean number of hymenopteran flower visitor species calculated for all locations. b. Efficiency of transect methods of sampling In each of the cropping periods and also in their pooled mean level, the number of species caught in both treatments with transect methods (treatment 8—ST, and treatment 9—VT) had significant differences between the three locations of the study (Table 4 (a)). The mean sampling record of hymenopteran flower visitor species numbers per site was significantly higher in Kutuha village (35.8) when compared to that in Panimirigaon (33.33) and very significantly higher compared to that recorded in Jajimukh (32.3).

**Table 3. a.** Number of hymenopteran flower visitor species recorded in different non-transect sampling methods at every site by cropping period and toria variety. **b.** Mean number of hymenopteran flower visitor species recorded by non-transect sampling methods, depending on site location.

**a**

| Treatment (Sampling Method) | Normal Cropping Period (TS 36 Variety) | | | | Late Cropping Period (TS 67 Variety) | | | |
|---|---|---|---|---|---|---|---|---|
| | Jajimukh | Panimirigaon | Kutuha | Mean | Jajimukh | Panimirigaon | Kutuha | Mean |
| Treatment 1 (YTG) | 4.01 | 4.13 | 4.51 | 4.21 b | 4.81 | 4.28 | 4.81 | 4.63 c |
| Treatment 2 (BTG) | 3.49 | 3.73 | 3.93 | 3.72 c | 4.12 | 4.21 | 4.35 | 4.23 d |
| Treatment 3 (WTG) | 3.51 | 3.53 | 3.88 | 3.64 c | 4.07 | 3.97 | 4.38 | 4.14 d |
| Treatment 4 (YTE) | 7.65 | 7.88 | 9.03 | 8.19 a | 7.49 | 7.98 | 9.00 | 8.16 b |
| Treatment 5 (BTE) | 2.19 | 2.38 | 2.50 | 2.36 d | 2.50 | 2.59 | 2.84 | 2.64 e |
| Treatment 6 (WTE) | 2.19 | 2.46 | 2.61 | 2.42 d | 2.46 | 2.71 | 2.83 | 2.67 e |
| Treatment 7 (OP) | 7.76 | 8.11 | 9.20 | 8.35 a | 8.13 | 8.58 | 9.84 | 8.85 a |
| Site Mean | 4.40 c | 4.60 b | 5.09 a | 4.70 | 4.80 b | 4.90 b | 5.44 a | 5.05 |
| Deviation (%) from OM (4.88) | (−) 9.84 | (−) 5.74 | (+) 4.30 | (−) 3.69 | (−) 1.64 | (+) 0.41 | (+) 11.48 | (+) 3.28 |

**b**

| Treatment (Sampling Method) | Jajimukh | Panimirigaon | Kutuha | Mean |
|---|---|---|---|---|
| Treatment 1 (YTG) | 4.41 | 4.21 | 4.66 | 4.43 b |
| Treatment 2 (BTG) | 3.80 | 3.97 | 4.14 | 3.97 bc |
| Treatment 3 (WTG) | 3.79 | 3.75 | 4.13 | 3.89 c |
| Treatment 4 (YTE) | 7.57 | 7.93 | 9.02 | 8.17 a |
| Treatment 5 (BTE) | 2.35 | 2.49 | 2.67 | 2.50 d |
| Treatment 6 (WTE) | 2.33 | 2.59 | 2.72 | 2.55 d |
| Treatment 7 (OP) | 7.95 | 8.35 | 9.52 | 8.61 a |
| Mean | 4.60 | 4.75 | 5.27 | OM = 4.88 |
| Deviation (%) from OM | (−) 5.74 | (−) 2.66 | (+) 7.99 | - |

a: Tabulated data correspond to mean values of eight assessments taken in two crops of normal and two crops of late cropping periods per locality. OM: overall mean; i.e., mean of all data under all treatments recorded in three locations in all cropping periods (=4.88). Values in mean column superscripted with different letters are statistically significantly different at $p = 0.01$ according to Tukey's test (SPSS 20). b: Tabulated data are pooled between the mean of normal and late cropping periods in different locations, observed in eight evaluations. Mean values superscripted with different letters are significantly different at $p = 0.01$. Values in mean column superscripted with different letters are statistically significantly different at $p = 0.01$ according to the Tukey HSD test (SPSS 20).

**Table 4. a.** Mean number of hymenopteran flower visitor species recorded by transect methods in two cropping periods of toria in three different locations of Assam, India. **b.** Number of hymenopteran flower visitor species sampled by transect methods in toria crops.

**a**

| Locations | Normal Cropping Period | | | Late Cropping Period | | | Pooled Mean (over Cropping Periods) |
|---|---|---|---|---|---|---|---|
| | T8 ST | T9 VT | Mean | T8 ST | T9 VT | Mean | |
| Jajimukh | 24.75 | 38.00 | 31.38 c | 26.75 | 39.69 | 33.22 b | 32.30 c |
| Panimirigaon | 25.69 | 39.75 | 32.72 b | 27.69 | 40.19 | 33.94 b | 33.33 b |
| Kutuha | 27.43 | 42.56 | 35.00 a | 29.23 | 41.06 | 35.15 a | 35.08 a |
| Mean | 25.96 | 40.10 | 33.03 | 27.89 | 40.31 | 34.10 | 33.57 |

**b**

| Treatments | Normal Cropping Period | Late Cropping Period | Mean |
|---|---|---|---|
| T8 (ST) | 25.96 | 27.89 | 26.93 |
| T9 (VT) | 40.10 | 40.31 | 40.21 |
| Mean | 33.03 | 34.10 | 33.57 |
| *t*-test | 3.408 ** | 2.767 ** | 11.585 ** |

a: Mean values in a column superscripted with the same letters do not differ significantly according to LSD ($p = 0.01$) and Tukey (SPSS 20) tests. b: Tabulated data (columns 2 and 3) are the mean of eight observations of three locations for 2 years. ** Significantly different at 1% level of significance.

Based on a *t*-test, it can be seen that the performance of VT was superior to ST in both the cropping periods individually as well as at their pooled mean level; highly significant differences between the number of hymenopteran flower visitor species were observed between them (Table 4 (b)).

In the normal cropping period, the efficiency of transect methods in terms of the hymenopteran flower visitor species caught was found to be higher than that of pan traps (Figure 4). The comparison of the nine sampling methods through confidence intervals (CI 95%) reveals that methods VT (T9) and ST (T8) recorded the highest values, showing a high variability as well.

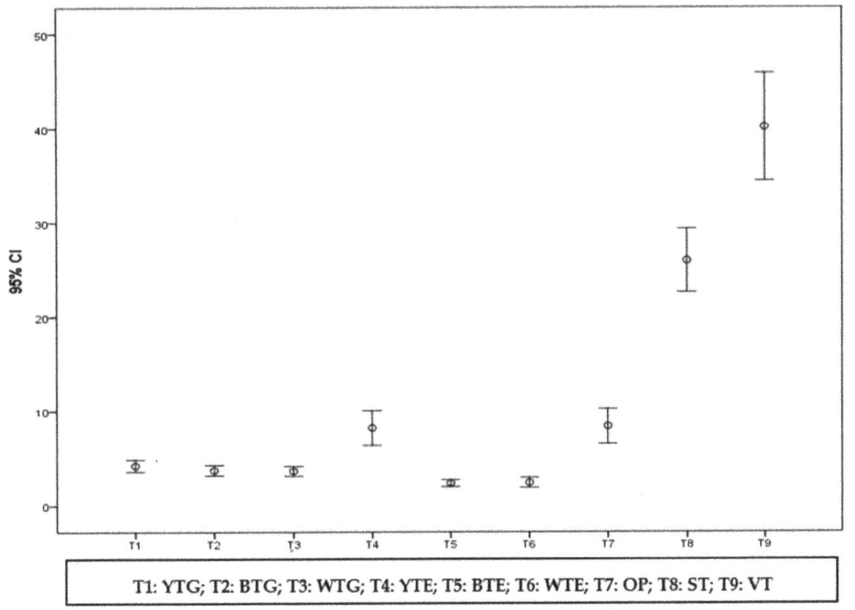


**Figure 4.** Error bar graphs on the efficiency of nine sampling methods (T1–T9) in a normal cropping period of toria crops.

Non-transect methods, BTG (T2), WTG (T3), BTE (T5), and WTE (T6), exhibited the lowest values in species collecting from toria crops, presenting thinner trust intervals and endorsing more accurate evaluations.

Using the nine sampling methods in the late-drilled period of toria cropping proved that their efficiency does not change, the ranking being similar to the one obtained in a normally drilled period of the crop (Figure 5). Transect methods VT (T9) and ST (T8) collected the highest number of hymenopteran flower visitor species, followed by OP (T7) and YTE (T4). Using the methods of coloured traps placed on the ground (T1, T2, T3) and elevated ones (T5, T6), a low number of hymenopteran flower visitor species were captured.

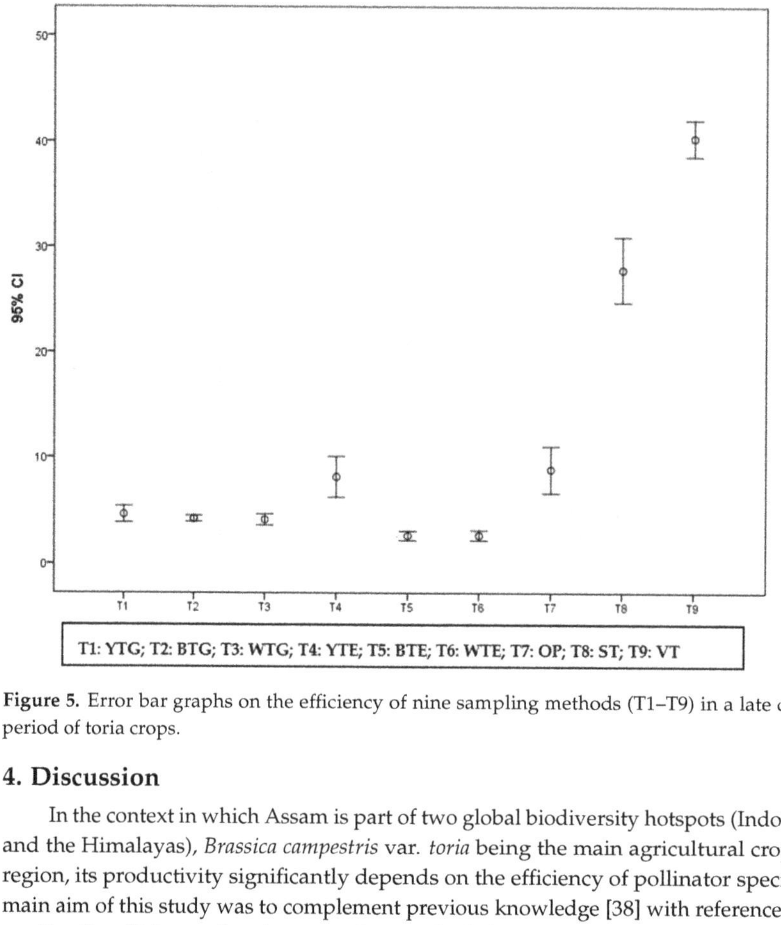


**Figure 5.** Error bar graphs on the efficiency of nine sampling methods (T1–T9) in a late cropping period of toria crops.

## 4. Discussion

In the context in which Assam is part of two global biodiversity hotspots (Indo–Burma and the Himalayas), *Brassica campestris* var. *toria* being the main agricultural crop in the region, its productivity significantly depends on the efficiency of pollinator species. The main aim of this study was to complement previous knowledge [38] with reference data regarding the efficiency of various sampling methods for hymenopteran floral visitor species.

### *4.1. Diversity of Floral Visitors Based on Sampling Methods*

In *Brassica campestris* var. *toria* (drilled in the optimal and later period), in all experimental sites dedicated to the study in three localities (Jajimukh, Panimirigaon, and Kutuha) using nine sampling methods, a very large number of hymenopteran floral visitor species were caught. The 64 species were assigned to nine Hymenoptera families (Apidae, Andrenidae, Megachilidae, Vespidae, Scoliidae, Halictidae, Crabronidae, Colletidae, and Sphecidae) (Table 1).

Active methods like VT (variable transect), ST (standardised transect walk), and OP (direct observations) proved to be the most effective, and through their use, important pollinators belonging to the Apidae and Halictidae families were collected (Table 1). The results align with those reported by Popic et al. and Roulston et al. [39,48]. As a group, the species of Apidae and Halictidae were detected in all the sampling methods, which indicates that species of these families visit multiple strata of crop plant vegetation.

Even though the number of species belonging to Vespidae was relatively larger (Table 1), most of them are predatory in habit and some species can alter the pollination dynamics; thus, vespids are less efficient as pollinators in toria crops as compared to many species of other families such as Apidae and Megachilidae since bees are considered important pollinators globally [64].

By using the OP method in the *Brassica campestris* var. *toria* ecosystem, we succeeded in collecting 32.8% of species belonging to seven families (Apidae, Andrenidae, Megachilidae, Vespidae, Halictidae, Crabronidae, and Sphecidae) out of the nine recorded.

By using the YTE method, we recorded the presence of 34.4% of toria crop hymenopteran floral visitor species belonging to six families (Apidae, Andrenidae, Megachilidae, Vespidae, Halictidae, and Crabronidae), while the use of passive methods placed at a ground level, BTE and WTE, recorded a very low percentage (9.4% and 10.9%, respectively) of the total toria hymenopteran floral visitor species (Table 1). YPT efficacy is also confirmed by studies conducted by Sounders and Luck [49], who found that this method caught the most pollinator insects across a great variety of habitats, although not all of the analysed ones, and concluded that pan trap colour attractiveness depends largely on habitat.

Therefore, the ability to sample more pollination-efficient species, as observed in YTE and OP, may increase the efficiency of the entire sampling module if these two are included along with VT and ST. These significant differences are supported by Figures 2 and 5, as well as Table 3 (a). Although transect methods (VT, ST) proved to be superior in collecting hymenopteran floral visitor species in a *Brassica campestris* var. *toria* ecosystem, some studies indicate that standardised methods may exclude essential pollinating species [65,66]. Likewise, traps installed at ground level, viz., YTG, BTG, and WTG, could sample some ground-dwelling species, even though their overall sampling efficiency is less (10.9–18.8%). The coloured traps placed at ground level (YTG, BTG, WTG) proved to be attractive for species belonging to the Halictidae family, but these were not efficient in capturing species belonging to the Colletidae family. Similar results were reported by Droege et al. [67], who also reported a taxonomic bias related to colour.

Therefore, ground traps may be complementary to other methods. In some small farms of Asian farmers, where non-crop vegetation and hedges are frequently present between two adjacent crop fields of small acreage, there are heterogeneous and homogeneous areas within the same crop landscape. In such a situation, the combination of complementary methods would always perform better.

### 4.2. Comparative Efficiency of Transect and Non-Transect Sampling Methods

The comparative analyses of transect sampling methods (VT and ST) and non-transect (OP, YTG, BTG, WTG, YTE, BTE, and WTE) revealed significant differences in their capacity to capture the diversity of hymenopteran floral visitor species in toria agroecosystem conditions in Assam.

Transect methods proved to be superior in capturing the number of species compared to non-transect methods across all three sites in both toria crops (normal cropping period and late cropping period) (Figures 4 and 5). Variable transect (VT), due to its flexibility, exhibits the most efficient active method, followed by direct observation (OP) and standard transect (ST). We found that VT had higher sampling records than ST. This may be due to the larger transecting area covered under VT treatment. We consider that species records in both ST and VT may be increased proportionately with pollinator diversity of a locality by intensifying the netting with proper temporal randomisation, particularly for annual crops with a shorter blooming period. Such flexibility in sampling intensity based on scientists' particular goals and specific knowledge about local diversity of pollinators has also been suggested in the case of traps by Shapiro et al. [65].

Analysing comparatively the efficiency of transect methods in toria crops drilled in regular and late periods, it was observed that their superiority was maintained, indicating a consistency regarding collecting the hymenopteran flower visitors. Temporal variations of species richness between two cropping periods under multiple sampling methods are a researchable issue.

Among the non-transect methods, the highest mean number of hymenopteran flower visitor species record was observed in OP followed by YTE (Table 3 (a)). The flower visitors responded poorly to other non-transect methods. Here lies the importance of incorporating the transect methods in a sampling module. Although non-transect methods YTG, BTG, WTG, BTE, and WTE constantly yielded lower results in terms of the number of hymenopteran flower visitor species, their complementary role cannot be ignored as they can provide more complete coverage of functional diversity.

Westphal et al. [34] recommend pan traps for pollinator monitoring schemes to provide reliable results when operated by many surveyors in different habitats, regions, and years, and also consider the transect walks and observation plots as the main methods in more detailed studies on plant–pollinator interactions. Despite its wide use in sampling pollinator diversity, the influence of colour on pan trap efficiency is not so clear; accordingly, the blue, yellow, and white pan traps are considered complementary in sampling the Hymenoptera community [66].

Transect methods have been shown to be remarkably superior to the non-transect ones in terms of sampling records. However, being contributory, both are required for such sampling studies, as discussed in the above sections. Multiple sampling methods have also been suggested in some recent studies [68,69].

The combination of transect methods with some selected passive methods (like OP and YTE) seems to be an optimal strategy for a broader characterisation of pollinator communities.

The results of this study highlight that variable transect (VT) recorded the highest values in the detection of floral visitor species, maintaining its efficiency even in cases where non-transect methods (YTG, BTG, WTG, YTE, BTE, and WTE) tend to fail in fully capturing local diversity.

The superiority of transect methods to pan traps is also supported by studies conducted by Berglund and Milberg [70] and by other subsequent research [71].

They claim that pan traps underestimate the number of species and individuals belonging to the Apidea family and overestimate Lepturinae and Cetoniidae. Similar differences were reported in studies performed by Cane et al. [46] and Roulston et al. [48] (2007), highlighting the limits of pan traps in fully characterising the bee fauna. Differences between passive and active methods, in terms of species abundance and the composition of collected bees, were also reported by Gibbs et al. [72] and McCravy and Ruholl [73].

### 4.3. Influence of Trap Colour and Position on Sampling Efficiency

Colour is one of the most important attractants for many flower-visiting insects, and their preferences for a specific colour is an important source of bias that needs to be considered in pan trap surveys [74].

Our data clearly indicate that the majority of the species sampled by pan traps had an affinity towards yellow (Table 2). Saunders and Luck [49] have also reported higher trap records in yellow pan traps. In our study, YTG had failed to trap some of the top surface visitors, which led to a lower sampling rate than YTE. On the other hand, possibly because of the affinity of some ground-dwelling and sub-surface species towards blue and white, the trap record in BTG and WTG was significantly superior to same-coloured elevated traps. This model of chromatic selection based on vertical layers was also documented by Nuttman et al. [75], underlining the fact that colour preferences vary among taxons, affecting feeding behaviour and activity level in a vegetal canopy. However, no colour was consistently found to be preferred if both placement positions (i.e., ground and elevated) were considered. Such facts have also been reported in some previous works, which emphasise that one colour cannot be considered more attractive than others when targeting a wide range of taxa in different ecological contexts [40,66,76]. The species diversity might be improved through (i) the simultaneous use of pan traps of different colours, which has been suggested for surveys targeting a wide range of taxa [47,74,77], or (ii) by increasing the number of traps based on particular goals and specific knowledge about local bee diversity, as reported by Shapiro et al. [65].

### *4.4. Implications for Future Research and Conservation*

As this present study took place near the Palearctic border, a higher number of Palearctic species may be recorded by similar diversity studies in localities in the upstream basin of the Burhidihing River. Identifying those wild pollinators may open new research avenues in relation to their conservation and colonisation for harnessing their pollination services in crop ecosystems of the Oriental region. The conservation of wild pollinators to support the toria production of small and marginal farmers in India and other Asian countries is an important issue. This is highlighted by Zou et al. [78] in smallholder oilseed farming systems of China. There is also the future scope of a study focusing on temporal dynamics, as demonstrated by Ludewig et al. [79] and Casiá-Achjé [80]. Millard et al. [59] and Tsang et al. [81] have demonstrated the impact of land use changes on biodiversity, particularly in human-modified areas like agricultural landscapes, and emphasised the effects of land use change on insect diversity at multiple scales.

A multisampling study on bee assemblage conducted in the traditional slash-and-burn agriculture (milpa) of the Yucatan Peninsula of Mexico (part of Mesoamerica, which is considered an important biodiversity hotspot), revealed that bee diversity was highest in forested areas and lower in cropped land; although, surprisingly, chilli pollination was enhanced by surrounding fallow, gardens, and pasture but reduced by surrounding forest cover [7]. Such studies on pollinator assemblage with due correlation with the surrounding fallow, non-crop plant community are another topic of interest in toria crop ecosystems in Assam state in India.

## 5. Conclusions

Overall, the data demonstrate that VT, followed by ST, consistently yielded higher species richness across diverse hymenopteran families, reinforcing their utility as core sampling methods for monitoring flower visitor diversity in toria crops. Their ability to capture both dominant and moderately represented taxa suggests that these methods may offer a more comprehensive assessment of the pollinator community structure, but this study reveals that no method alone is adequate to sample the whole hymenopteran flower visitor diversity, including pollinator species in toria crop ecosystems. The most deployed method of sampling, i.e., coloured pan traps filled with soapy water, failed to justify its efficiency as the sole method of sampling. Another conclusion was that active methods (VT, ST, and OP) exerted a higher efficacy in capturing a higher number of species belonging to Apidae and Halictidae. The YTE method proved to be the most effective passive method for capturing flower visitors in toria crops. On the other hand, the lowest efficiency in catching hymenopterans was shown by blue-coloured traps.

A cluster of coloured pan traps and netting with transect walk were found to be complementary to each other; hence, both transect and non-transect methods as a module should be used in order to record a wider species composition in toria crop ecosystems.

Such a complementary combination of sampling techniques will be particularly useful in areas with rich flower visitor diversity, with implications for long-term monitoring and conservation planning.

Such modules will also be fruitful for benchmark surveys to see the depletion of pollinator species richness, if any, due to climate change over time. In continuation of this present study, there is the scope of identifying wild Palearctic hymenopteran pollinators in the upstream basin of the Burhidihing River of Assam, India, which may open new research avenues on their conservation and colonisation for harnessing their pollination services in crop ecosystems in the Oriental region.

**Author Contributions:** Conceptualisation, A.K.S., B.N. and M.K.D.; methodology, A.K.S., B.N. and M.K.D.; formal analysis, B.N. and A.K.S.; investigation, A.K.S.; resources, A.K.S. and M.K.D.; writing—original draft preparation, A.K.S. and R.S.; writing—review and editing, A.K.S., A.C. and R.S.; funding acquisition, A.C. and R.S. This work is part of the doctoral research program of the first author. All authors have read and agreed to the published version of the manuscript.

**Funding:** This research received no external funding.

**Data Availability Statement:** This article is based on the PhD research work of the first author; no raw data will be shared.

**Acknowledgments:** The authors are highly indebted to Neil Anderson, Former Researcher-cum-Project leader, Department of Biosciences, University of Oslo, Norway, for his encouragement to take up the field research and for sharing the methodology of the sampling techniques. The farmers who allowed us to conduct research work in their crop fields and provided accommodation during the field studies are duly acknowledged. The authors are also thankful to the scientists of (i) the National Bureau of Agricultural Insect Resource, Bengaluru, India, and (ii) the University of Agricultural Science, Bengaluru, India, for the taxonomic identification of the insects.

**Conflicts of Interest:** The authors declare no conflicts of interest.

## References

1. Klein, A.M.; Vaissiere, B.E.; Cane, J.H.; Steffan-Dewenter, I.; Cunningham, S.A.; Kremen, C.; Tscharntke, T. Importance of pollinators in changing landscapes for world crops. *Proc. R. Soc. B* **2007**, *274*, 303–313. [CrossRef] [PubMed]
2. Nicholls, C.; Altieri, M. Plant biodiversity enhances bees and other insect pollinators in agro-ecosystems: A review. *Agron. Sustain. Dev* **2013**, *33*, 257–274. [CrossRef]
3. Mcgregor, S.E. *Insect Pollination of Cultivated Crop Plants*; USDA/ARS Agriculture Handbook 496; Agricultural Research Service, US Department of Agriculture: Washington, DC, USA, 1976; p. 139.
4. Tepedino, V.J. The importance of bees and other insect pollinators in maintaining floral species composition. *Great Basin Nat. Mem.* **1979**, *3*, 17.
5. Free, J.B. *Insect Pollination of Crops*; Academic Press: London, UK, 1993; p. 684.
6. Gallai, N.; Salles, J.M.; Settele, J.; Vaissiere, B.E. Economic valuation of the vulnerability of world agriculture confronted with pollinator decline. *Ecol. Econ.* **2009**, *68*, 810–821. [CrossRef]
7. Landaverde-González, P.; Quezada-Euán, J.J.G.; Theodorou, P.; Murray, T.E.; Husemann, M.; Ayala, R.; Moo-Valle, H.; Vandame, R.; Paxton, R.J. Sweat bees on hot chilies: Provision of pollination services by native bees in traditional slash-and-burn agriculture in the Yucatán Peninsula of tropical Mexico. *J. Appl. Ecol.* **2017**, *54*, 1814–1824. [CrossRef]
8. Intergovernmental Science-Policy Platform on Biodiversity and Ecosystem Services. *The Assessment Report on Pollinators, Pollination and Food Production*; IPBES Secretariat: Bonn, Germany, 2016; Available online: https://ipbes.net (accessed on 3 March 2025).
9. Faegri, K.; Van Der Pijl, L. *Principales of Pollination Ecology*, 3rd ed.; Elsevier: Amsterdam, The Netherlands, 2013; p. 256.
10. Klein, A.M.; Steffan-Dewenter, I.; Tscharntke, T. Pollination of Coffea canephora in relation to local and regional agroforestry management. *Oecologia* **2003**, *134*, 607–615. [CrossRef]
11. Culley, T.M.; Weller, S.G.; Sakai, A.K. The evolution of wind pollination in angiosperms. *Trends Ecol. Evol.* **2002**, *17*, 361–369. [CrossRef]

12. Abrol, D.P. *Pollination Biology: Biodiversity Conservation and Agricultural Production*; Springer: Dordrecht, The Netherlands, 2012; p. 792.
13. Stanley, J.; Sah, K.; Subbanna, A.R.N.S. How efficient is the Asian honey bee, *Apis cerana* in pollinating mustard, *Brassica campestris* var. *toria*? Pollination behavior, pollinator efficiency, pollinator requirements and impact of pollination. *J. Agric. Res.* **2017**, *56*, 439–451. [CrossRef]
14. Sarma, A.K.; Chauhan, J.S. Pollinator diversity in *Brassica* crops and significance of pollinators in improving productivity: A review. *J. Agric. Res.* **2015**, *1*, 35–40.
15. Taba, N.; Sharma, P.; Devadas, V.S.; Hazarika, G.N.; Monlai, S. Performance of Toria (*Brassica campestris* L.) Varieties under Namsai Conditions. *Int. J. Curr. Microbiol. Appl. Sci.* **2020**, *9*, 2101–2103.
16. Williams, I.H. The pollination requirements of swede rape (*Brassica napus* L.) and of turnip rape (*Brassica campestris* L.). *J. Agric. Sci.* **1978**, *91*, 343–348. [CrossRef]
17. Badenes-Pérez, F.R. Benefits of Insect Pollination in Brassicaceae: A Meta-Analysis of Self-Compatible and Self-Incompatible Crop Species. *Agriculture* **2022**, *12*, 446. [CrossRef]
18. Brittain, C.; Kremen, C.; Klein, A.M. Biodiversity buffers pollination from changes in environmental conditions. *Glob. Chang. Biol.* **2013**, *19*, 540–547. [CrossRef]
19. Woodcock, B.A.; Garratt, M.P.D.; Powney, G.D.; Shaw, R.F.; Osborne, J.L.; Soroka, J.; Lindström, S.A.M.; Stanley, D.; Ouvrard, P.; Edwards, M.E.; et al. Meta-Analysis Reveals That Pollinator Functional Diversity and Abundance Enhance Crop Pollination and Yield. *Nat. Commun.* **2019**, *10*, 1481. [CrossRef] [PubMed]
20. Rader, R.; Howlett, B.G.; Cunningham, S.A.; Westcott, D.A.; Newstrom-Lloyd, L.E.; Walker, M.K.; Teulon, D.A.J.; Edwards, W. Alternative Pollinator Taxa Are Equally Efficient but Not as Effective as the Honeybee in a Mass Flowering Crop. *J. Appl. Ecol.* **2009**, *46*, 1080–1087. [CrossRef]
21. Junqueira, C.N.; Pereira, R.A.S.; da Silva, R.C.; Alves Cardoso Kobal, R.O.; Araújo, T.N.; Prato, A.; Pedrosa, J.; Martínez-Martínez, C.A.; Castrillon, K.P.; Felício, D.T.; et al. Do Apis and Non-Apis Bees Provide a Similar Contribution to Crop Production with Different Levels of Pollination Dependency? A Review Using Meta-Analysis. *Ecol. Entomol.* **2021**, *47*, 76–83. [CrossRef]
22. Garibaldi, L.A.; Steffan-Dewenter, I.; Winfree, R.; Aizen, M.A.; Bommarco, R.; Cunningham, S.A.; Kremen, C.; Carvalheiro, L.G.; Harder, L.D.; Afik, O.; et al. Wild Pollinators Enhance Fruit Set of Crops Regardless of Honey Bee Abundance. *Science* **2013**, *339*, 1608–1611. [CrossRef]
23. Rader, R.; Bartomeus, I.; Garibaldi, L.A.; Garratt, M.P.D.; Howlett, B.G.; Winfree, R.; Cunningham, S.A.; Mayfield, M.M.; Arthur, A.D.; Andersson, G.K.S.; et al. Non-Bee Insects Are Important Contributors to Global Crop Pollination. *Proc. Natl. Acad. Sci. USA* **2016**, *113*, 146–151. [CrossRef]
24. Földesi, R.; Howlett, B.G.; Grass, I.; Batáry, P. Larger Pollinators Deposit More Pollen on Stigmas across Multiple Plant Species—A Meta-Analysis. *J. Appl. Ecol.* **2021**, *58*, 699–707. [CrossRef]
25. Rader, R.; Howlett, B.G.; Cunningham, S.A.; Westcott, D.A.; Edwards, W. Spatial and temporal variation in pollinator effectiveness: Do unmanaged insects provide consistent pollination services to mass flowering crops? *J. Appl. Ecol.* **2012**, *49*, 126–134. [CrossRef]
26. Kleijn, D.; Winfree, R.; Bartomeus, I.; Carvalheiro, L.G.; Henry, M.; Isaacs, R.; Klein, A.-M.; Kremen, C.; M'Gonigle, L.K.; Rader, R.; et al. Delivery of crop pollination services is an insufficient argument for wild pollinator conservation. *Nat. Commun.* **2015**, *6*, 7414. [CrossRef]
27. Winfree, R.; Fox, W.; Williams, J.; Reilly, N.M.; Cariveau, J.R.D.P. Abundance of common species, not species richness, drives delivery of a realworld ecosystem service. *Ecol. Lett.* **2015**, *18*, 626–635. [CrossRef] [PubMed]
28. Winfree, R.; Reilly, J.R.; Bartomeus, I.; Cariveau, D.P.; Williams, N.M.; Gibbs, J. Species turnover promotes the importance of bee diversity for crop pollination at regional scales. *Science* **2018**, *359*, 791–793. [CrossRef] [PubMed]
29. Oliver, T.H.; Heard, M.S.; Isaac, N.J.B.; Roy, D.B.; Procter, D.; Eigenbrod, F.; Freckleton, R.; Hector, A.; Orme, C.D.L.; Petchey, O.L.; et al. Biodiversity and Resilience of Ecosystem Functions. *Trends Ecol. Evol.* **2015**, *30*, 673–684. [CrossRef] [PubMed]
30. Jauker, F.; Bondarenko, B.; Becker, H.C.; Steffan-Dewenter, I. Pollination efficiency of wild bees and hoverflies provided to oilseed rape. *Agric. For. Entomol.* **2012**, *14*, 81–87. [CrossRef]
31. Vaissiere, B.E.; Breno, M.F.; Barbara, G.H. *Protocol to Detect and Assess Pollination Deficits in Crops: A Handbook for Its Use*; FAO, UN: Rome, Italy, 2011; pp. 1–70.
32. Wood, T.J.; Holland, J.M.; Goulson, D. Providing foraging resources for solitary bees on farmland: Current schemes for pollinators benefit a limited suite of species. *J. Appl. Ecol.* **2016**, *54*, 323–333. [CrossRef]
33. Thompson, A.; Frenzel, M.; Schweiger, O.; Musche, M.; Groth, T.; Roberts, S.P.M.; Kuhlmann, M.; Knight, T.M. Pollinator sampling methods influence community patterns assessments by capturing species with different traits and at different abundances. *Ecol. Indic.* **2021**, *132*, 108284. [CrossRef]

34. Westphal, C.; Bommarco, R.; Carre, G.; Lamborn, E.; Morison, N.; Petanidou, T.; Potts, S.G.; Roberts, S.P.M.; Szentgyo Rgyi, H.; Tscheulin, T.; et al. Measuring bee biodiversity in different European habitats and bio-geographical regions. *Ecol. Monogr.* **2008**, *78*, 653–671. [CrossRef]
35. Potts, S.G.; Petanidou, T.; Roberts, S.; O'Toole, C.; Hulbert, A.; Willmer, P. Assessing pollinator biodiversity: Standardized methods for monitoring. *J. Appl. Ecol.* **2008**, *45*, 9–14.
36. Potts, S.G.; Vulliamy, B.; Roberts, S.; O'toole, C.; Dafni, A.; Ne'eman, G.; Willmer, P. Role of nesting resources in organising diverse bee communities in a mediterranean landscape. *Ecol. Entomol.* **2005**, *30*, 78–85. [CrossRef]
37. Hutchinson, L.A.; Oliver, T.H.; Breeze, T.D.; O'Connor, R.S.; Potts, S.G.; Roberts, S.P.M.; Garratt, M.P.D. Inventorying and monitoring crop pollinating bees: Evaluating the effectiveness of common sampling methods. *Insects Conserv. Divers.* **2022**, *15*, 299–311. [CrossRef]
38. Sarma, A.K.; Deka, M.K.; Neog, B. Species richness of Hymenopteran flower visitors in *Brassica campestris* var. *toria* in Assam, India: A comparison of five sampling methods. *AtaXE* **2024**, *70*, 314–328. [CrossRef]
39. Popic, T.J.; Davila, Y.C.; Wardle, G.M. Evaluation of common method for sampling invertebrate pollinator assemblages: Net samplings outperform Pan traps. *PLoS ONE* **2013**, *8*, e66665. [CrossRef] [PubMed]
40. Campbell, J.W.; Hanula, J.L. Efficiency of Malaise traps and coloured pan traps for collecting flower visiting insects from three forested ecosystems. *J. Insect Conserv.* **2007**, *11*, 399–408. [CrossRef]
41. Geroff, R.K.; Gibbs, J.; McCravy, K.W. Assessing bee (Hymenoptera: Apoidea) diversity of an Illinois restored tallgrass prairie: Methodology and conservation considerations. *J. Insect Conserv.* **2014**, *18*, 951–964. [CrossRef]
42. Biesmeijer, J.C.; Roberts, S.P.M.; Reemer, M.; Ohlemüller, R.; Edwards, M.; Peeters, T.; Kunin, W.E. Parallel declines in pollinators and insect-pollinated plants in Britain and the Netherlands. *Science* **2006**, *313*, 351–354. [CrossRef]
43. Giovanetti, M.; Albertazzi, S.; Flaminio, S.; Ranalli, R.; Bortolotti, L.; Quaranta, M. Pollination in Agroecosystems: A Review of the Conceptual Framework with a View to Sound Monitoring. *Land* **2021**, *10*, 540. [CrossRef]
44. Nielsen, A.; Steffan-Dewenter, I.; Westphal, C.; Messinger, O.; Potts, S.G.; Roberts, S.P.M.; Settele, J.; Szentgyörgyi, H.; Vaissière, B.E.; Vaitis, M.; et al. Assessing bee species richness in two Mediterranean communities: Importance of habitat type and sampling techniques. *Ecol. Res.* **2011**, *26*, 969–983. [CrossRef]
45. Leong, J.M.; Thorp, R.W. Colour-coded sampling: Pan trap colour preferences oligolectic and non-oligolectic bees associated with a vernal pool plant. *Ecol. Entomol.* **1999**, *24*, 329–335. [CrossRef]
46. Cane, J.H.; Minckley, R.L.; Kervin, R.J. Sampling bees (Hymenoptera: Apiformes) for pollinator community studies: Pitfalls of pan trapping. *J. Kans. Entomol. Soc.* **2000**, *73*, 225–231. Available online: http://www.jstor.org/stable/25085973 (accessed on 5 March 2025).
47. Toler, T.R.; Evans, E.W.; Tepedino, V.J. Pan-trapping for Bee (Hymenoptera: Apiformes) in Utah's West Desert: The importance of colour diversity. *Pan-Pac Entomol.* **2005**, *81*, 103–113.
48. Roulston, T.H.; Smith, S.A.; Brewster, A.L. A comparison of Pan Trap and Intensive Net Sampling Techniques for documenting a bee (Hymenoptera: Apiformes) Fauna. *J. Kans. Entomol. Soc.* **2007**, *80*, 179–181. [CrossRef]
49. Saunders, M.E.; Luck, G.W. Pan trap catches of pollinator insects vary with habitat. *Aus. J. Entomol.* **2013**, *52*, 106–113. [CrossRef]
50. Lebuhn, G.; Droege, S.; Connor, E.F.; Gemmill-Herren, B.; Potts, S.G.; Minckley, R.L.; Griswold, T.; Jean, R.; Kula, E.; Roubik, E.W.; et al. Detecting insect pollinator declines on regional and global scales. *Conserv. Biol.* **2013**, *27*, 113–120. [CrossRef] [PubMed]
51. Aguiar, A.P.; Sharkov, A. Blue Pan Traps for collecting Stephanidae (Hymenoptera). *J. Hymenopt. Res.* **1997**, *6*, 422–423.
52. Gollan, J.R.; Ashcroft, M.B.; Batley, M. Comparison of yellow and white pan traps in surveys of bee fauna in New South Wales, Australia (Hymenoptera: Apoidea: Anthophila). *Aust. J. Entomol.* **2011**, *50*, 174–178. [CrossRef]
53. Prado, S.G.; Ngo, H.T.; Florez, J.A.; Collazo, J.A. Sampling bees in tropical forests and agroecosystems: A review. *J. Insect Conserv.* **2017**, *21*, 753–770. [CrossRef]
54. Mayer, C.; Adler, L.; Armbruster, W.S.; Dafni, A.; Eardle, C.; Huan, S.Q.; Kevan, P.G.; Ollerton, J.; Packer, L.; Ssymank, A.; et al. Pollination ecology in the 21st century: Key questions for future research. *J. Pollinat. Ecol.* **2011**, *3*, 8–23. [CrossRef]
55. Ollerton, J.; Winfree, R.; Tarrant, S. How many flowering plants are pollinated by animals? *Oikos* **2011**, *120*, 321–326. [CrossRef]
56. Scheper, J.; Reemer, M.; van Kats, R.; Ozinga, W.A.; van der Linden, G.T.J.; Schaminée, J.H.J.; Siepel, H.; Kleijn, D. Museum specimens reveal loss of pollen host plants as key factor driving wild bee decline in The Netherlands. *Proc. Natl. Acad. Sci. USA* **2014**, *111*, 17552–17557. [CrossRef]
57. Winfree, R.; Aguilar, R.; Vázquez, D.P.; LeBuhn, G.; Aizen, M.A. A meta-analysis of bees' responses to anthropogenic disturbance. *Ecology* **2009**, *90*, 2068–2076. [CrossRef] [PubMed]
58. Newbold, T.; Hudson, L.N.; Hill, S.L.L.; Contu, S.; Lysenko, I.; Senior, R.A.; Börger, L.; Bennett, D.J.; Choimes, A.; Collen, B.; et al. Global effects of land use on local terrestrial biodiversity. *Nature* **2015**, *520*, 45–50. [CrossRef] [PubMed]

59. Millard, J.; Outhwaite, C.L.; Kinnersley, R.; Freeman, R.; Gregory, R.D.; Adedoja, O.; Gavini, S.; Kioko, E.; Kuhlmann, M.; Ollerton, J.; et al. Global effects of land-use intensity on local pollinator biodiversity. *Nat. Commun.* **2021**, *12*, 2902. [CrossRef] [PubMed]
60. Regional Meteorological Department, Kolkata. 2020. Available online: https://mausam.imd.gov.in/kolkata/ (accessed on 6 March 2025).
61. Thakuria, C. Yield assessment of Indian mustard variety NRCHB101 with toria varieties TS 36 and TS 38 in Dibrugarh district of Assam. *Pharma Innov. J.* **2023**, *12*, 1502–1503.
62. Deka, B.C.; Parisa, D.; Singha, A.K.; Siangshai, R.; Massar, D.A. (Eds.) *Impact of Technologies on Oilseeds Production in North Eastern Region*; ICAR-Agricultural Technology Application Research Institute (ATARI): Zone-VII: Umiam, Meghalaya, India, 2018; p. 28.
63. Belavadi, V.V.; Ganeshaiah, K.N. *Insect Pollination Manual*; Indian Council of Agricultural Research: New Delhi, India, 2013; pp. 1–44.
64. Willmer, P. *Pollination and Floral Ecology*; Princeton University Press: Princeton, NJ, USA, 2011; pp. 1–832.
65. Shapiro, L.H.; Tepedino, V.J.; Minckley, R.L. Bowling for bees: Optimal sample number for "bee bowl" sampling transects. *J. Insect Conserv.* **2014**, *18*, 1105–1113. [CrossRef]
66. Moreira, E.F.; Silva Santos, R.L.S.; Penna, U.L.; Coca, C.A.; Oliveira, F.F.; Blandina Felipe Viana, B.F. Are pan traps colors complementary to sample community of potential pollinator insects? *J. Insect Conserv.* **2016**, *20*, 583–596. [CrossRef]
67. Droege, S.; Tepedino, V.J.; LeBuhn, G.; Link, W.; Minckley, R.L.; Chen, Q.; Conrad, C. Spatial scale and sampling interval effects on abundance and species richness of native bees. *Biol. Conserv.* **2010**, *143*, 1068–1074.
68. Rogers, S.R.; Tarpy, D.R.; Burrack, H.J. Bee Species Diversity Enhances Productivity and Stability in a Perennial Crop. *PLoS ONE* **2014**, *9*, e97307. [CrossRef]
69. Devi, M.; Sharma, H.K.; Thakur, R.K.; Bhardwaj, S.K.; Rana, K.; Thakur, M.; Ram, B. Diversity of insect pollinators in reference to seed set of mustard (*Brassica juncea* L.). *Int. J. Curr. Microbiol. Appl. Sci.* **2017**, *6*, 2131–2144. [CrossRef]
70. Berglund, H.L.; Milberg, P. Sampling of fower-visiting insects: Poor correspondence between the catches of colour pan−trap and sweep netting. *Eur. J. Entomol.* **2019**, *116*, 425–431. [CrossRef]
71. Lezzeri, M.; Lozano, V.; Brundu, G.; Floris, I.; Pusceddu, M.; Quaranta, M.; Satta, A. Standardized transect walks outperform pan traps in assessing wild bee community in a Mediterranean protected area (Asinara National Park, Italy). *Biodivers. Conserv.* **2024**, *33*, 2329–2344. [CrossRef]
72. Gibbs, J.; Joshi, N.K.; Wilson, J.K.; Rothwell, N.L.; Powers, K.; Haas, M.; Gut, L.; Biddinger, D.J.; Isaacs, R. Does passive sampling accurately reflect the bee (Apoidea:Anthophila) communities pollinating apple and sour cherry orchards? *Environ. Entomol.* **2017**, *46*, 579–588. [CrossRef] [PubMed]
73. McCravy, K.W.; Ruholl, J.D. Bee (Hymenoptera: Apoidea) diversity and sampling methodology in a Midwestern USA deciduous forest. *Insects* **2017**, *8*, 81. [CrossRef] [PubMed]
74. Vrdoljak, S.M.; Samways, M.J. Optimising coloured pan traps to survey flower visiting insects. *J. Insect Conserv.* **2012**, *16*, 345–354. [CrossRef]
75. Nuttman, C.V.; Otieno, M.; Kwapong, P.K.; Combey, R.; Willmer, P.; Potts, S.G. The utility of aerial pan-trapping for assessing insect pollinators across vertical strata. *J. Kans. Entomol. Soc.* **2011**, *84*, 260–270. [CrossRef]
76. Grundel, R.; Frohnapple, K.J.; Jean, R.P.; Pavlovic, N.B. Effectiveness of bowl trapping and netting for inventory of a bee community. *Environ. Entomol.* **2011**, *40*, 374–380. [CrossRef]
77. Kirk, W.D.J. Ecologically selective coloured traps. *Ecol. Entomol.* **1984**, *9*, 35–41. [CrossRef]
78. Zou, Y.; Xiao, H.; Felix, J.J.; Bianchi, A.; Jauker, F.; Luo, S.; Werf, W. Wild pollinators enhance oilseed rape yield in small-holder farming systems in China. *BMC Ecol.* **2017**, *17*, 6. [CrossRef]
79. Ludewig, M.J.; Landaverde-González, P.; Götz, K.P.; Chmielewski, F.-M. Initial assessment to understand the effect of air temperature on bees as floral visitors in urban orchards. *J. Insect Conserv.* **2023**, *27*, 1013–1022. [CrossRef]
80. Casiá-Ajché, Q.B.; Escobedo-Kenefic, N.; Escobar-González, D.; Cardona, E.; Mejía-Coroy, A.; Morales-Siná, J.; Enríquez, E.; Landaverde-González, P. Unveiling the effects of land use and intra-seasonal variation on bee and plant diversity and their ecological interactions in vegetation surrounding coffee plantations. *Front. Bee Sci.* **2024**, *2*, 1408854. [CrossRef]
81. Tsang, T.P.; De Santis, A.A.; Armas-Quiñonez, G.; Ascher, J.S.; Ávila-Gómez, E.S.; Báldi, A.; Ballare, K.; Balzan, M.V.; Banaszak-Cibicka, W.; Bänsch, S.; et al. Land use change consistently reduces $\alpha$-but not $\beta$-and $\gamma$-diversity of bees. *Glob. Chang. Biol.* **2025**, *31*, e70006. [CrossRef] [PubMed]

*agronomy*

MDPI

*Article*

# Entomopathogenic Fungus Treatment Affects Trophic Interactions by Altering Volatile Emissions in Tomato

**Asim Munawar [1,†], Haonan Zhang [1,†], Jinyi Zhang [1,†], Xiangfen Zhang [1], Xiao-Xiao Shi [2], Xuan Chen [1], Zicheng Li [1], Xiaoli He [1,3], Jian Zhong [1], Zengrong Zhu [1,3], Yaqiang Zheng [4,*] and Wenwu Zhou [1,*]**

[1] State Key Laboratory of Rice Biology, Ministry of Agricultural and Rural Affairs Key Laboratory of Molecular Biology of Crop Pathogens and Insect Pests, Institute of Insect Sciences, Zhejiang University, Hangzhou 310058, China; asim_munawar@zju.edu.cn (A.M.); jamehowlnen@zju.edu.cn (H.Z.); 22016107@zju.edu.com (J.Z.); 22316277@zju.edu.cn (X.Z.); 12116104@zju.edu.cn (X.C.); lizicheng386@zju.edu.cn (Z.L.); xiaolihe@zju.edu.cn (X.H.); 11816067@zju.edu.cn (J.Z.); zrzhu@zju.edu.cn (Z.Z.)

[2] Zhejiang Academy of Forestry, Hangzhou 310023, China; shixiao6656@163.com

[3] Hainan Institute, Zhejiang University, Sanya 572000, China

[4] Resource and Utilization Research Center of Medicinal Cordyceps, Guizhou University of Traditional Chinese Medicine, Guiyang 550025, China

* Correspondence: zhengyaqiang131@gzy.edu.cn (Y.Z.); wenwuzhou@zju.edu.cn (W.Z.)

† The authors contributed equally to this study.

**Abstract:** Entomopathogenic fungi (EPFs) can influence plant–insect interactions through complex molecular and chemical mechanisms. This study investigates how EPF treatment of tomato plants modulates volatile organic compound (VOC) emissions and subsequent trophic interactions between tomato plants, the herbivorous pest *Phthorimaea absoluta*, and the parasitic wasp, *Trichogramma chilonis*. Our results demonstrate that EPF-treated plants exhibited reduced attractiveness to adult *P. absoluta* moths, which were actively repelled by EPF-induced VOCs. Conversely, these same plants showed enhanced recruitment of the parasitoid *T. chilonis*, which demonstrated positive chemotaxis toward the modified VOC profile. Chemical analysis revealed significantly elevated emissions of key VOCs in EPF-treated plants, particularly (*E*)-β-Caryophyllene, β-phellandrene, and α-Phellandrene. This increase is correlated with enhanced production of defense-related phytohormones, including JA, SA, and JA-Ile, which may regulate VOC biosynthesis pathways. Behavioral response studies using synthetic VOCs and electroantennogram (EAG) measurements confirmed that these EPF-induced VOCs elicited strong olfactory responses in both insect species. To summarize, EPF treatment reshapes multitrophic interactions by strategically modulating plant VOC emissions and activating defense signaling pathways in tomato plants, providing new insights for potential applications in sustainable pest management strategies.

**Keywords:** entomopathogenic fungus; *Phthorimaea absoluta; Trichogramma chilonis*; olfactory preferences; volatile emissions

## 1. Introduction

Entomopathogenic fungi (EPFs) play a significant role in the regulation of insect populations [1]. These EPFs serve as promising biological control agents that can serve as environmentally sustainable alternatives to chemical insecticides [2]. EPFs exhibit insecticidal properties against numerous insect pests, including *Myzus persicae*, *Aleurodicus rugioperculatus*, and *Leucoptera coffeella* [3,4]. In addition to its direct effects on insects, EPF colonization can also lead to indirect effects on herbivores via affecting plant defenses [5] and recently has gained increasing research attention [6].

EPFs have emerged as multi-functional biocontrol agents in agricultural systems, extending beyond their direct pathogenic effects on insect pests [7]. When colonizing plants as endophytes, these EPFs establish complex symbiotic relationships that significantly alter plant physiology, growth patterns, and defense capabilities [8,9]. Endophytic colonization by EPFs induces significant changes in the host plant's physiology and metabolism, influencing growth, development, and disease resistance by modulating the metabolic pathways that enhance resilience to various biotic and abiotic stresses [10]. These growth-promoting effects are mediated through the fungal production of phytohormones and enzymes that facilitate nutrient solubilization and uptake [11]. During plant development, EPFs enhance the immune system by triggering the synthesis of key secondary metabolites—including phenolics, terpenoids, and alkaloids—that function as natural defenses against herbivore attacks [12–14]. These plant-derived metabolites play vital roles in regulating plant–insect interactions by influencing various aspects of plant defense and signaling mechanisms [15]. However, the specific mechanisms by which EPF treatment alters plant defense and how these changes impact trophic interactions remain poorly characterized despite their potential significance for pest management strategies.

VOCs are ubiquitous substances in plants that regulate insect–plant interactions and have been elucidated as a pivotal component of plant defense mechanisms against herbivory [15,16]. When subjected to external stressors, plants activate their direct and indirect defense mechanisms [17]. They synthesize and release intricate blends of VOCs that play a crucial role in mediating indirect resistance against herbivores and their natural enemies [18,19]. Changes in plant physiology and biochemistry induced by EPF colonization can modulate VOC emissions [20–22]. EPF colonization has been demonstrated to affect plant VOC emissions [23–25], likely due to the shared molecular synthesis pathway between VOCs and jasmonic acid (JA), which serves as a regulatory factor for VOC production [19,26]. Phytohormones, whose levels may be altered by EPF colonization, also influence VOC release. Ultimately, EPF colonization-induced changes in plant VOC emissions can change the behavior of herbivorous insects and their associated natural enemies.

Tomato (*Solanum lycopersicum*), originating from South America, is the second-largest vegetable crop grown globally. *P. absoluta* is one of the most devastating pests of Solanaceous plants and is recorded as a major invasive pest worldwide due to its high reproductive capacity, substantial food consumption, and extensive migration behaviors [27–29]. Adults preferentially oviposit on stems or leaf blades and the larvae directly consume leaves or fruits upon hatching. Larvae typically mine leaves and consume substantial amounts of mesophyll, leading to reduced photosynthesis. In severe infestations, tomato yields can be reduced by 80–100% [30]. Insecticides, such as pyrethroids and abamectin, are primarily used for pest control in tomatoes, but *P. absoluta* has developed resistance [31,32]. Concurrently, the injudicious application of pesticides has adverse effects on human health, the agricultural environment, and non-target insects. In these situations, using EPF for the indirect management of herbivore pests provides a better alternative. Trichogramma wasps are widely used as biocontrol agents [33–35]. Among these, *T. chilonis* is a widespread egg parasitoid employed to manage various lepidopteran herbivores [36].

EPF colonization has been shown to induce physiological and biochemical changes in plants, particularly altering their VOC emissions [36,37]. These plant-emitted VOCs serve as important chemical cues for *T. chilonis* wasps to locate their hosts [36]. Such changes may consequently affect the behavior of both *P. absoluta* and *T. chilonis*. In this study, we tested two hypotheses: (I) the EPF colonization of tomato plants may influence their attractiveness to adult herbivorous insects and parasitoids, altering trophic interactions; (II) these altered trophic interactions under EPF colonization may be mediated by changes in VOC emissions. This study may provide insights for understanding the molecular mechanisms of EPF treatment on insect–plant interactions.

## 2. Materials and Methods

### 2.1. Plant and Insect Culture

Tomato seeds (*S. lycopersicum*, cv. 'Heinz') were used in the study. The seeds underwent surface sterilization by immersion in 1% NaClO for 5 min, followed by three thorough rinses with sterile water. The sterilized seeds were then placed on sterile moist filter paper to germinate. Once small plantlets developed, they were transferred to 250 mL pots. All the plants were randomly distributed in an environmental chamber maintained at 28 °C, with a 16 h/8 h light/dark cycle and 80% relative humidity.

*P. absoluta* specimens were obtained from Yuxi City, Yunnan Province, China, and their population was reared for more than 10 generations on tomato plants under controlled conditions: 28 °C, 16 h:8 h light/dark cycle, and 80% relative humidity. The egg parasitoid, *T. chilonis*, was sourced commercially from Keyun Biocontrol, China, and subsequently reared for more than 4 generations on *P. absoluta* eggs in laboratory conditions prior to use in experiments. All the experiments were conducted between [06/2020] and [03/2025].

### 2.2. Fungal Culture and Colonization on Tomato Plants

All the EPF strains used in this study were sourced from the Resource and Utilization Research Center of Medicinal Cordyceps, Guizhou University of Traditional Chinese Medicine, Guiyang, China. The obtained EPFs were activated and cultured on Potato Dextrose Agar (PDA; Haibo Biology, Qingdao, Shandong, China) at 27 °C. To prepare the fungal inoculum, conidia were scraped from the culture using a sterile scraper and suspended in sterile distilled water containing 0.05% TWEEN® 80. The conidia concentration was adjusted to $1 \times 10^8$ conidia $mL^{-1}$ using a hemocytometer. A control solution was prepared using sterile distilled water containing 0.05% TWEEN® 80 without fungal conidia. The suspension was uniformly sprayed on tomato leaves until runoff. To promote conidia germination and colonization, inoculated plants were enclosed in plastic bags for 24 h post-treatment. Among all the strains, Cordyceps fumosorosea (YNKM210801) was selected for an in-depth analysis based on preliminary assessments, as it demonstrated the most pronounced effects on herbivore behavior.

To confirm the endophytic colonization of tomato plants, leaves from both the *Cordyceps fumosorosea* (YNKM210801)-inoculated and control plants were surface-sterilized in 1% hypochlorite for 5 min. After sterilization, the leaves were cut into $0.5 \times 0.5$ $cm^2$ pieces and placed on Potato Dextrose Agar (PDA) supplemented with dodine and chloramphenicol following the method described previously [38]. Plates were incubated at $28 \pm 1$ °C with a 16:8 h light/dark cycle and 85% relative humidity for 15 days. After the incubation period, the presence of *C. fumosorosea* (YNKM210801) was assessed. The colonization rate was determined by counting the number of leaf pieces showing EPF growth. Final colonization assessment was conducted at 28 days post inoculation (dpi) following soil drenching treatments. The experiment was repeated three times. The visual evidence of EPF colonization is shown in Figure S1.

### 2.3. P. absoluta Oviposition and T. chilonis Parasitism Assays

The oviposition preference of *P. absoluta* and the parasitism choice of *T. chilonis* were assessed using our previously established bioassay [36] (Figure S2A). In detail, individual cages (65 cm L × 55 cm H × 60 cm W) were set up, each containing one EPF-inoculated plant and one control plant. Pots were covered with aluminum foil to eliminate soil VOC interference. Two newly mated *P. absoluta* females (24 h post-emergence) were introduced into each cage for a 24 h oviposition period, after which the total number of eggs on each plant was quantified.

For the *T. chilonis* parasitism assay, standardized egg masses of *P. absoluta* (N = 150 eggs) were prepared on 30 mm × 30 mm filter paper pieces. These were obtained by allowing gravid *P. absoluta* females to oviposit on 120 mm × 120 mm filter paper in a 3 L transparent plastic container for 24 h. The oviposited filter papers were then sectioned and eggs counted microscopically, with excess eggs removed to maintain the standard count. These egg masses were affixed to the abaxial surface of apical leaflets on both the inoculated and control tomato plants. Eight pairs of newly emerged *T. chilonis* were released into each cage. After a 12 h exposure period, the egg-bearing filter papers were transferred to ventilated Petri dishes (BeyoGold, 100 × 20 mm), sealed with Parafilm, and incubated at 28 °C for subsequent parasitism assessment.

### 2.4. Sampling of Foliar VOC Blends

Foliar VOCs were sampled using Polydimethylsiloxane (PDMS) tubes (Carl Roth, Karlsruhe, Germany) following the methodology detailed in our previous studies [15,36]. Briefly, two clean PDMS tubes and a single leaflet from each plant were enclosed in a sealed plastic chamber for VOC sampling. For constitutive VOC analysis, intact leaves were used. To assess herbivory-induced plant volatiles (HIPVs), the plants were first infested with *P. absoluta* larvae. A blank treatment consisting of an empty plastic chamber with PDMS tubes was included as a control. The VOC analysis was performed using a thermal desorption-gas chromatography/mass spectrometry (TD-GC-MS) system (TD-100XR, Markes International; Trace 1300 GC, ISQ 7000 single quadrupole MS, Thermo Fisher Scientific, Milan, Italy). The resulting chromatographic peaks were normalized by fresh leaf mass and integrated. Compound identification was conducted using the NIST v and rep libraries integrated within the Chromeleon software v7.2.8 (Thermo Scientific), complemented by comparison with authentic standards.

### 2.5. Phytohormone Extraction and Quantification

Two fully expanded compound leaves were harvested from the treatment plants and immediately flash-frozen in liquid nitrogen until analysis. Phytohormone extraction and quantification procedures were conducted using an ultra-performance liquid chromatography-tandem mass spectrometry (UPLC-MS/MS) system (LCMS-8040 system, Shimadzu Corporation, Kyoto, Japan) following previously established protocols without modifications [39,40].

### 2.6. Insect Choice Assay with Y-Tube Olfactometer

The behavioral responses of *P. absoluta* and *T. chilonis* to synthetic VOCs were assessed using a glass Y-tube olfactometer system adapted from our previous study [36]. The Y-tube consisted of a 16 cm stem and two 10 cm arms set at a 60° angle, with an inner diameter of 10 mm. Purified air was supplied by an air pump (0.02 MPa) through activated charcoal, distilled water, and silica gel. Airflow rates were set at 0.5 L $min^{-1}$ for *P. absoluta* and 0.1 L $min^{-1}$ for *T. chilonis*, regulated through Teflon tubing connected to the Y-tube

(Figure S2B). Prior to testing, newly emerged females of both species were paired with males for 24 h. Synthetic VOCs were diluted in n-hexane (≥98.0%, Shanghai Aladdin Biochemical Technology Co.Ltd. Xinjinqiao, Pudong, Shanghai) to concentrations of 0.01, 0.05, 0.1, 0.5, and 1.0 μL mL$^{-1}$. Aliquots were applied to filter paper (4 cm × 2 cm) using a glass micropipette. The treated filter paper was placed in a glass tube (10 cm × 2 cm) connected to one arm of the Y-tube, while a hexane-treated control was placed in the opposite arm. Individual insects were introduced into the stem and allowed to make a choice. Non-responsive individuals were excluded from the analysis. To mitigate positional bias, odor source positions were alternated every five trials and fresh VOC samples were used. The apparatus was cleaned and sterilized after every 10 trials.

For the whole-plant preference tests, a similar setup was employed. The plants were enclosed in transparent plastic sheets, with the upper ends connected to the air source and choice arm via Teflon tubes. The plant pots were covered with aluminum foil to minimize soil VOC interference.

### 2.7. Electroantennogram (EAG) and Odor Delivery

The EAG responses of *P. absoluta* and *T. chilonis* antennae were recorded following the methodology described previously [36]. The apical tip of each antenna was excised with micro-scissors and inserted into a recording electrode containing 0.2 M KCl solution. A similar reference electrode was attached to a micromanipulator and connected to the distal end of the antenna, completing the circuit. The EAG signals were amplified using a high-impedance (>$10^{12}$ Ω) pre-amplifier coupled to an EAG amplifier (Syntech, AM-02, Kirchzarten, Germany). The amplified signals were then processed and digitized using a signal acquisition interface board (Syntech, IDAC-02).

Odor stimuli were delivered using an air stimulus controller (Syntech, CS-05) with a constant flow (2 L min$^{-1}$) of humidified air passing over the antenna through a glass tube (20 × 0.5 cm) positioned 20 mm from the preparation. VOCs were introduced via a pipette tip (Corning Life Sciences, Jiangsu, China) inserted 5 cm from the end of the glass tube. During stimulation, 0.4 L min$^{-1}$ of air was pulsed through the pipette tip into the main airflow for 0.40 s. The test compounds were serially diluted in n-hexane to concentrations of 0.01, 0.05, 0.1, 0.5, and 1.0 μL mL$^{-1}$ for dose–response studies. Aliquots of 7 μL of each concentration were applied to filter paper strips and inserted into the pipette tip. Odors were presented in order of increasing concentration, with 30 s intervals between applications to allow for antennal recovery. A pipette tip containing 7 μL of n-hexane on filter paper served as a control. The EAG responses were quantified as the amplitude of deflection relative to the control stimulus. Peak amplitudes were marked and expressed as relative amplitudes (mV) compared to the standard response.

### 2.8. Synthetic VOC Applications and *P. absoluta* Behavior Assays

To investigate the influence of plant VOCs on *P. absoluta* behavioral responses, we applied synthetic VOCs to the plant leaf surfaces following the protocols established [15]. (*E*)-β-Caryophyllene, at a biologically relevant concentration of 1.5 μL mL$^{-1}$, was precisely dissolved in liquefied lanolin (BBI Life Sciences) to create a standardized treatment solution. The control plants received applications of pure lanolin, with the lanolin aliquot prepared according to methods previously described [41]. Following the (*E*)-β-Caryophyllene treatment, the plants were immediately subjected to *P. absoluta* oviposition behavioral assays using the experimental protocols detailed in the preceding sections.

### 2.9. Statistical Analyses

The total number of eggs oviposited and parasitized was analyzed using a generalized linear model (GLM) with Poisson distribution [42]. Pearson's chi-square ($\chi^2$) tests were used to evaluate differences in the responses of *P. absoluta* and *T. chilonis* to odor stimuli. The EAG responses of both species were analyzed using one-way analysis of variance (ANOVA) and means were separated by post hoc comparisons (Tukey's HSD; $p \leq 0.05$). Phytohormone data were analyzed using *T*-tests. Hierarchical cluster analysis using Euclidean distances and Ward's method was performed to generate VOC heatmaps; utilizing the 'pheatmap' package in R., VOC data were further analyzed using analysis of variance (ANOVA) and principal component analysis (PCA). Statistical analyses were conducted using the R software version 4.0.2.

## 3. Results

### 3.1. Effects of EPF Treatment on Herbivore and Parasitoid Behavioral Preferences

We investigated the effects of EPF treatment on plant–insect interactions through oviposition and parasitism choice assays using tomato plants with EPF-colonized roots. Tomato plants treated by *Beauveria bassiana* (HXBA202002, HX200801, M190901), *Metarhizium fkavoviride* (HXMF211004), *Metarhizium anisopliae* (QLSMA201102), *Cordyceps cateniobliqua* (BMZ2075841), *Cordyceps cateinannulata* (BNCC150028), and *Cordyceps fumosorosea* (YNKM210801) received fewer eggs from *P. absoluta* females compared to control plants (Figure S3A–I). Tomato treated by *Metarhizium rileyi* (QLSMR211004) received a similar number of eggs as those of untreated plants. Among these EPF strains, *C. fumosorosea* (YNKM210801) demonstrated the most potent deterrent effect on oviposition (Figure S3G–I). The exogenous application of *C. fumosorosea* (YNKM210801) to tomato plants also resulted in reduced oviposition by *P. absoluta* females (GLM: T = 2.22, $p = 0.015$), confirming the effectiveness of this EPF when applied externally (Figure S4). Additionally, the root application of *C. fumosorosea* (YNKM210801) significantly altered plant attractiveness to both herbivores and parasitoid wasps. In choice experiments, *P. absoluta* showed a strong preference for control plants over *C. fumosorosea* (YNKM210801)-treated plants for oviposition (GLM: T = 2.13, $p = 0.019$). In contrast, the egg parasitoid *T. chilonis* significantly preferred *C. fumosorosea* (YNKM210801)-treated plants for parasitism (GLM: T = 2.39, $p = 0.01$) (Figure 1A,B). These preference patterns persisted even in plants subjected to the larval feeding of *P. absoluta*, indicating that herbivore damage did not substantially alter the effects of EPF treatment on plant attractiveness to either herbivores or parasitoids.

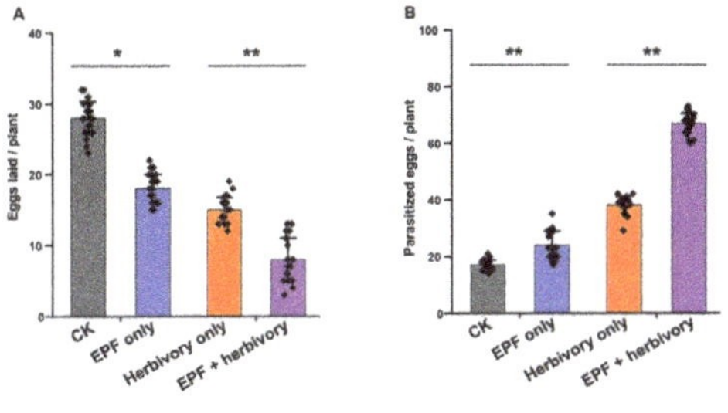


**Figure 1.** *C. fumosorosea* (YNKM210801) treatment reduced herbivore moth attraction while increasing parasitoid preference for tomato plants: (**A**) Oviposition preference of female *P. absoluta* ($n = 24$ tests per treatment). (**B**) Parasitism preference of *T. chilonis* ($n = 25$ tests per treatment). Each replicate utilized a new set of plants. The intact and herbivore-induced plants were assessed on separate days. Asterisks (*) denote significant differences between means within the treatments (Tukey HSD following GLM ANOVA, *: $p < 0.05$; **: $p < 0.01$). The data are mean ± SE.

To further investigate the olfactory responses of both insect species, we conducted Y-tube olfactometer experiments (Figure 2A). The *P. absoluta* females exhibited a significant preference for the control plants compared to the *C. fumosorosea* (YNKM210801)-treated plants ($\chi^2$ = 5.58, $p$ = 0.01). Conversely, *T. chilonis* displayed stronger attraction to the *C. fumosorosea* (YNKM210801)-treated plants than to the control plants ($\chi^2$ = 6.33, $p$ = 0.01). The olfactory responses of both insect species intensified when the plants (both control and *C. fumosorosea* (YNKM210801)-treated) were subjected to larval feeding (Figure 2B,C). These results suggest that the *C. fumosorosea* (YNKM210801) treatment modifies the plant VOC profile, thereby influencing plant interactions with herbivores and their natural enemies.

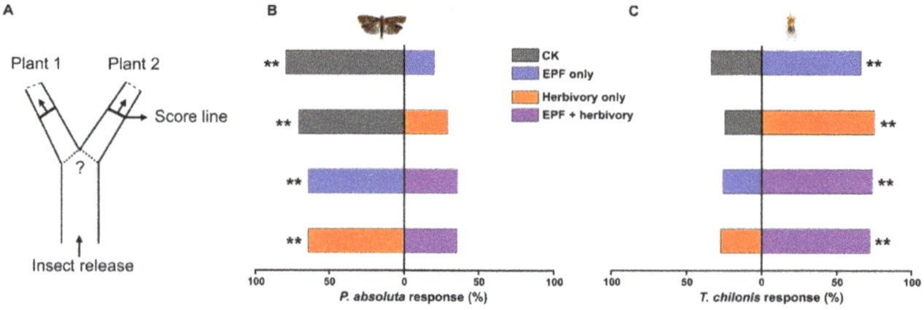


**Figure 2.** *C. fumosorosea* (YNKM210801) treatment altered olfactory behaviors of herbivores and parasitoids: (**A**) Schematic illustration of Y-tube olfactometer used to test insect olfactory responses. (**B**,**C**) Olfactory responses of *P. absoluta* females and *T. chilonis* to control or *C. fumosorosea* (YNKM210801)-treated plants ($n$ = 100 individuals released per test). Non-responding insects were excluded from analysis. Asterisks (*) indicate significant differences in preference between control and EPF-treated plants ($\chi^2$ test; **: $p < 0.01$).

### 3.2. EPF Treatment Modulates VOC Emissions and Induces Phytohormone Accumulation

To elucidate the mechanisms underlying the altered plant attraction to herbivores and parasitoid wasps following the EPF root treatment, we analyzed the VOC emissions in tomato plants (Figure 3A). The *C. fumosorosea* (YNKM210801) treatment induced a distinct VOC profile compared to the control plants, characterized by enhanced emissions of several compounds. This effect remained consistent in both the undamaged and herbivory-damaged plants, indicating that herbivore damage did not significantly alter the *C. fumosorosea* (YNKM210801)-induced VOC emissions. Generally, the VOC levels detected from the undamaged leaves were lower than those from the herbivory-induced leaves. Among the *C. fumosorosea* (YNKM210801)-enhanced VOCs, (*E*)-β-caryophyllene and β-phellandrene exhibited higher emissions in the undamaged leaves, with further enhancement observed in both the *C. fumosorosea* (YNKM210801)-treated and herbivory-damaged leaves. Notably, α-phellandrene was the only VOC significantly enhanced by the combination of the *C. fumosorosea* (YNKM210801) treatment and herbivory damage, while α-humulene was uniquely enhanced in both the *C. fumosorosea* (YNKM210801)-treated leaves and *C. fumosorosea* (YNKM210801) + herbivory damaged leaves (Figure 3A). Additionally, the emissions of several other VOCs including α-pinene, 3-carene, and 2-carene were significantly affected across all the treatments. The principal component analysis (PCA) of the VOC profiles revealed clear treatment-specific signatures, with the first two principal components explaining 94.26% of the total variance (PC1: 84.3%; PC2: 9.96%) (Figure 3B). Furthermore, the *C. fumosorosea* (YNKM210801)-treated plants exhibited distinct temporal patterns in their VOC emissions. Among the most abundant VOC in tomato, (*E*)-β-caryophyllene, α-phellandrene, and β-phellandrene demonstrated time-dependent emission profiles (Figure 4A–C). Notably, (*E*)-β-caryophyllene was the only volatile compound that maintained significant emission levels across all the time points examined (Figure 4A). In contrast, α-phellandrene showed no significant emission at the 6 h time point (Figure 4B), while β-phellandrene emissions were not significant at the 3 h collection period (Figure 4C). These temporal variations in VOC emission patterns suggest that the *C. fumosorosea* (YNKM210801) treatment induces a complex and dynamic reprogramming of plant secondary metabolism, potentially as part of the plant's adaptive response to fungal presence.

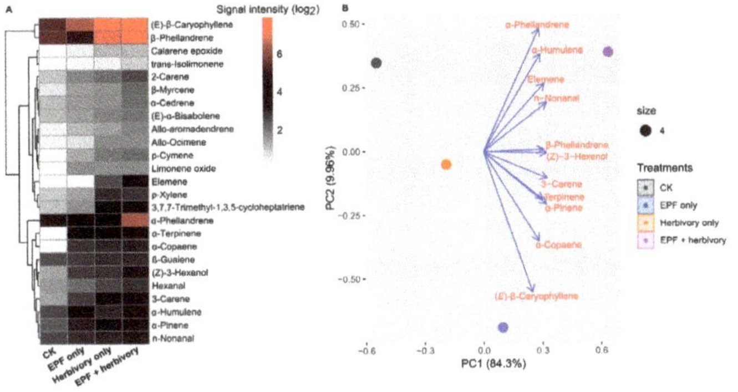


**Figure 3.** *C. fumosorosea* (YNKM210801) treatment enhances VOC emission in tomato plants: (**A**) Heatmap showing relative abundance of VOCs from control and *C. fumosorosea* (YNKM210801)-treated tomato plants ($n$ = 6 plants/treatment, 6 h post-treatment). (**B**) PCA of VOC profiles demonstrating treatment-specific clustering of compound signatures.

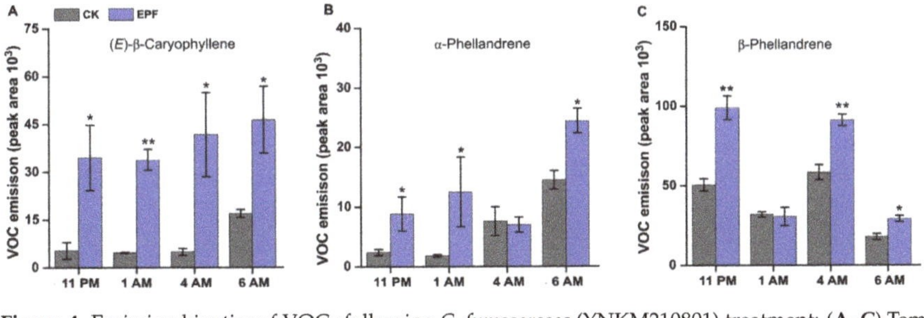


**Figure 4.** Emission kinetics of VOCs following *C. fumosorosea* (YNKM210801) treatment: (**A**–**C**) Temporal patterns of highly abundant VOCs emitted from tomato plants treated with *C. fumosorosea* (YNKM210801) compared to control plants. Asterisks (*) denote significant differences between control and *C. fumosorosea* (YNKM210801)-treated plants at each time point ($t$-test, *: $p < 0.05$ and **: $p < 0.01$). Data are presented as means ± SE.

EPF colonization also significantly influenced phytohormone accumulation in tomato leaves. In the undamaged plants, JA (T8 = 2.70, $p = 0.027$) and JA-Ile (T8 = 2.84, $p = 0.022$) levels were significantly higher in the *C. fumosorosea* (YNKM210801)-treated plants compared to controls (Figure 5A,B). This pattern persisted in the herbivory-damaged plants as well. Interestingly, the SA levels were higher in the undamaged plants than in the damaged plants following the *C. fumosorosea* (YNKM210801) treatment (T8 = 2.96, $p = 0.039$) (Figure 5C). These differential responses in phytohormone accumulation suggest that various phytohormones may play distinct roles in mediating plant indirect defense responses to *C. fumosorosea* (YNKM210801) colonization and herbivory.

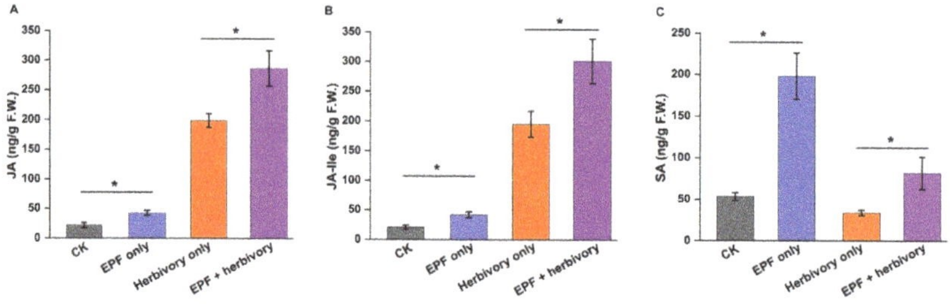


**Figure 5.** *C. fumosorosea* (YNKM210801) treatment induces phytohormone accumulation in tomato plants: (**A–C**) Phytohormone concentrations in control and *C. fumosorosea* (YNKM210801)-treated tomato plants ($n$ = 5 plants per treatment). Means between treatment groups were compared using *T*-tests. Asterisks (*) indicate significant differences between control and treated plants within each treatment (*t*-test; *: $p < 0.05$). Bars represent means ± standard error.

### 3.3. Behavioral and Electrophysiological Responses of Insects to Synthetic VOC Odors

To verify whether the altered VOC profiles are responsible for the observed changes in plant–herbivore–parasitoid interactions, we conducted Y-tube olfactometer and EAG studies using pure VOCs. The Y-tube olfactometer assays revealed distinct behavioral responses of *P. absoluta* and *T. chilonis* female adults to the tested VOCs (Figure 6A–F). The olfactory response of *T. chilonis* demonstrated high sensitivity and attraction to all three tested VOCs with concentration-dependent patterns. For (*E*)-β-caryophyllene, *T. chilonis* consistently showed stronger responses at lower concentrations, with a significant decrease in response at 1.0 μL mL$^{-1}$ ($\chi^2$ = 4.79, $p$ = 0.028) (Figure 6A). α-Phellandrene elicited reduced responses at both 0.01 μL mL$^{-1}$ ($\chi^2$ = 4.29, $p$ = 0.038) and 0.5 μL mL$^{-1}$ ($\chi^2$ = 5.23, $p$ = 0.022), with enhanced responses observed at 0.05, 0.1 and 1.0 μL mL$^{-1}$ concentrations (Figure 6B). For β-phellandrene, *T. chilonis* exhibited decreased response at 0.1 μL mL$^{-1}$ ($\chi^2$ = 4.93, $p$ = 0.026) while showing stronger responses at the other tested concentrations (Figure 6C). Conversely, compared to the control (n-hexane), all three VOCs exhibited repellent effects on *P. absoluta* (Figure 6D–F). The olfactory responses to (*E*)-β-caryophyllene were significant across all the concentrations except 0.05 μL mL$^{-1}$ ($\chi^2$ = 3.59, $p$ = 0.058) (Figure 6D). Responses to α-phellandrene were non-significant at higher concentrations of 0.5 μL mL$^{-1}$ ($\chi^2$ = 3.42, $p$ = 0.064) and 1.0 μL mL$^{-1}$ ($\chi^2$ = 3.47, $p$ = 0.062) while showing the strongest repellent effect at the medium concentration of 0.1 μL mL$^{-1}$ ($\chi^2$ = 6.23, $p$ = 0.01) (Figure 6E). The repellent response of *P. absoluta* to β-phellandrene was significant across most concentrations, with a decreased effect at 0.5 μL mL$^{-1}$ ($\chi^2$ = 5.16, $p$ = 0.023) and a non-significant response at 0.1 μL mL$^{-1}$ ($\chi^2$ = 3.53, $p$ = 0.060) (Figure 6F). These findings indicate that VOCs differentially modulate the behavior of both the parasitoid and pest species, suggesting their potential application in integrated pest management strategies.

We also investigated whether the direct application of pure VOCs to tomato leaves could alter *P. absoluta* oviposition behavior (Figure S5A). Notably, the tomato plants treated with (*E*)-β-caryophyllene exhibited significantly reduced oviposition, with the *P. absoluta* females laying fewer eggs on these plants compared to the control plants treated with lanolin only (GLM: T = 13.59, $p$ = 0.001; Figure S5B). This finding further supports the role of (*E*)-β-caryophyllene as an important mediator of plant–insect interactions. It is worth noting that in our previous study, we demonstrated that lanolin application alone does not induce changes in VOC emission patterns in potato plants, confirming that the observed effects were specifically due to the applied VOC.

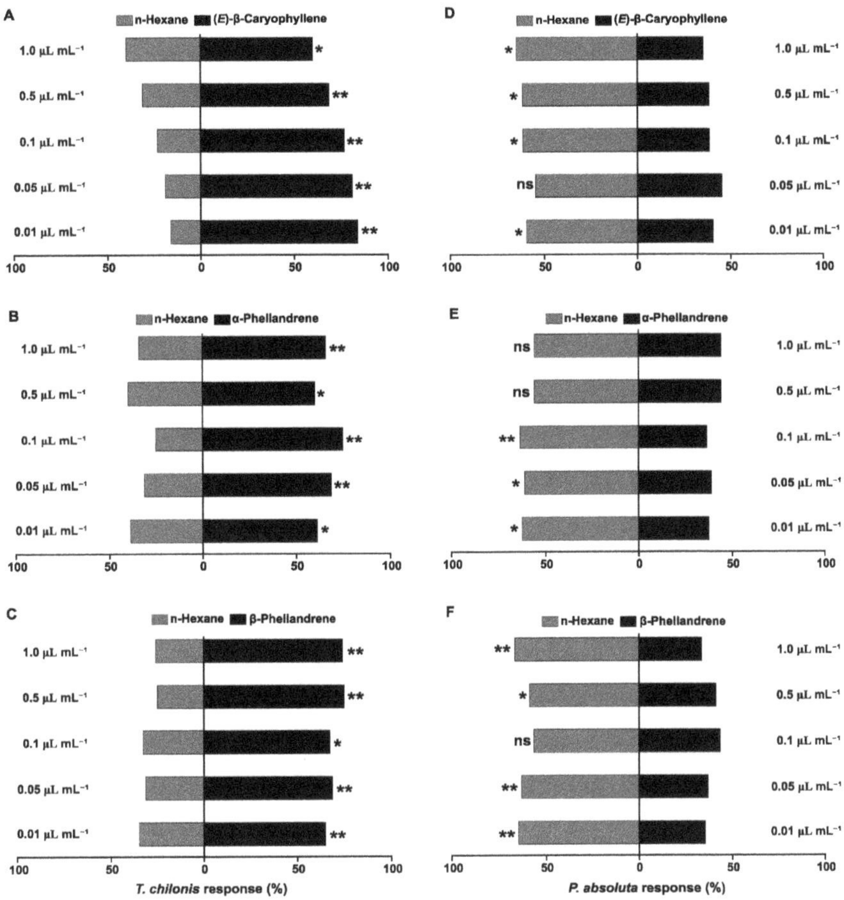


**Figure 6.** Herbivore female moths and their egg parasitoids show differential responses to tomato plant VOCs. (**A–C**) Choice responses of *P. absoluta* and (**D–F**) *T. chilonis* females to varying concentrations of key synthetic tomato VOCs ($n$ = 100 insects tested/concentration). n-Hexane served as control. Non-responding insects were excluded from analysis. Asterisks (*) indicate significant differences in choice response ($\chi^2$ test; *: $p < 0.05$, **: $p < 0.01$). ns; non-significant.

The EAG studies revealed that both insect species, *P. absoluta* and *T. chilonis*, exhibited distinct response patterns to pure VOCs, specifically (*E*)-β-caryophyllene and β-phellandrene (Figure 7A,B). These findings suggest that these plant-emitted VOCs are detectable and recognized by the olfactory systems of both insects. Notably, *T. chilonis* displayed a significantly stronger concentration-dependent EAG response to (*E*)-β-caryophyllene (F3,11 = 61.6, $p$ = 0.001), whereas *P. absoluta* showed a weaker response at lower concentrations, with a marked increase only at higher concentrations (F3,11 = 110.9, $p$ = 0.001; Figure 7A). Also, both species exhibited EAG responses to β-phellandrene, with *T. chilonis* demonstrating the highest sensitivity (F3,11 = 101.4, $p$ = 0.001) compared to *P. absoluta* (F3,11 = 134.7, $p$ = 0.001; Figure 7B). These results indicate that the antennae of both insects are capable of detecting and responding to these specific plant VOCs, albeit with varying sensitivity and intensity.

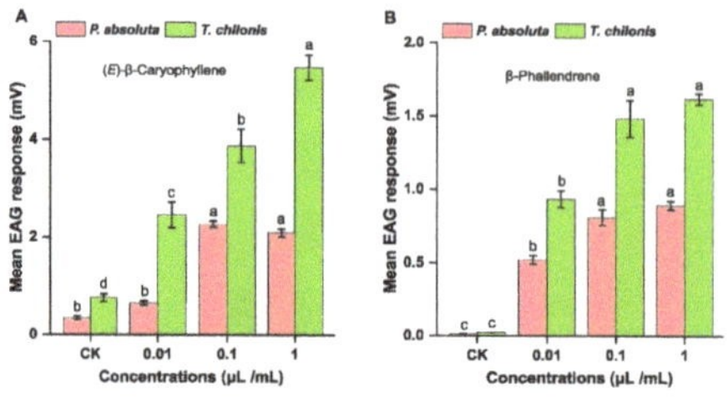


**Figure 7.** EAG responses of herbivore female moth and egg parasitoid antennae to tomato plant VOCs. (**A**,**B**) EAG responses of *T. chilonis* and *P. absoluta* females to varying concentrations of (*E*)-β-caryophyllene and β-phellandrene ($n$ = 3 insects tested/concentration). CK indicates antennal response to n-hexane. Different letters denote significant differences between concentrations within each insect species ($p \leq 0.05$, Tukey's HSD following ANOVA). Bars represent means ± standard error.

## 4. Discussion

EPFs play important roles in crop protection by simultaneously controlling insect pests and altering plant immune responses [43]. These EPFs have the capacity to colonize plant tissues endophytically and can elicit systemic defense responses [44]. EPF colonization triggers a cascade of biochemical and physiological changes within the plant, including the up-regulation of defense-related genes, altered production of secondary metabolites, and priming of the immune system for enhanced resistance against a broad spectrum of pathogens and herbivores [45]. This study investigated the effects of EPF treatment on tomato plant defenses and its consequent impact on plant–insect interactions.

Through choice assay tests and olfactory studies, our findings reveal that EPF treatment significantly alters the plant attractiveness to both herbivores and parasitoid wasps. The EPF treatment of tomato plants significantly reduced the attraction of the female herbivore *P. absoluta* while enhancing the attraction of the parasitoid wasp *T. chilonis*. Importantly, these effects persisted and were stronger even when the plants were subjected to herbivory to produce more VOC. These alterations in plant chemotype and subsequent insect behavior suggest that EPF treatment may serve as an effective strategy for enhancing plant indirect defenses against herbivores while promoting the attraction of beneficial insects [1]. Previous studies have explored the effect of EPFs on insect behavior, such as *Beauveria bassiana* treatment to faba bean seeds which altered the choice preferences and development of the aphid parasitoid, *Aphidius colemani* [46]. Similarly, EPF applications have been found to affect the behavior and life history of peach aphid, *Myzus persicae* [47], and the population growth of two-spotted spider mites, *Tetranychus urticae* [48]. However, the role of plant VOCs mediating insect behavior under such conditions remained largely unexplored. Moreover, it has been discovered that the release of constitutive plant VOC plays an important role in mediating the attraction/repellence of insect species [49,50]. This study found that EPF treatment induces changes in the responses of herbivorous insects and parasitoids for tomato plants, which was likely attained via the alteration of the VOCs by the host, which is involved in the interaction of these organisms. Also, the observed changes in insect preference were consistent in both choice assays and olfactometer experiments, indicating that the effect is primarily mediated through plant VOC emissions rather than visual or tactile cues.

Plant-released VOCs play important roles in regulating plant–insect interactions in agroecosystem [16]. EPF-treated tomato plants emitted a distinct VOC blend compared to control plants, with several compounds showing enhanced emissions. Interestingly, the VOC profile alterations persisted even under herbivore attack. Notably, (*E*)-β-caryophyllene and β-phellandrene exhibited increased emissions in both the undamaged and herbivore-damaged EPF-treated plants. These sesquiterpenes have been previously associated with indirect plant defenses, attracting natural enemies and repelling herbivores [15,36,51]. Our results showed that (*E*)-β-caryophyllene and β-phellandrene were repellent to *P. absoluta* while attractive to *T. chilonis*. Moreover, both insects also responded differently to several other key tomato leaf VOCs, indicating that these VOCs might also be involved in promoting or deterring natural enemies and herbivores [52]. Our olfaction studies provided crucial evidence linking the observed changes in plant VOC profiles to specific insect behavioral and physiological responses. *P. absoluta* and *T. chilonis* exhibited distinct preferences for different VOCs, with responses often showing dose dependency [36]. Most tested VOCs were repellent to *P. absoluta* but attractive to *T. chilonis*, aligning with the overall pattern of herbivore deterrence and parasitoid attraction observed in whole-plant assays.

The altered plant–insect interactions observed in our study can be largely attributed to the significant changes in VOC profiles induced by EPF treatment. The alterations in VOC profiles and plant–insect interactions were accompanied by significant changes in phytohormone levels following EPF treatment. JA and SA are key regulators of plant responses to biotic stresses [53,54], particularly in mediating defenses against herbivores and necrotrophic pathogens [36,55]. Our results showed that the levels of JA, JA-Ile, and SA were increased in both undamaged and herbivore-damaged EPF-treated plants. This simultaneous activation of both the JA and SA pathways is noteworthy, as these hormones often exhibit antagonistic interactions [56]. However, recent studies have shown that certain beneficial microbes can circumvent this antagonism, leading to a synergistic activation of multiple defense pathways [57]. These changed phytohormone levels likely play a crucial role in modulating the plant VOC emissions, as both JA and SA are known to regulate the biosynthesis of various defense-related VOC [15]. Furthermore, the sustained increase in these defense hormones in the EPF-treated plants, even prior to the herbivore attack, suggests that the fungus may be priming the plant defense responses, potentially enabling a more rapid and robust response to subsequent stresses.

Our findings reveal a significant relationship between (*E*)-β-Caryophyllene treatment and *P. absoluta* oviposition behavior. When the tomato plants were treated with (*E*)-β-Caryophyllene, we observed a marked reduction in oviposition by adult female *P. absoluta*. This observation aligns with our Y-tube olfactometer assays, which demonstrated that this VOC effectively repels *P. absoluta*, thus confirming the involvement of (*E*)-β-Caryophyllene in modulating host preference behaviors. Previous research demonstrated that application of (*E*)-β-Caryophyllene to potato leaves significantly altered the movement patterns of predatory mites across plant organs [15]. Taken together with our current results, these findings suggest that EPFs mediate an increase in VOC emission in tomato plants, which may serve to prime both the direct and indirect plant defense responses [58]. Unlike traditional pesticides that directly target pests, EPFs appear to enhance the plant's own defense mechanisms, creating a less favorable environment for herbivores while simultaneously attracting their natural enemies. The use of EPFs as a biocontrol agent could reduce reliance on chemical pesticides, thereby minimizing environmental impacts and slowing the development of pesticide resistance in pest populations [59,60]. This research was conducted under controlled laboratory conditions, which may not fully reflect the complexity of field environments [61]. Future studies should validate these findings in diverse field conditions to assess the robustness and applicability of EPF-induced effects across different environmental contexts. Also, further investigation into the molecular mechanisms underlying the observed changes, particularly the signaling pathways mediating the interaction between EPFs and plant defense responses, could provide deeper insights and potentially identify targets for enhancing these beneficial effects.

## 5. Conclusions

Our study provides evidence that EPF treatment significantly alters plant–insect interactions in tomato plants through multiple mechanisms. Choice assays and olfactometer experiments consistently demonstrated that the EPF-treated plants repelled the herbivore *P. absoluta* while attracting the parasitoid *T. chilonis*, with these effects intensifying under herbivory conditions. A chemical analysis revealed that the EPF treatment induced distinct VOC profile changes, particularly elevating emissions of (*E*)-β-caryophyllene and β-phellandrene, compounds associated with plant defensive functions. Our olfaction studies directly linked these specific VOCs to the observed behavioral responses, with (*E*)-β-caryophyllene demonstrably repelling *P. absoluta* while attracting *T. chilonis*. Furthermore, oviposition experiments confirmed that the (*E*)-β-caryophyllene treatment significantly reduced egg-laying by female *P. absoluta*, providing a mechanistic explanation for the observed herbivore deterrence. These altered plant–insect interactions coincided with significant phytohormonal changes, as the EPF treatment elevated the levels of JA, JA-Ile, and SA in both the undamaged and herbivore-damaged plants. This research highlights the potential of EPFs as a biocontrol strategy that strategically modifies plant defense responses to create less favorable conditions for herbivores while promoting natural enemy attraction (Figure 8).

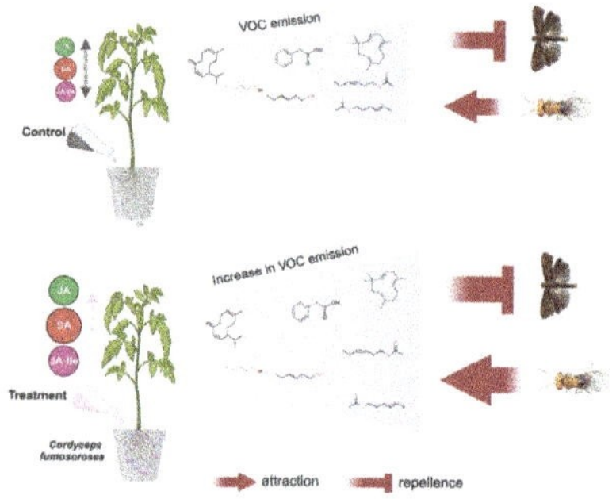


**Figure 8.** Graphical abstract. Endophytic fungal (EPF) treatment reduces herbivore preference and enhances the attraction of their egg parasitoid, mediated by the increased emissions of VOCs and phytohormone inductions. The size of the arrows represents the relative intensity of attraction or repellence.

**Supplementary Materials:** The following supporting information can be downloaded at https://www.mdpi.com/article/10.3390/agronomy15051161/s1, Figure S1: Detection of *Cordyceps fumosorosea* (YNKM210801) colonization on leaves. Figure S2: Schematic drawings of the setup used for the choice experiments. Figure S3: Effect of EPF treatment on the oviposition selection behavior of *P. absoluta*. Figure S4: Effect of *Cordyceps fumosorosea* (YNKM210801) spray application on the oviposition selection behavior of *P. absoluta* for tomato plants. Figure S5: Exogenous application of (*E*)-β-caryophyllene reduces herbivore oviposition preference in tomato plants.

**Author Contributions:** Conceptualization, W.Z. and Y.Z.; methodology, A.M., J.Z. (Jinyi Zhang), X.Z., X.-X.S., X.C., X.H., Y.Z., and W.Z.; software, A.M. and J.Z. (Jian Zhong); validation, A.M., J.Z. (Jinyi Zhang), J.Z. (Jian Zhong), X.C., and W.Z.; formal analysis, A.M., Z.L., J.Z. (Jinyi Zhang), X.Z., X.-X.S., X.C., X.H., J.Z. (Jian Zhong), Y.Z., and W.Z.; investigation, A.M., H.Z., and J.Z. (Jinyi Zhang); resources, W.Z. and Z.Z.; data curation, A.M., H.Z., J.Z. (Jinyi Zhang), and Z.L.; writing—original draft preparation, A.M., H.Z., J.Z. (Jinyi Zhang), and W.Z.; writing—review and editing, A.M., H.Z., Z.L., J.Z. (Jinyi Zhang), X.Z., X.-X.S., X.C., X.H., Y.Z., and W.Z.; visualization, A.M. and H.Z.; supervision, W.Z., Z.Z., and Y.Z.; project administration, W.Z. and Z.Z.; funding acquisition, W.Z., A.M., and Z.Z. All authors have read and agreed to the published version of the manuscript.

**Funding:** This work was funded by the National Key R&D Program (2023YFC2604500), National Nature Science Foundation of China (Grant Nos. W2433069 and 32272636), and the Yunnan FVF-IPM Joint Lab (no. 202303AP140018).

**Data Availability Statement:** The original contributions presented in this study are included in the article/Supplementary Materials. Further inquiries can be directed to the corresponding authors.

**Acknowledgments:** The authors are grateful to their colleagues for their help in the experiments and the Yaqiang Zheng for providing EPF strains.

**Conflicts of Interest:** The authors declare that they have no conflict of interest.

## References


1. Sharma, A.; Shaurav, S.; Yadav, P.K. Entomopathogenic fungi and their relevance in sustainable agriculture: A review. *Cogent Food Agric.* **2023**, *9*, 2180857. [CrossRef]
2. Shang, S.-Q.; Wei-Zhen, L.I.; Chen, Y.-N.; Tong, Z.H.U. The influence of Ahy1 strain of *Acremonium hansfodii* on functional response of *Neoseiulus barkeri* to *Tetranychus urticae. Syst. Appl. Acarol.* **2023**, *28*, 903–911. [CrossRef]
3. Thakur, A.; Singh, V.; Kaur, A.; Kaur, S. Insecticidal potential of an endophytic fungus, *Cladosporium uredinicola*, against *Spodoptera litura. Phytoparasitica* **2013**, *41*, 373–382. [CrossRef]
4. García-Estrada, C.; Cat, E.; Santamarta, I. Beauveria bassiana as Biocontrol Agent: Formulation and Commercialization for Pest Management. In *Agriculturally Important Microorganisms: Commercialization and Regulatory Requirements in Asia*; Singh, H.B., Sarma, B.K., Keswani, C., Eds.; Springer: Singapore, 2016; pp. 81–96.
5. Eilenberg, J.; Keller, S.; Humber, R.A.; Jensen, A.H.; Jensen, A.B.; Görg, L.M.; Muskat, L.C.; Kais, B.; Gross, J.; Patel, A.V. *Pandora cacopsyllae* Eilenberg, Keller & Humber (Entomophthorales: Entomophthoraceae), a new species infecting pear psyllid *Cacopsylla pyri* L. (Hemiptera: Psyllidae). *J. Invertebr. Pathol.* **2023**, *200*, 107954. [CrossRef]
6. Bamisile, B.S.; Dash, C.K.; Akutse, K.S.; Keppanan, R.; Afolabi, O.G.; Hussain, M.; Qasim, M.; Wang, L. Prospects of endophytic fungal entomopathogens as biocontrol and plant growth promoting agents: An insight on how artificial inoculation methods affect endophytic colonization of host plants. *Microbiol. Res.* **2018**, *217*, 34–50. [CrossRef] [PubMed]
7. Dutta, R.; Santra, H.K.; Banerjee, D. Multifunctional Efficacy and Eco-friendly Applications of Fungal Endophytes. In *Fungal Endophytes Volume II: Applications in Agroecosystems and Plant Protection*; Abd-Elsalam, K.A., Hashem, A.H., Eds.; Springer Nature: Singapore, 2025; pp. 33–60.
8. Hong, L.; Wang, Q.; Zhang, J.; Chen, X.; Liu, Y.; Asiegbu, F.O.; Wu, P.; Ma, X.; Wang, K. Advances in the beneficial endophytic fungi for the growth and health of woody plants. *For. Res.* **2024**, *4*, e028. [CrossRef]
9. Baron, N.C.; Rigobelo, E.C. Endophytic fungi: A tool for plant growth promotion and sustainable agriculture. *Mycology* **2022**, *13*, 39–55. [CrossRef]
10. Ji, X.; Xia, Y.; Zhang, H.; Cui, J.L. The microscopic mechanism between endophytic fungi and host plants: From recognition to building stable mutually beneficial relationships. *Microbiol. Res.* **2022**, *261*, 127056. [CrossRef]
11. Taheri, P.; Puopolo, G.; Santoyo, G. Plant growth-promoting microorganisms: New insights and the way forward. *Microbiol. Res.* **2025**, *297*, 128168. [CrossRef]
12. Anjali; Kumar, S.; Korra, T.; Thakur, R.; Arutselvan, R.; Kashyap, A.S.; Nehela, Y.; Chaplygin, V.; Minkina, T.; Keswani, C. Role of plant secondary metabolites in defence and transcriptional regulation in response to biotic stress. *Plant Stress* **2023**, *8*, 100154. [CrossRef]
13. Saini, H.P.; Meena, M.; Sahoo, A.; Mehta, T. A review on fungal endophytes of the family Fabaceae, their metabolic diversity and biological applications. *Heliyon* **2025**, *11*, e42153. [CrossRef] [PubMed]
14. Toppo, P.; Kagatay, L.L.; Gurung, A.; Singla, P.; Chakraborty, R.; Roy, S.; Mathur, P. Endophytic fungi mediates production of bioactive secondary metabolites via modulation of genes involved in key metabolic pathways and their contribution in different biotechnological sector. *3 Biotech* **2023**, *13*, 191. [CrossRef] [PubMed]
15. Munawar, A.; Xu, Y.; Abou El-Ela, A.S.; Zhang, Y.; Zhong, J.; Mao, Z.; Chen, X.; Guo, H.; Zhang, C.; Sun, Y.; et al. Tissue-specific regulation of volatile emissions moves predators from flowers to attacked leaves. *Curr. Biol.* **2023**, *33*, 2321–2329.e2325. [CrossRef]
16. Munawar, A.; Zhu, Z.; Machado, R.A.R.; Zhou, W. Beyond 'push–pull': Unraveling the ecological pleiotropy of plant volatile organic compounds for sustainable crop pest management. *Crop Health* **2023**, *1*, 18. [CrossRef]
17. He, X.; Wang, Y.; Munawar, A.; Zhu, J.; Zhong, J.; Zhang, Y.; Guo, H.; Zhu, Z.; Baldwin, I.T.; Zhou, W. Manipulating stomatal aperture by silencing StSLAC1 affects potato plant–herbivore–parasitoid tritrophic interactions under drought stress. *New Phytol.* **2025**, *245*, 2133–2149. [CrossRef]
18. Nawrocka, J.; Szymczak, K.; Skwarek-Fadecka, M.; Małolepsza, U. Toward the Analysis of Volatile Organic Compounds from Tomato Plants (*Solanum lycopersicum* L.) Treated with *Trichoderma virens* or/and Botrytis cinerea. *Cells* **2023**, *12*, 1271. [CrossRef]

19. Rubio, M.B.; Monti, M.M.; Gualtieri, L.; Ruocco, M.; Hermosa, R.; Monte, E. *Trichoderma harzianum* Volatile Organic Compounds Regulated by the THCTF1 Transcription Factor Are Involved in Antifungal Activity and Beneficial Plant Responses. *J. Fungi* **2023**, *9*, 654. [CrossRef]
20. Střîbrská, B.; Hradecký, J.; Čepl, J.; Modlinger, R.; Tomášková, I.; Jirošová, A. Physiological and biochemical indicators in Norway spruces freshly infested by Ips typographus: Potential for early detection methods. *Front. For. Glob. Change* **2023**, *6*, 1197229. [CrossRef]
21. van Zijll de Jong, E.; Kandula, J.; Rostás, M.; Kandula, D.; Hampton, J.; Mendoza-Mendoza, A. Fungistatic Activity Mediated by Volatile Organic Compounds Is Isolate-Dependent in *Trichoderma* sp. *"atroviride* B". *J. Fungi* **2023**, *9*, 238. [CrossRef]
22. Sarkar, A.K.; Sadhukhan, S. Unearthing the alteration in plant volatiles induced by mycorrhizal fungi: A shield against plant pathogens. *Physiol. Plant* **2023**, *175*, e13845. [CrossRef]
23. Mitina, G.V.; Stepanycheva, E.A.; Choglokova, A.A.; Cherepanova, M.A. Features of Behavioral Reactions of the Peach Aphid *Myzus persicae* (Sulzer, 1776) (Hemiptera, Aphididae) to Volatile Organic Compounds of Entomopathogenic Fungi of the Genus Lecanicillium. *Entomol. Rev.* **2021**, *101*, 1015–1023. [CrossRef]
24. Ballot, A.; Dore, J.; Rey, M.; Meiffren, G.; Langin, T.; Joly, P.; Dreux-Zigha, A.; Taibi, A.; Prigent-Combaret, C. Dimethylpolysulfides production as the major mechanism behind wheat fungal pathogen biocontrol, by Arthrobacter and Microbacterium actinomycetes. *Microbiol. Spectr.* **2023**, *11*, e0529222. [CrossRef] [PubMed]
25. Guo, H.; Shi, X.; Han, J.; Ren, Q.; Gao, Z.; Zhang, A.; Wang, H.; Du, Y. VOCs from fungi-infected apples attract and increase the oviposition of yellow peach moth *Conogethes punctiferalis*. *Pest Manag. Sci.* **2023**, *79*, 5208–5219. [CrossRef] [PubMed]
26. Eckert, S.; Eilers, E.J.; Jakobs, R.; Anaia, R.A.; Aragam, K.S.; Bloss, T.; Popp, M.; Sasidharan, R.; Schnitzler, J.P.; Stein, F.; et al. Inter-laboratory comparison of plant volatile analyses in the light of intra-specific chemodiversity. *Metabolomics* **2023**, *19*, 62. [CrossRef]
27. Biondi, A.; Guedes, R.N.C.; Wan, F.H.; Desneux, N. Ecology, Worldwide Spread, and Management of the Invasive South American Tomato Pinworm, *Tuta absoluta*: Past, Present, and Future. *Annu. Rev. Entomol.* **2018**, *63*, 239–258. [CrossRef]
28. Sánchez-Hernández, C.; López, M.G.; Délano-Frier, J.P. Reduced levels of volatile emissions in jasmonate-deficient spr2 tomato mutants favour oviposition by insect herbivores. *Plant Cell Environ.* **2006**, *29*, 546–557. [CrossRef]
29. Contreras, E.; Rodriguez-Herva, J.J.; Isabel, D.; Emilia, L.-S.; Martinez, M. Previous interaction with phytopathogenic bacteria alters the response of Arabidopsis against *Tetranychus urticae* herbivory. *J. Plant Interact.* **2023**, *18*, 2144651. [CrossRef]
30. Desneux, N.; Han, P.; Mansour, R.; Arnó, J.; Brévault, T.; Campos, M.R.; Chailleux, A.; Guedes, R.N.C.; Karimi, J.; Konan, K.A.J.; et al. Integrated pest management of *Tuta absoluta*: Practical implementations across different world regions. *J. Pest Sci.* **2022**, *95*, 17–39. [CrossRef]
31. Depenbusch, L.; Teresa, S.; Pepijn, S.; Mahin, S.; Krishnadas, M.; Nasir, U.; Hanson, P. Tomato pests and diseases in Bangladesh and India: Farmers' management and potential economic gains from insect resistant varieties and integrated pest management. *Int. J. Pest Manag.* **2023**, *69*, 1–15. [CrossRef]
32. Guedes, R.N.C.; Roditakis, E.; Campos, M.R.; Haddi, K.; Bielza, P.; Siqueira, H.A.A.; Tsagkarakou, A.; Vontas, J.; Nauen, R. Insecticide resistance in the tomato pinworm *Tuta absoluta*: Patterns, spread, mechanisms, management and outlook. *J. Pest Sci.* **2019**, *92*, 1329–1342. [CrossRef]
33. Koller, J.; Sutter, L.; Gonthier, J.; Collatz, J.; Norgrove, L. Entomopathogens and Parasitoids Allied in Biocontrol: A Systematic Review. *Pathogens* **2023**, *12*, 957. [CrossRef]
34. Fortes, A.D.R.; Coelho, A.; Amorim, D.J.; Demetrio, C.G.B.; Parra, J.R.P. Biology and quality assessment of *Telenomus remus* (Hymenoptera: Scelionidae) and *Trichogramma* spp. (Hymenoptera: Trichogrammatidae) in eggs of *Spodoptera* spp. for augmentative biological control programs. *J. Insect Sci.* **2023**, *23*, 1–10. [CrossRef]
35. Shi, R.; Yu, J.; Chang, X.; Qiao, L.; Liu, X.; Lu, L. Recent Advances in Research into Jasmonate Biosynthesis and Signaling Pathways in Agricultural Crops and Products. *Processes* **2023**, *11*, 736. [CrossRef]
36. Munawar, A.; Zhang, Y.; Zhong, J.; Ge, Y.; Abou El-Ela, A.S.; Mao, Z.; Ntiri, E.S.; Mao, L.-J.; Zhu, Z.; Zhou, W. Heat stress affects potato's volatile emissions that mediate agronomically important trophic interactions. *Plant Cell Environ.* **2022**, *45*, 3036–3051. [CrossRef] [PubMed]
37. Duan, C.; Liao, H.; Wang, K.; Ren, Y. The research hotspots and trends of volatile organic compound emissions from anthropogenic and natural sources: A systematic quantitative review. *Environ. Res.* **2023**, *216*, 114386. [CrossRef]
38. Correa, T.A.; Santos, F.S.; Camargo, M.G.; Quinelato, S.; Bittencourt, V.; Golo, P.S. Comparison of Methods for Isolating Entomopathogenic Fungi from Soil Samples. *J. Vis. Exp.* **2022**, *179*, e63353. [CrossRef]
39. Wu, J.; Hettenhausen, C.; Meldau, S.; Baldwin, I.T. Herbivory rapidly activates MAPK signaling in attacked and unattacked leaf regions but not between leaves of *Nicotiana attenuata*. *Plant Cell* **2007**, *19*, 1096–1122. [CrossRef]
40. Song, J.; Bian, J.; Xue, N.; Xu, Y.; Wu, J. Inter-species mRNA transfer among green peach aphids, dodder parasites, and cucumber host plants. *Plant Divers.* **2022**, *44*, 1–10. [CrossRef]

41. Baldwin, I.T.; Schmelz, E.A.; Zhang, Z.P. Effects of octadecanoid metabolites and inhibitors on induced nicotine accumulation in *Nicotiana sylvestris*. *J. Chem. Ecol.* **1996**, *22*, 61–74. [CrossRef]
42. Michael, J. Crawley, Generalized Linear Models. In *The R Book*; Crawley, M.J., Ed.; Weily: Hoboken, NJ, USA, 2012; pp. 557–578. [CrossRef]
43. Islam, W.; Adnan, M.; Shabbir, A.; Naveed, H.; Abubakar, Y.S.; Qasim, M.; Tayyab, M.; Noman, A.; Nisar, M.S.; Khan, K.A.; et al. Insect-fungal-interactions: A detailed review on entomopathogenic fungi pathogenicity to combat insect pests. *Microb. Pathog.* **2021**, *159*, 105122. [CrossRef]
44. Sani, I.; Jamian, S.; Saad, N.; Abdullah, S.; Mohd Hata, E.; Jalinas, J.; Ismail, S.I. Inoculation and colonization of the entomopathogenic fungi, *Isaria javanica* and *Purpureocillium lilacinum*, in tomato plants, and their effect on seedling growth, mortality and adult emergence of *Bemisia tabaci* (Gennadius). *PLoS ONE* **2023**, *18*, e0285666. [CrossRef]
45. Bamisile, B.S.; Afolabi, O.G.; Siddiqui, J.A.; Xu, Y. Endophytic insect pathogenic fungi-host plant-herbivore mutualism: Elucidating the mechanisms involved in the tripartite interactions. *World J. Microbiol. Biotechnol.* **2023**, *39*, 326. [CrossRef] [PubMed]
46. Jensen, R.E.; Cabral, C.; Enkegaard, A.; Steenberg, T. Influence of the plant interacting entomopathogenic fungus *Beauveria bassiana* on parasitoid host choice-behavior, development, and plant defense pathways. *PLoS ONE* **2020**, *15*, e0238943. [CrossRef] [PubMed]
47. Wilberts, L.; Vuts, J.; Caulfield, J.C.; Thomas, G.; Birkett, M.A.; Herrera-Malaver, B.; Verstrepen, K.J.; Sobhy, I.S.; Jacquemyn, H.; Lievens, B. Impact of endophytic colonization by entomopathogenic fungi on the behavior and life history of the tobacco peach aphid *Myzus persicae* var. *nicotianae*. *PLoS ONE* **2022**, *17*, e0273791. [CrossRef] [PubMed]
48. Rasool, S.; Cárdenas, P.D.; Pattison, D.I.; Jensen, B.; Meyling, N.V. Isolate-Specific Effect of Entomopathogenic Endophytic Fungi on Population Growth of Two-Spotted Spider Mite (*Tetranychus urticae* Koch) and Levels of Steroidal Glycoalkaloids in Tomato. *J. Chem. Ecol.* **2021**, *47*, 476–488. [CrossRef]
49. Huang, X.Z.; Xiao, Y.T.; Köllner, T.G.; Jing, W.X.; Kou, J.F.; Chen, J.Y.; Liu, D.F.; Gu, S.H.; Wu, J.X.; Zhang, Y.J.; et al. The terpene synthase gene family in *Gossypium hirsutum* harbors a linalool synthase GhTPS12 implicated in direct defence responses against herbivores. *Plant Cell Environ.* **2018**, *41*, 261–274. [CrossRef]
50. McCallum, E.J.; Cunningham, J.P.; Lücker, J.; Zalucki, M.P.; De Voss, J.J.; Botella, J.R. Increased plant volatile production affects oviposition, but not larval development, in the moth *Helicoverpa armigera*. *J. Exp. Biol.* **2011**, *214*, 3672–3677. [CrossRef]
51. Proffit, M.; Birgersson, G.; Bengtsson, M.; Reis, R., Jr.; Witzgall, P.; Lima, E. Attraction and oviposition of *Tuta absoluta* females in response to tomato leaf volatiles. *J. Chem. Ecol.* **2011**, *37*, 565–574. [CrossRef]
52. Ninkovic, V.; Markovic, D.; Rensing, M. Plant volatiles as cues and signals in plant communication. *Plant Cell Environ.* **2021**, *44*, 1030–1043. [CrossRef]
53. Santino, A.; Taurino, M.; De Domenico, S.; Bonsegna, S.; Poltronieri, P.; Pastor, V.; Flors, V. Jasmonate signaling in plant development and defense response to multiple (a)biotic stresses. *Plant Cell Rep.* **2013**, *32*, 1085–1098. [CrossRef]
54. Xin, K.; Wu, Y.; Ikram, A.U.; Jing, Y.; Liu, S.; Zhang, Y.; Chen, J. Salicylic acid cooperates with different small molecules to control biotic and abiotic stress responses. *J. Plant Physiol.* **2025**, *304*, 154406. [CrossRef]
55. Benjamin, G.; Pandharikar, G.; Frendo, P. Salicylic Acid in Plant Symbioses: Beyond Plant Pathogen Interactions. *Biology* **2022**, *11*, 861. [CrossRef] [PubMed]
56. Caarls, L.; Pieterse, C.M.; Van Wees, S.C. How salicylic acid takes transcriptional control over jasmonic acid signaling. *Front. Plant Sci.* **2015**, *6*, 170. [CrossRef]
57. Nishad, R.; Ahmed, T.; Rahman, V.J.; Kareem, A. Modulation of Plant Defense System in Response to Microbial Interactions. *Front. Microbiol.* **2020**, *11*, 1298. [CrossRef] [PubMed]
58. Hossain, M.M. Chapter 4—Upscaling plant defense system through the application of plant growth-promoting fungi (PGPF). In *Microbial Technology for Agro-Ecosystems*; Kumar, V., Iram, S., Eds.; Academic Press: Cambridge, MA, USA, 2024; pp. 61–95.
59. Pathak, V.M.; Verma, V.K.; Rawat, B.S.; Kaur, B.; Babu, N.; Sharma, A.; Dewali, S.; Yadav, M.; Kumari, R.; Singh, S.; et al. Current status of pesticide effects on environment, human health and it's eco-friendly management as bioremediation: A comprehensive review. *Front. Microbiol.* **2022**, *13*, 962619. [CrossRef] [PubMed]
60. Siddiqui, J.A.; Fan, R.; Naz, H.; Bamisile, B.S.; Hafeez, M.; Ghani, M.I.; Wei, Y.; Xu, Y.; Chen, X. Insights into insecticide-resistance mechanisms in invasive species: Challenges and control strategies. *Front. Physiol.* **2022**, *13*, 1112278. [CrossRef]
61. Boldorini, G.X.; McCary, M.A.; Romero, G.Q.; Mills, K.L.; Sanders, N.J.; Reich, P.B.; Michalko, R.; Gonçalves-Souza, T. Predators control pests and increase yield across crop types and climates: A meta-analysis. *Proc. Biol. Sci.* **2024**, *291*, 20232522. [CrossRef]

*agronomy*

MDPI

*Article*

# *Satureja kitaibelii* Essential Oil and Extracts: Bioactive Compounds and Pesticide Properties

**Milena Nikolova [1], Aneta Lyubenova [2,*], Elina Yankova-Tsvetkova [1], Borislav Georgiev [1], Genadi Gavrilov [3] and Anna Gavrilova [3]**

1 Institute of Biodiversity and Ecosystem Research, Bulgarian Academy of Sciences, 23 Acad. G. Bonchev Street, 1113 Sofia, Bulgaria; mtihomirova@gmail.com (M.N.); e_jankova@abv.bg (E.Y.-T.); bobogeorgiev5@gmail.com (B.G.)

2 Department of Agrobiotechnology, AgroBioInstitute, Agricultural Academy, 8 Dragan Tsankov Blvd, 1164 Sofia, Bulgaria

3 Faculty of Pharmacy, Medical University—Pleven, 1 Sv. Kliment Ohridski Street, 5800 Pleven, Bulgaria; genadi.gavrilov@mu-pleven.bg (G.G.); anna.gavrilova@mu-pleven.bg (A.G.)

* Correspondence: anetalyubenova@abi.bg

**Abstract:** In recent years, the essential oil of *Satureja* species has been studied as a source of biocidal activity with potential applications in organic farming such as bio-pesticides. The present study aims to determine the potential of essential oil (EO), exudate fraction (EF) and methanolic extract (ME) of *Satureja kitaibelii* Wierzb. ex Heuff. to inhibit the mycelial growth of phytopathogenic fungi and acetylcholinesterase (AChE). Additionally, ME was tested for inhibitory activity on seed germination and root elongation. Phytochemical analysis was conducted using gas chromatography–mass spectrometry (GC–MS) and thin-layer chromatography (TLC). Biological activities were studied using in vitro methods. *p*-Cymene, limonene, geraniol, carvacrol and borneol were identified as the main components of EO. Oleanolic and ursolic acid, carvacrol and flavonoid aglycones were determined as the most abundant bioactive compounds of EF, whereas rosmarinic acid and flavonoid glycosides were found in ME. EO reduced the growth of all tested plant pathogens, indicated by 40% to 84% inhibition of mycelial growth (IMG). The growth rates of oomycetes *Phytophthora cryptogea* Pethybr. & Laff. and *Phytophthora nicotianae* Breda de Haan were affected to the greatest extent with 84% and 68% IMG. EF showed the most potent AChE inhibitory activity with $IC_{50}$ value of 0.18 mg/mL. Aqueous solutions of the ME with a concentration above 5 mg/mL were found to inhibit seed germination by more than 90%, whereas a reduction in root elongation was observed at 3 mg/mL. The present study provides for the first time data for the pesticidal properties of EO, EF and ME of *S. kitaibelii*.

**Keywords:** acetylcholinesterase; seed germination; exudate; flavonoids; *p*-cymene; *Phytophthora* spp.; antifungal bioassay; agar disk-diffusion method

## 1. Introduction

Pest control by natural remedies is an important characteristic of organic agriculture that is following the world's desire for an environmentally friendly way of life. Synthetic biocides negatively impact the environment due to their poor biodegradability, which is a precondition for the loss of biodiversity and human and animal health problems [1,2]. Another disadvantage is the development of resistance to synthetic biocides in invertebrates, as observed in over 500 insect and mite species, including cross-resistance (to pesticides from more than one chemical class) [3]. Alternatively, the use of broad-spectrum insecticides leads to a reduction in the populations of beneficial insects [4]. Natural biocides are safer because of a more rapid degradation and the lack of accumulation in the environment [2]. Thus, the need for novel natural sources of biocidal activity is increasing significantly [5].

In recent decades, substantial data have been accumulated on the potential of EOs as pesticidal agents [6–9]. Essential oil is a complex mixture of volatile compounds—mono- and sesquiterpenes, phenylpropanoids, phenols, etc.—that are localized in the cytoplasm of plant cells, components of various secretory structures—glands, secretory hairs, resin, ducts and secretory cavities. Strong phytotoxic effects and antifungal and insecticidal activities have been previously reported for essential oils [10–13]. Not all plant species have a high yield of essential oils, and this is a challenge when it comes to the large number of requirements in agriculture.

An exudate is a mixture of lipophilic compounds accumulated on the surface of the plants. Examples of such substances include terpenoids, flavonoid aglycones, lipids and waxes. Such compounds may be of key biological importance. Several studies show their external location to be associated with a protective role against different biotic and abiotic factors, including herbivores and pathogens [14–16]. Plant exudate fractions (EFs) are readily and quickly extracted; no special equipment is required. It has been previously reported that EFs of *Origanum vulagre* ssp. *hirtum* inhibit the mycelial growth of *Phytophthora* isolates by about 80% [17]. Furthermore, exudates have been found to exhibit seed germination inhibitory activity and antimicrobial properties [18–20].

Compared to EOs and EFs, alcoholic and water extracts often demonstrate weaker biocidal activities [21,22]. Despite that, when satisfactory levels of activity are achieved, extracts do have a high potential for agricultural application because they are obtained in much larger quantities [23].

Species of the genus *Satureja* are plants traditionally used in the food and pharmaceutical industries. In recent years, the EOs of *S. hortensis* L. and other species such as *S. cuneifolia* Ten. have been studied as sources of biocidal activity with possible applications in organic farming such as bio-pesticides. Carvacrol, thymol, *p*-cymene, γ-terpinene, α-pinene, β-ocimene, camphene and camphor are the dominant EO compounds that are most likely responsible for the pronounced biocidal effects of *Satureja* spp. [24–26]. Great antimicrobial properties have been found for the EO of *S. hortensis* against 23 bacteria and 15 fungal and yeast species; however, in the same study, the ME did not exhibit any antimicrobial activities [27]. Studies on various *Satureja* species reported notable inhibitory effects of their EOs on viruses, *Leishmania* spp., other protozoa, insects, acari, nematodes, helminths, mollusks, etc. [28,29]. In particular, *Satureja montana*, a notable close relative to *S. kitaibelii*, possesses insecticidal and nematocidal effects [28].

The genus *Satureja* is represented in the Bulgarian flora by five native and one introduced species (*S. hortensis* L.). *S. kitaibelii* Wierzb. ex Heuff., previously assumed to be a subspecies of *S. montana* L., is a Balkan endemic plant with wide distribution across the Balkan Peninsula [30]. It is one of the most characteristic species of the petrophyte steppes, which are found in many karst low-hilly areas and canyons in Bulgaria and Eastern Serbia [31,32]. *S. kitaibelii* is a perennial subshrub, 30–70 cm in height, with a well-established root system. The stems are square in cross section and glabrous or sericeous on two opposite sides, with opposite, coriaceous and narrow lanceolate leaves. The inflorescence is an apical, laxous verticillaster with pink labiate flowers. All aerial parts of the plant emit a strong pleasant aroma [30,33].

Only a few studies have been performed on the biological activities of *S. kitaibelii*. The ethanolic extract has been found to exhibit significant antioxidant potential and antibacterial properties against *Micrococcus luteus* and *Pseudomonas aeruginosa* [34]. The EO of this species has been reported to possess antimicrobial activity against human pathogens such as *Candida albicans*, *Pseudomonas aeruginosa*, *Staphylococcus aureus*, *Escherichia coli*, etc. [35–38].

Geraniol, *p*-cymene, limonene and borneol have been determined as the main components of *S. kitaibelii*'s EO. Their ratio varies greatly according to the origin of the population, developmental stage and the plant parts used for the extraction. Ten chemotypes are distinguished for this species [39–41]. The presence of such variability means that researchers need to examine as many origins as possible in the search for the most effective EO profile for the needs of their respective studies. Previous analyses determined phenolic compounds such as rosmarinic acid, clinopodic acid, flavonoids and jasmonic acid derivatives as the main bioactive compounds of the ethanolic extract of *S. kitaibelii* [34]. Triterpene acids, flavonoid aglycones and phenolic acids have been found in this species' exudate [42].

The present study aims to determine the potential of essential oil, exudate fraction and methanolic extract of *S. kitaibelii* to inhibit the mycelial growth of phytopathogenic fungi and acetylcholinesterase. Additionally, the species' methanolic extract was tested for inhibitory activity on seed germination. Acetylcholinesterase inhibitory activity provides data on the potential repellent (insecticidal) activity.

## 2. Materials and Methods

### 2.1. Plant Material

Aerial parts of *S. kitaibelii* plants in full flowering stage were harvested in October 2022 in the protected area "Kailaka" near the town of Pleven, Middle Danube plain, Bulgaria (Figure 1). The bedrock in this area is limestone, and the altitude is 200 m a.s.l. The sample material was air-dried without exposure to direct sunlight.

**Figure 1.** *Satureja kitaibelii* in its natural locality—The Danube Plain, Bulgaria (photos: Genadi Gavrilov).

### 2.2. Phytochemical Analysis

#### 2.2.1. Extraction Procedure

Essential oil (EO). EO was obtained by a standard extraction procedure [43]. Air-dried, not ground, aerial parts were subjected to hydrodistillation for 4 h using a conventional glass Clevenger-type apparatus. The EO was dried over anhydrous sodium sulfate and stored at 4 °C until experimental use.

Exudate fraction (EF). Exudate was obtained according to the method introduced by Prof. Eckhart Wollenweber to study exudate flavonoids but without removing substances with a terpenoid structure in the present study [44]. Dry, not ground, aerial parts of *S. kitaibelii* were dipped into acetone for 4–5 min. Afterwards, the fraction was filtered and evaporated to dryness using a rotary vacuum evaporator at 40 °C. The dry EF was stored at 4 °C before analyses.

Methanolic extract (ME). Air-dried and powdered aerial parts were extracted with methanol by a classic process of maceration for 24 h at room temperature. After filtration, the organic solvent was evaporated, and the dry extract was stored at 4 °C before analyses.

2.2.2. Derivatization of Methanolic Extract and Exudate Fraction

The EF and ME were derivatized before GC–MS analysis as described by Berkov et al. [45]. Fifty milligrams of methanolic extract and exudate fraction were silylated with 50 μL of *N,O*-bis(trimethylsilyl)trifluoroacetamide (BSTFA) in 50 μL of pyridine for 2 h at 70 °C. Compounds of the ME and EF were identified as TMS derivatives using NIST 05 database (NIST Mass Spectral Database, PC-Version 5.0, 2005), Golm Metabolome Database and home-made MS databases. Amounts of all identified compounds are presented as response ratios calculated for each compound relative to the internal standard (3,5-dichloro-4-hydroxybenzoic acid (1 mg/mL)) using the calculated areas for both components. The internal standard was added (50 μL) at the beginning of the extraction process.

2.2.3. Gas Chromatography–Mass Spectrometry

The GC–MS spectra were recorded on a Thermo Scientific Focus GC coupled with Thermo Scientific (Waltham, MA, USA) DSQ mass detector operating in EI mode at 70 eV. ADB-5MS column (30 m × 0.25 mm × 0.25 mm) was used.

The chromatographic conditions used for the EO were as follows: column temperature was 60 °C for 10 min, then programmed at the rate of 3 °C $min^{-1}$ to 200 °C, and finally, held isothermally for 10 min. The injector temperature was 220 °C. The flow rate of carrier gas (Helium) was 1 mL $min^{-1}$. The split ratio was 1:50, and 1 μL of the solution was injected. Significant quadrupole MS operating parameters: interface temperature 240 °C; scan mass range of 40 to 400 $m/z$ at a sampling rate of 1.0 scan $s^{-1}$.

The chromatographic conditions used for the EF and ME were as follows: 100–180 °C at 15 °C $min^{-1}$, 180–300 °C at 5 °C $min^{-1}$ and 10 min hold at 300 °C. The injector temperature was 250 °C. The flow rate of the carrier gas (Helium) was 0.8 mL $min^{-1}$. The split ratio was 1:10 and 1 μL of the solution was injected.

The individual components were identified by their retention times (RTs), retention indices (RIs), relative to C5-C28 n-alkanes, referring to known compounds from the literature, and also by comparison with those of NIST 14 Library [46]. The percentage of composition of the essential oil was computed from the GC peaks areas.

2.2.4. Thin-Layer Chromatography

TLC analyses were performed on 20 × 10 cm silica gel Kiselgel 60 F254 plates (Merck, Rahway, NJ, USA) and DC Alufolien Polyamide 11 F254 (5555) plates (Merck). Five microliters of methanolic extract were applied to the TLC plates. The mobile phases used on silica gel were as follows: $S_1$ (toluene–dioxin–acetic acid = 90:25:4), $S_2$ (ethyl acetate–formic acid–acetic acid–methylethylketone–methanol–water = 50:7:3:30:10) and $S_3$ (ethyl acetate–acetic acid–water = 20:2:1). The mobile phase used on polyamide was $S_4$ (toluene–methylethylketone–methanol = 60:30:15). Chromatograms were observed under UV light at 336 nm before and after spraying with "Naturstoffreagenz A", 1% solution of diphenylboric acid ethanolamine complex in methanol. Identification of compounds was carried out by co-chromatography of the EF and ME with authentic standards and comparing their retardation factors ($R_f$) and fluorescence emission under UV before and after spraying with reagent.

### 2.3. *Antifungal/Anti-Oomycete Bioassay*

The three *S. kitaibelii* samples (EO, EF, ME) were tested against three fungal and two oomycete species to determine their pesticidal properties. These isolates were previously obtained from different ecosystems in Bulgaria, except for *Phytophthora nicotianae* originating from potted ornamental plants. Their identification was based on morphology and ITS enzyme restriction for *P. cryptogea* as well as morphology and ITS sequences for the rest.

Namely, these are *Alternaria alternata* (Fr.) Keissl., 1912 (NCBI GenBank accession number PQ803669), *Botrytis cinerea* Pers. (PQ345538), *Fusarium oxysporum* Schltdl. (PQ345539), and *Phytophthora nicotianae* (PQ460002). The two *Phytophthora* spp. were maintained on V8 Agar Media (16 g agar, 100 mL Campbell's (Camden, NJ, USA) V8 Juice, and 900 mL distilled water), and fungal isolates on PDA (BD Difco™, Franklin Lakes, NJ, USA). In the conducted bioassay, the antifungal properties of EO, EF and ME derived from *S. kitaibelii* were assessed through a modified agar disk-diffusion technique [47]. A concentration of 100 mg/mL was achieved for the EF and ME by dissolving the dry samples in DMSO and methanol, respectively. The bioassay consisted of six variants for each isolate: EO, EF, ME, two control treatments with pure solvents (100% DMSO and methanol) and control without treatment. Three replicates were conducted for each variant. One day prior to the introduction of *S. kitaibelii* samples, the isolates were re-cultured in 9 mm Petri dishes, each containing 20 mL of a suitable nutrient agar medium. To achieve simultaneous growth, they were incubated overnight. The following day, 2 × 15 µL of either EF or ME were dripped directly onto the agar medium in the Petri dish, equidistant from the center. Similarly, the EO was administered using a volume of 2 × 2 µL. The concentrations were selected based on results from previous similar studies [17,48]. The Petri dishes were cultivated in a climatic chamber in darkness at 25 °C. The results were documented when the mycelial colonies in the controls reached the periphery of the Petri dishes. Photographs (Canon, Tokyo, Japan EOS 4000D) of all mycelial colonies were taken, and their mycelial growth areas were measured using the image analysis program ImageJ 1. 54g [49]. Based on the collected information (mean mycelial growth area for each treatment/isolate variant), the inhibition percentage was determined [50] using the following equation:

$$\%\,\mathrm{IMG} = 100(\mathrm{C} - \mathrm{T})\mathrm{C}^{-1},$$

where % IMG is the percentage of inhibition of mycelial growth, C is the area of the fungal colony without treatment (control) and T is the area of the treated fungal colony.

### 2.4. Acetylcholinesterase (AChE) Inhibition Assay

The microplate assay used for measuring AChE inhibitory activity was performed in 96-well plates using a modified method of Ellman et al. [51], as described by López et al. [52]. Acetylthiocholine iodide (ATCI) in solution with 5,5′-dithiobis(2-nitrobenzoic acid) (DTNB) was used as a substrate for the acetylcholinesterase from *Electrophorus electricus* (Sigma-Aldrich, Darmstadt, Germany). *S. kitaibelii* sample solutions with concentrations between 0.08 and 5 mg/mL were tested. Fifty microliters of AChE (0.25 U/mL) dissolved in phosphate buffer (8 mM $K_2HPO_4$, 2.3 mM $NaH_2PO_4$, 0.15 M NaCl, pH 7.5) and 50 µL of the samples dissolved in the same buffer were added to the wells. The plates were incubated for 30 min at room temperature before the addition of 100 µL of the substrate solution (0.04 M $Na_2HPO_4$, 0.2 mM DTNB, 0.24 mM ATCI, pH 7.5). The absorbances were read in a microplate reader (BIOBASE, ELISA-EL10A, Jinan, China) at 405 nm after 5 min. Enzyme activity was calculated as inhibition percentage compared to an assay using a buffer instead of an inhibitor. Galanthamine was used as a positive control. The AChE inhibitory data were then analyzed with the software package Prism 3 (Graph Pad Inc., San Diego, CA, USA). The $IC_{50}$ values were measured in triplicate, and the results are presented as means ± SD.

### 2.5. Seed Germination Inhibition Bioassay

A hundred seeds of *Lolium perenne* L. (a common crop weed) were placed per Petri dish on filter papers moistened with the tested solutions. The methanolic extract, in a water–

methanol mixture (99.5:0.5), was assayed at a concentration of 1, 3, 5 and 8 mg/mL. The control consisted only of the water–methanol mixture. The samples were incubated at room temperature for seven days. At the end of the incubation period, the rate of germination inhibition and root length inhibition were calculated using the following equations [53]:

$$GI = [(GC - TG) \div GC] \times 100,$$

where GI is the rate of germination inhibition (%); GC is the germination rate of seeds treated with control solutions; TG is the germination rate of seeds treated with ME solution;

$$REI = [(REC - TRE) \div REC] \times 100$$

where REI is the rate of root elongation inhibition (%); REC is the root length of the control; TRE is the root length of the treated plant.

The lengths of the roots were measured using a ruler in millimeters.

Seed germination assays were performed in three independent experiments.

### 2.6. Data Analysis

Statistical analyses were performed using Microsoft Excel software. Results are presented as means with standard deviation (SD) and standard error (SE). The AChE inhibitory data were analyzed using the software package Prism 3 (Graph Pad Inc., San Diego, CA, USA). The statistical significance of the differences between mean values for the treated and untreated variants from antifungal/anti-oomycete bioassay tests was determined by a one-way ANOVA test, accepting $p \leq 0.05$ to be significant.

## 3. Results

### 3.1. Phytochemical Analysis

#### 3.1.1. Essential Oil

The EO was obtained with a yield of 0.1% (*w*/*v*). The chemical composition of the EO was determined using GC/MS. *p*-Cymene (23.94%), limonene (8.57%), geraniol (8.63%), carvacrol (7.22%) and endo-borneol (7.15%) were identified as the main components (Table 1).

**Table 1.** Identified compounds in EO of the aerial parts of *S. kitaibelii*.

| RT | RI | Compound | Area, % |
|---|---|---|---|
| 16.04 | 927 | β-Thujene | 0.60 ± 0.1 |
| 16.41 | 935 | α-Pinene | 3.68 ± 0.4 |
| 16.76 | 952 | Camphene | 1.20 ± 0.7 |
| 18.14 | 991 | β-Myrcene | 0.96 ± 0.1 |
| 18.50 | 1017 | α-Terpinene | 0.38 ± 0.1 |
| 19.71 | 1022 | *p*-Cymene | 23.94 ± 1.8 |
| 19.82 | 1031 | D-Limonene | 8.57 ± 0.9 |
| 20.00 | 1032 | Eucalyptol | 0.84 ± 0.2 |
| 20.14 | 1042 | β-Ocimene | 0.67 ± 0.1 |
| 20.77 | 1060 | γ-Terpinene | 2.70 ± 0.5 |
| 21.28 | 1070 | Sabinene hydrate | 1.67 ± 0.7 |
| 21.98 | 1097 | Linalool | 0.53 ± 0.1 |
| 24.89 | 1166 | endo-Borneol | 7.15 ± 0.6 |
| 25.02 | 1177 | Terpinen-4-ol | 2.67 ± 0.4 |
| 26.51 | 1244 | Carvacrol methyl ether | 6.97 ± 0.5 |
| 26.75 | 1253 | Geraniol | 8.63 ± 0.4 |

Table 1. Cont.

| RT | RI | Compound | Area, % |
|---|---|---|---|
| 28.22 | 1299 | Carvacrol | 7.22 ± 0.6 |
| 31.01 | 1388 | (−)-β-Bourbonene | 1.45 ± 0.1 |
| 32.08 | 1419 | Caryophyllene | 1.29 ± 0.3 |
| 33.59 | 1432 | β-Copaene | 0.53 ± 0.1 |
| 33.84 | 1509 | β-Bisabolene | 1.28 ± 0.2 |
| 35.58 | 1578 | (+)-Spathulenol | 1.20 ± 0.2 |
| 35.71 | 1581 | Caryophyllene oxide | 1.57 ± 0.1 |

3.1.2. Exudate Fraction

The EF is composed mainly of lipophilic compounds dissolved on the surface of the plant tissue. The EF from *S. kitaibelii* was obtained with a yield of 0.7% (*w*/*w*). Ursolic and oleanolic acids, carvacrol and n-alkanes were found as the most abundant components (Table 2). Phenolic acids, monoterpenoids, flavonoids, sterols, fatty acids and organic acid were also identified. Flavonoid aglycones such as scutellarein-6,7,8-trimethyl ether (xanthomicrol) and scutellarein-6,7-dimethyl ether were determined by TLC (Table 3).

Table 2. Components of *S. kitaibelii* exudate fraction and methanolic extract as determined by GC–MS.

| RI | Compound | Amount, µg * | |
|---|---|---|---|
| | | Exudate Fraction | Methanolic Extract |
| 1220 | Borneol | 163.9 ± 26 | 16.5 ± 3 |
| 1249 | Benzoic Acid | – | 2.9 ± 0.7 |
| 1289 | Glycerol | 479.9 ± 31 | 755.3 ± 24 |
| 1321 | Succinic acid | 19.0 ± 8 | 79.0 ± 11 |
| 1339 | Carvacrol | 634.7 ± 17 | 259.8 ± 19 |
| 1396 | Hydroquinone | 3.1 ± 0.9 | – |
| 1497 | Malic acid | 33.7 ± 4 | 22.1 ± 6 |
| 1624 | Ribofuranose | – | 156.6 ± 10 |
| 1635 | 4-Hydroxybenzoic acid | 11.4 ± 0.5 | 5.2 ± 2 |
| 1776 | Vanillic acid | 1.1 ± 0.6 | – |
| 1803 | Fructose 1 | 273.1 ± 33 | 661.0 ± 25 |
| 1811 | Fructose 2 | 450.7 ± 45 | 1510.4 ± 67 |
| 1835 | Protocatechuic acid | 10.2 ± 3 | 10.4 ± 8 |
| 1842 | Quinic acid | 18.6 ± 5 | 167.4 ± 12 |
| 1855 | Syringic acid | – | 8.5 ± 0.9 |
| 1890 | D-Glucopyranose | 130.5 ± 6 | 1122.2 ± 68 |
| 1946 | 4-Hydroxycinnamic acid | 4.01 ± 1.4 | 2.5 ± 0.7 |
| 1996 | Methyl caffeate | 10.6 ± 1.2 | – |
| 2040 | Hexadecanoic acid | 646.4 ± 49 | 134.6 ± 28 |
| 2080 | Catechollactate | 16.4 ± 2.9 | 222.6 ± 31 |
| 2104 | Ferulic acid | 15.1 ± 3.2 | 0.8 ± 0.1 |
| 2129 | Myo-Inositol | 43.4 ± 9 | 2876.9 ± 87 |
| 2155 | Caffeic acid | 16.1 ± 2.4 | 56.1 ± 16 |
| 2212 | Octadienoic acid | 161.5 ± 11 | 170.9 ± 21 |
| 2218 | Octatrienoic acid | 302.4 ± 24 | 241.86 ± 28 |
| 2246 | Stearic acid | 40.7 ± 3 | 29.9 ± 12 |
| 2628 | Sucrose | 410.0 ± 32 | 6597.0 ± 167 |
| 2838 | Tetracosanoic acid | 69.4 ± 10 | – |
| 2872 | Naringenin | 74.7 ± 12 | – |
| 2900 | Nonacosane $C_{29}H_{60}$ | 486.2 ± 56 | – |
| 2942 | Taxifolin | 8.5 ± 2 | – |
| 3100 | Hentriacontane $C_{31}H_{64}$ | 340.9 ± 44 | – |
| 3122 | Methylated flavone | 50.2 ± 10 | – |
| 3194 | β-Sitosterol | 137.0 ± 13 | 96.2 ± 33 |
| 3335 | β-Amyrin | 16.2 ± 8 | 3.8 ± 2 |

**Table 2.** *Cont.*

| RI | Compound | Amount, µg * | |
|---|---|---|---|
| | | Exudate Fraction | Methanolic Extract |
| 3455 | Rosmarinic acid | 7.5 ± 2 | 469.2 ± 43 |
| 3540 | Uvaol | 107.3 ± 17 | 20.3 ± 8 |
| 3570 | Oleanolic acid | 3505.55 ± 204 | 128.6 ± 12 |
| 3580 | Betulinic acid | 67.5 ± 21 | 2.3 ± 0.9 |
| 3620 | Ursolic acid | 4457.9 ± 267 | 244.5 ± 21 |
| 3632 | Micromeric acid | 47.3 ± 12 | - |

* amounts (µg) are the mean of three independent experiments and represent the response ratios calculated for each compound relative to the internal standard

**Table 3.** Phenolic compounds identified in the exudate fraction and methanolic extract using TLC.

| Compounds | $R_f$ Values in Different TLC Conditions * | | | |
|---|---|---|---|---|
| | SG, $S_1$ | SG, $S_2$ | SG, $S_3$ | PA, $S_4$ |
| Xanthomicrol | 0.62 | | | 0.95 |
| Scutellarein 6 methyl ether | 0.50 | | | 0.71 |
| Rosmarinic acid | | 0.98 | 0.73 | |
| Luteolin-7-glucoronide | | 0.50 | | |
| Rutin | | 0.31 | 0.06 | |

* SG and PA correspond to silica gel and polyamide; $S_{1-4}$ are the mobile phases used as described in Section 2.2.4.

#### 3.1.3. Methanolic Extract

The ME was obtained with a yield of 8.5% (*w*/*w*). Its composition was examined using GC–MS and TLC. Some of the compounds reported in the EF (carvacrol, borneol, oleanolic acid, ursolic acid, sterols, terpenes) were also found in the ME, but here, they are in much lower quantities. The GC–MS analysis revealed sucrose, fructose and glucose as the most abundant compounds (Table 2). TLC screening with authentic flavonoid compounds showed the presence of flavonoid glycosides (rutin and luteolin-7-glucuronide) (Table 3). Additionally, rosmarinic acid was found to be the most abundant phenolic acid during TLC with $S_3$ mobile phase.

### 3.2. *Fungal and Oomycete Growth Inhibition Properties*

Mycelial colonies in the controls reached the periphery of the Petri dishes in 4–7 days depending on the species. The results from data analysis showed a lack of significant difference between the three types of controls and significant differences ($p < 0.05$) in the mycelial growth rate inhibition in the EO, EF and ME variants with the Fungi. EF and ME of *S. kitaibelii* exerted no inhibitory effect on the mycelial growth rate of *P. cryptogea* and *P. nicotianae* at the administered doses of 2 × 15 µL of 100 mg/mL stock solution. Contrastingly, the mycelial growth of two of the fungal pathogens investigated was suppressed, albeit to a low degree. For *B. cinerea*, an IMG between 55% (ME) and 59% (EF) was recorded, and for *F. oxysporum*, the IMG was between 37% (EF) and 39% (ME) (Figures 1 and 2).

Generally, the essential oil reduced the growth of all tested pathogens, indicated by an IMG between 40 and 84%. *Phytophthora* spp. were affected the strongest—84% IMG for *P. cryptogea* and 68% IMG for *P. nicotianae*. Out of the three fungal species, *F. oxysporum* and *A. alternata* were largely inhibited—65% and 50% IMG, respectively. The growth of *B. cinerea* was reduced by 40% (Figures 2 and 3).

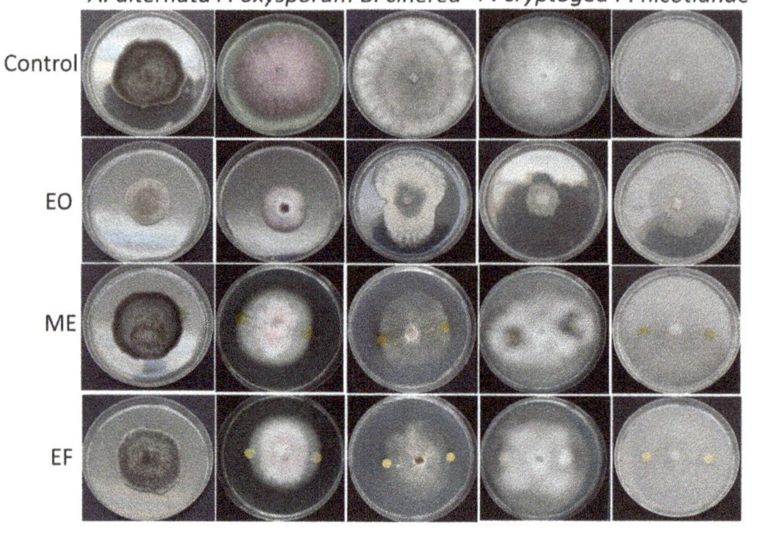


**Figure 2.** In vitro evaluation of the fungal and oomycete growth inhibition properties of essential oil (EO), exudate fraction (EF) and methanolic extract (ME) of *S. kitaibelii*.

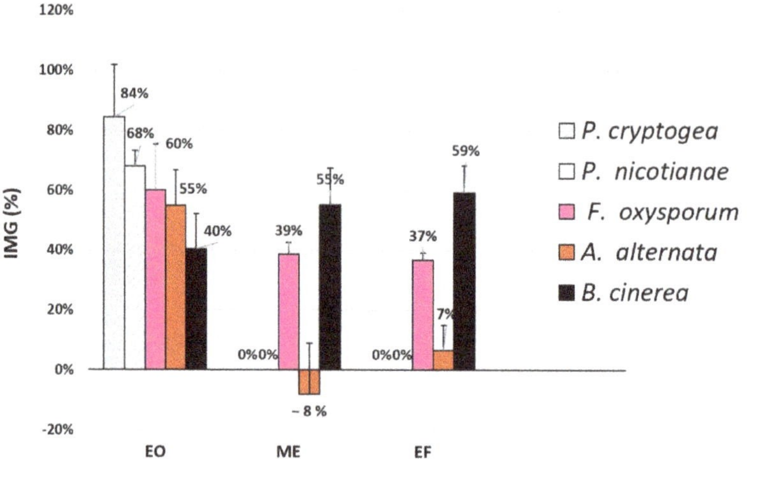


**Figure 3.** Inhibition of mycelial growth (%) in fungi and oomycetes by essential oil (EO), exudate fraction (EF) and methanolic extract (ME) of *S. kitaibelii*. Bars represent the standard deviation ($n$ = 4).

### 3.3. Acetylcholinesterase (AChE) Inhibitory Activity

EO, EF and ME from *S. kitaibelii* were studied for inhibitory effects on AChE by an in vitro assay. EO and ME exhibited similar AChE activity with $IC_{50}$ values of 3.68 ± 0.17 and 3.76 ± 0.25 mg/mL, respectively. EF exhibited the most potent inhibition on AChE with $IC_{50}$ value of 0.18 ± 0.03 mg/mL. Galanthamine (positive control) achieved an $IC_{50}$ value of 0.35 ± 0.01 μg/mL (1.22 ± 0.04 μM).

### 3.4. Seed Germination Inhibition Bioassay

The inhibitory activity of the ME from *S. kitaibelii* on *L. perenne* seed germination and root elongation was studied in the concentration range of 1–8 mg/mL. Aqueous solutions of the extract with a concentration of 5 and 8 mg/mL were found to inhibit seed germination and root growth by more than 90% (Table 4). A substantial reduction in root elongation was also observed at 3 mg/mL.

**Table 4.** Inhibition of seed germination and root growth of *L. perenne* by methanolic extract (ME) of *S. kitaibelii*.

| ME Concentration, mg/mL | Inhibition of Seed Germination, % | Inhibition of Root Growth, % |
|---|---|---|
| 1 | 13.4 ± 9 | 1 ± 1 |
| 3 | 21.8 ± 8 | 56 ± 6 |
| 5 | 95.8 ± 3 | 97.5 ± 1 |
| 8 | 99.4 ± 1 | 100 ± 0 |

## 4. Discussion

### 4.1. Phytochemical Constituents

The main component of the EO of the studied *S. kitaibelii* population was determined as *p*-cymene. Limonene, borneol and carvacrol were found as the next most abundant compounds. The established EO profile corresponds to the previously reported data for the chemical composition of this species [41,54,55] and refers to the *p*-cymene/limonene chemotype [40]. A variety of biological activities have been already published for the monoterpenes identified in the EO. Antimicrobial, antiparasitic, antiviral, antitumor, anti-inflammatory, antinociceptive, neuroprotective and other activities have been established for *p*-cymene [56,57]. A lot of pharmacological activities have been reported for the remaining well-represented monoterpenes in the EO—borneol, carvacrol and limonene—defining them as strongly bioactive molecules [58,59]. It is important to emphasize that carvacrol is a lead molecule for pest control [60,61]. In the present study of the EO profile of *S. kitaibelii, however,* carvacrol content is limited to 7.22% only.

The EF consisting of compounds located on the surface of plant tissues is rich in secondary metabolites with allelopathic potential [14,15]. Ursolic and oleanolic acids were found as the main bioactive compounds in the EF of *S. kitaibelii*. Various activities have been reported for these acids, but in the context of the present study, the anticholinesterase and antimicrobial properties are of importance [62–64]. The composition of *S. kitaibelii* EF we described follows previously reported data [41,42].

*S. kitaibelii*'s ME contained mainly polar primary and secondary metabolites. The established metabolite profile is coherent with previous research [34,65]. Rosmarinic and other phenolic acids, as well as flavonoid glycosides, were found as substances with known biological activities [66].

### 4.2. Fungal and Oomycete Growth Inhibition Properties

Fungi and fungal-like organisms such as oomycetes are causative agents of one of the most devastating diseases in plants. Equally severe harm is caused by such plant pathogens during post-harvest, affecting between 25% and 50% of the total production.

It has been found that essential oils rich in carvacrol demonstrate the most extensive and potent antifungal effects at minimal active doses (0.05–5 μg/mL) [20]. Plant extracts containing caffeic acid and rosmarinic acid inhibited zoospore germination of *Phytophthora* spp. [67]. The key components identified in the EO from *S. kitaibelii* (*p*-cymene, limonene, geraniol, carvacrol, borneol) are known to be effective antimicrobial agents, which is why this species keeps attracting the attention of researchers in medicine, food technology and agriculture. Several studies demonstrated the lack of antifungal activity of *p*-cymene against pathogenic fungi responsible for human diseases, such as *Rhizopus oryzae* and *Aspergillus niger*. However, the evaluation of essential oils rich in this monoterpene against *Aspergillus flavus*, *Saccharomyces cerevisiae* and *A. niger* has shown promising results [56]. Jian et al. demonstrated the ability of limonene to cause significant damage to the mycelium and conidia of *Fusarium graminearum* [68], revealing the potential of a limonene-formulated product as an alternative to synthetic fungicides. The essential oil obtained from *Cinnamomum camphora* chvar., rich in endo-borneol, demonstrated significant suppressive effects on five *Fusarium* species responsible for potato dry rot [69].

As shown in our results, the EO from *S. kitaibelii* is rich in bioactive components with antifungal and anti-oomycete properties and could find application in biological plant protection.

### 4.3. Acetylcholinesterase (AChE) Inhibitory Activity

The acetylcholinesterase inhibitory activity test was conducted to assess the possible presence of insecticidal potential because the mechanism of action of organophosphorus insecticides is acetylcholinesterase inhibition [70]. Several previously published articles highlight some of the main components of the EO, EF and ME from *S. kitaibelii* as AChE inhibitors: oleanolic and ursolic acids [63,64], *p*-cymene [71] and rosmarinic acid [72,73]. Rosmarinic acid has also been reported as a potent insecticide against important pests like *Acyrthosiphon pisum* [74]. Since oleanolic and ursolic acids are the most abundant components of the EF and with carvacrol also being present, it is not surprising that this fraction achieved the best AChE inhibition out of all three samples. Regarding the EO and MF, bioactive compounds such as *p*-cymene, rosmarinic acid and carvacrol appear to be in insufficient quantities in order to demonstrate good results in AChE inhibition.

The $IC_{50}$ value determined for the EF ($0.18 \pm 0.03$ mg/mL) suggests a good to strong AChE inhibition considering the fact that this is not a pure compound. This value is approximately 500 times larger than that of galanthamine (positive control). An assay including the purified forms of some putatively bioactive components could showcase the most important ones for AChE inhibition.

The $IC_{50}$ values for the EO and ME from *S. kitaibelii* are approximately 10 000 times larger than that of galanthamine (positive control), meaning these samples are practically inactive towards the enzyme.

Similar results for the AChE inhibitory potential of extracts and essential oils from the closely related species *Satureja montana* have been previously reported. Stronger AChE inhibition has been observed probably due to much larger quantities of carvacrol and the presence of its isomer thymol in the samples from *S. montana* [75–77].

### 4.4. Seed Germination Inhibition Bioassay

When working with alkaloid-bearing plants, the suitable concentration at which the species are screened is 1 mg/mL, while for non-alkaloid-containing plant species, the concentration at which the screening studies are performed is usually higher [78–80]. Seed germination inhibition activity was tested within the concentration range of 1–8 mg/mL for the ME of *S. kitaibelii*. This range is consistent with similar experiments conducted by other research groups [10,81]. In this assay, only ME was used due to low yields of EO and EF.

The achieved inhibition of seed germination by the ME at concentrations 5 and 8 mg/mL is good and promising compared to data already reported in the literature. Caffeic acid and its derivatives showed an inhibitory effect on the growth and germination of *Lantana indica* seeds [66]. Hernández and Munné-Bosch [82] showed that naringenin, a compound found in our study of the EF, is a strong seed germination inhibitor. Muñoz et al., 2020 considered carvacrol to be a good candidate for bioherbicide formulations [61]. These compounds are present in the ME from *S. kitaibelii* and could contribute to the phytotoxic properties observed in our study.

*Nerium oleander* L. flower extract suppressed the growth of *Lolium multiflorum* Lam. (Italian ryegrass) at 40 g $L^{-1}$ [78], which is a much higher concentration. In a screening study including six plant extracts, the one from *Tamarix mannifera* Ehrenb. ex Bunge completely inhibited the seed germination and seedling growth of *Phalaris minor* Retz., also at 40 g $L^{-1}$ [79]. Other plant extracts at this concentration showed weaker inhibition. The application of *Cardus cardunculus* (L.) Baill. crude extract at a concentration of 10 g $L^{-1}$ has led to the inhibition of roots and hypocotyl growth by 97% and 91%, respectively [80].

## 5. Conclusions

The present study was aimed at determining the pesticidal potential of the essential oil, exudate fraction and methanolic extract of *Satureja kitaibelii*. Results showed that the EO has fungicide potential, inhibiting the growth of all studied phytopathogens. Of importance is its inhibitory activity against *Phytophthora* spp. The ME exhibits weed suppression by inhibiting the seed germination of ryegrass. The EF displayed inhibitory activity against acetylcholinesterase, which is a good base for further analysis of its insecticidal potential. Our results showed that choosing the right approach (extraction method) is crucial to obtaining an extract or a fraction rich in substances with a particular biological activity. Here, we describe for the first time the pesticidal capacity of *S. kitaibelii* and provide a good direction for further studies.

**Author Contributions:** Conceptualization, M.N.; collection and identification of plant material, A.G. and G.G.; formal analysis, M.N., A.L., B.G. and E.Y.-T.; methodology and data curation, M.N., A.L., B.G. and E.Y.-T.; writing—original draft preparation, M.N and A.L.; writing—review and editing, M.N., A.L., B.G. and A.G. All authors have read and agreed to the published version of the manuscript.

**Funding:** This study was financially supported by Medical University—Pleven research Grant №18/2024 under the project "Biological and phytochemical characteristic of *Satureja kitaibelii* Wierzb. ex Heuff. (Lamiaceae) near Pleven, Bulgaria "and Bulgarian Operational Programme Science and Education for Smart Growth 2014–2020, co-financed by the European Union through the European Structural and Investment Funds (Grant BG05M2OP001-1.002-0012).

**Data Availability Statement:** The raw data supporting the conclusions of this article will be made available by the authors on request.

**Conflicts of Interest:** The authors declare no conflicts of interest.

## References

1. Schrader, K.; Andolfi, A.; Cantrell, C.; Cimmino, A.; Duke, S.; Osbrink, W.; Wedge, D.; Evidente, A. A survey of phytotoxic microbial and plant metabolites as potential natural products for pest management. *Chem. Biodivers.* **2010**, *7*, 2261. [CrossRef] [PubMed]
2. Okwute, S.K. Plants as Potential Sources of Pesticidal Agents: A Review. In *Pesticides—Advances in Chemical and Botanical Pesticides*; Soundararajan, R.P., Ed.; IntechOpen: London, UK, 2012; pp. 207–232. [CrossRef]
3. Mann, R.S.; Kaufman, E. Natural product pesticides: Their development, delivery and use against insect vectors. *MiniRev. Org. Chem.* **2012**, *9*, 185. [CrossRef]
4. Musa, F.M.; Carli, C.; Susuri, L.; Pireva, I. Monitoring of *Myzus persicae* (Sluzer) in potato fields in Kosovo. *Acta Agric. Slov.* **2004**, *83*, 379. [CrossRef]
5. Cai, X.; Gu, M. Bioherbicides in Organic Horticulture. *Horticulturae* **2016**, *2*, 2. [CrossRef]

6. Raveau, R.; Fontaine, J.; Lounès-Hadj Sahraoui, A. Essential Oils as Potential Alternative Biocontrol Products against Plant Pathogens and Weeds: A Review. *Foods* **2020**, *9*, 3. [CrossRef]
7. Jouini, A.; Verdeguer, M.; Pinton, S.; Araniti, F.; Palazzolo, E.; Badalucco, L.; Laudicina, V.A. Potential Effects of Essential Oils Extracted from Mediterranean Aromatic Plants on Target Weeds and Soil Microorganisms. *Plants* **2020**, *9*, 10. [CrossRef]
8. Verdeguer, M.; Sánchez-Moreiras, A.M.; Araniti, F. Phytotoxic Effects and Mechanism of Action of Essential Oils and Terpenoids. *Plants* **2020**, *9*, 11. [CrossRef]
9. Garrido-Miranda, K.A.; Giraldo, J.D.; Schoebitz, M. Essential Oils and Their Formulations for the Control of *Curculionidae* Pests. *Fron. Agron.* **2022**, *4*, 876687. [CrossRef]
10. Nikolova, M.; Berkov, S. Use of essential oils as natural herbicides. *Ecol. Balk.* **2018**, *10*, 259–265.
11. Werrie, P.-Y.; Durenne, B.; Delaplace, P.; Fauconnier, M.L. Phytotoxicity of essential oils: Opportunities and constraints for the development of biopesticides. A review. *Foods* **2020**, *9*, 1291. [CrossRef]
12. Lengai, G.; Muthomi, J.; Mbega, E. Phytochemical activity and role of botanical pesticides in pest management for sustainable agricultural crop production. *Sci. Afr.* **2020**, *7*, e00239. [CrossRef]
13. Nikolova, M.; Yovkova, M.; Yankova-Tsvetkova, E.; Traykova, B.; Stefanova, T.; Aneva, I.; Berkov, S. Biocidal Activity of *Origanum vulgare* subsp. hirtum Essential Oil. *Acta Univ. Agric. Silvic. Mendel. Brun.* **2021**, *69*, 569–578. [CrossRef]
14. Chaves, N.; Escudero, J.C. Allelopathic Effect of *Cistus ladanifer* on Seed Germination. *Funct. Ecol.* **1997**, *11*, 432–440. Available online: http://www.jstor.org/stable/2390378 (accessed on 29 January 2025). [CrossRef]
15. Onyilagha, J.; Grotewold, E. The biology and structural distribution of surface flavonoids. In *Recent Research Developments in Plant Science*; Pandalai, S.G., Ed.; Research Signpost: Trivandrum, India, 2004; pp. 53–71.
16. Karamanoli, K.; Menkissoglu-Spiroudi, U.; Bosabalidis, A.M.; Vokou, D.; Constantinidou, H.-I.A. Bacterial colonization of the phyllosphere of nineteen plant species and antimicrobial activity of their leaf secondary metabolites against leaf associated bacteria. *Chemoecology* **2005**, *15*, 59–67. [CrossRef]
17. Lyubenova, A.; Nikolova, M.; Slavov, S.B. Impact of *Origanum vulgare* subsp. *hirtum* (Link) Ietswaart derived extracts and essential oil on plant pathogens from genus *Phytophthora*. *Bulg. J. Agric. Sci.* **2024**, *30*, 833–838.
18. Sotirova, A.; Mutafova, B.; Berkov, S.; Nikolova, M. Antibacterial Activity of Methanol Extract and Acetone Exudates from Bulgarian Plants. *Acta Microbiol. Bulg.* **2022**, *38*, 48–51.
19. Nikolova, M.; Yankova-Tsvetkova, E.; Stefanova, T.; Stoyanov, S.; Berkov, S. Evaluation of *Helichrysum arenarium* flower exudate as an inhibitor on *Lolium perenne* seed germination under laboratory conditions. *Acta Agrob.* **2023**, *76*, 761. [CrossRef]
20. Lyubenova, A.; Rusanova, M.; Nikolova, M.; Slavov, S.B. Plant extracts and *Trichoderma* spp.: Possibilities for implementation in agriculture as biopesticides. *Biotechnol. Biotechnol. Equip.* **2023**, *37*, 159–166. [CrossRef]
21. Şahin, F.; Güllüce, M.; Daferera, D.; Sökmen, A.; Sökmen, M.; Polissiou, M.; Agar, G.; Özer, H. Biological activities of the essential oils and methanol extract of *Origanum vulgare* ssp. vulgare in the Eastern Anatolia region of Turkey. *Food Control* **2004**, *15*, 549–557. [CrossRef]
22. Kordali, S.; Kabaagac, G.; Sen, İ.; Yilmaz, F.; Najda, A. Phytotoxic Effects of Three *Origanum* Species Extracts and Essential Oil on Seed Germinations and Seedling Growths of Four Weed Species. *Agronomy* **2022**, *12*, 2581. [CrossRef]
23. Srikrishnah, S.; Begam, U.J. Review on Use of Plant Extracts in Weed Control. *Curr. Trends Biomedical. Eng. Biosci.* **2019**, *18*, 309–317. [CrossRef]
24. Garcia-Rellán, D.; Verdeguer, M.; Salamone, A.; Blázquez, M.A.; Boira, H. Chemical Composition, Herbicidal and Antifungal Activity of *Satureja cuneifolia* Essential Oils from Spain. *Nat. Prod. Commun.* **2016**, *11*, 1934578X1601100636. [CrossRef]
25. Hazrati, H.; Saharkhiz, M.J.; Niakousari, M.; Moein, M. Natural herbicide activity of *Satureja hortensis* L. essential oil nanoemulsion on the seed germination and morphophysiological features of two important weed species. *Ecotoxicol. Environ. Saf.* **2017**, *142*, 423–430. [CrossRef] [PubMed]
26. Fierascu, I.; Dinu-Pirvu, C.E.; Fierascu, R.C.; Velescu, B.S.; Anuta, V.; Ortan, A.; Jinga, V. Phytochemical Profile and Biological Activities of *Satureja hortensis* L.: A Review of the Last Decade. *Molecules* **2018**, *23*, 10. [CrossRef]
27. Güllüce, M.; Sökmen, M.; Daferera, D.; Ağar, G.; Ozkan, H.; Kartal, N.; Polissiou, M.; Sökmen, A.; Sahin, F. In vitro antibacterial, antifungal, and antioxidant activities of the essential oil and methanol extracts of herbal parts and callus cultures of *Satureja hortensis* L. *J. Agric. Food Chem.* **2003**, *51*, 3958–3965. [CrossRef]
28. Ebadollahi, A.; Jalali Sendi, J.; Ziaee, M.; Krutmuang, P. Acaricidal, Insecticidal, and Nematicidal Efficiency of Essential Oils Isolated from the *Satureja* Genus. *Int. J. Environ. Res. Public Health* **2021**, *18*, 6050. [CrossRef]
29. Tepe, B. Inhibitory Effect of *Satureja* on Certain Types of Organisms. *Rec. Nat. Prod.* **2015**, *9*, 1–18.
30. Anchev, M. Genus *Satureja* L. In *Flora Reipublicae Popularis Bulgaricae*; Velčev, V., Ed.; Aedibus Academiae Scientiarum Bulgaria: Sofia, Bulgaia, 1989; pp. 336–338. (In Bulgarian)

31. Matevski, V.; Čarni, A.; Ćušterevska, R.; Kostadinovski, M.; Mucina, L. Syntaxonomy and biogeography of dry grasslands on calcareous substrates in the central and southern Balkans. *Appl. Veg. Sci.* **2018**, *21*, 488–513. [CrossRef]
32. Todorova, S.; Tzonev, R. Bromo Moesiacae-Stipetum Epilosae—A New Association from the Relict Mountain Steppe Vegetation in South-Western Bulgaria. *Hacquetia* **2010**, *9*, 185–206. [CrossRef]
33. Šilić, C. Monographie der Gattungen *Satureja* L., *Calamintha* Miller *Micromeria* Bentham, *Acinos* Miller und *Clinopodium* L. In *Flora Jugoslaviens*; Svjetlost: Sarajevo, Bosnia and Herzegovina, 1979.
34. Gopčević, K.; Grujić, S.; Arsenijević, J.; Karadžić, I.; Izrael-Živković, L.; Maksimović, Z. Phytochemical Properties of *Satureja kitaibelii*, Potential Natural Antioxidants: A New Insight. *Plant Foods Hum. Nut.* **2019**, *74*, 179–184. [CrossRef]
35. Kundaković, T.; Milenković, M.; Zlatković, S.; Kovacević, N.; Goran, N. Composition of *Satureja kitaibelii* essential oil and its antimicrobial activity. *Nat. Prod. Commun.* **2011**, *6*, 1353–1356. [CrossRef] [PubMed]
36. Mihajilov-Krstev, T.; Kitić, D.; Radnović, D.; Ristić, M.; Mihajlović-Ukropina, M.; Zlatković, B. Chemical composition and antimicrobial activity of Satureja kitaibelii essential oil against pathogenic microbial strains. *Nat. Prod. Commun.* **2011**, *6*, 1167–1172. [CrossRef] [PubMed]
37. Aćimović, M.; Zorić, M.; Zheljazkov, V.D.; Pezo, L.; Čabarkapa, I.; Stanković Jeremić, J.; Cvetković, M. Chemical Characterization and Antibacterial Activity of Essential Oil of Medicinal Plants from Eastern Serbia. *Molecules* **2020**, *25*, 5482. [CrossRef] [PubMed]
38. Dimitrijević, M.V.; Miladinović, L.C.; Marković, M.S.; Arsić, B.; Mihajilov-Krstev, T.M.; Miladinović, D.L. New Facts on the Antimicrobial Essential Oil of *Satureja kitaibelii*. *Chem. Biodivers.* **2024**, *21*, e202301418. [CrossRef]
39. Zheljazkov, V.D.; Semerdjieva, I.B.; Cantrell, C.L.; Astatkie, T.; Aćimović, M. Phytochemical Variability of Essential Oils of Two Balkan Endemic Species: *Satureja pilosa* Velen. and *S. kitaibelii* Wierzb. ex Heuff. (Lamiaceae). *Molecules* **2022**, *27*, 3153. [CrossRef]
40. Miladinović, D.L.; Dimitrijević, M.V.; Miladinović, L.C.; Marković, M.S.; Stojanović, G.S. Seasonal variation in the essential oil of *Satureja kitaibelii* determines chemotypes. *J. Essent. Oil Res.* **2022**, *34*, 567–575. [CrossRef]
41. Dodoš, T.; Rajčević, N.; Janaćković, P.; Vujisić, L.; Marin, P.D. Essential oil profile in relation to geographic origin and plant organ of *Satureja kitaibelii* Wierzb. Ex Heuff. *Ind. Crop Prod.* **2019**, *139*, 111549. [CrossRef]
42. Gavrilova, A.; Nikolova, M.; Gavrilov, G. Phytochemical screening of *Satureja kitaibelii* Wierzb. Ex Heuff. Extracts by GC/MS and TLC. *Farmacia* **2023**, *71*, 91–96. [CrossRef]
43. Souiy, Z. Essential Oil Extraction Process. In *Essential Oils—Recent Advances, New Perspectives and Applications*; Viskelis, J., Ed.; Intech Open: Rijeka, Croatia, 2024; pp. 1–15. [CrossRef]
44. Wollenweber, E.; Dörr, M.; Rustaiyan, A.; Roitman, J.; Graven, E. Notes: Exudate Flavonoids of Some Salvia and a *Trichostema* Species. *Z. Naturforsch. C J. Biosci.* **1992**, *47*, 782–784. [CrossRef]
45. Berkov, S.; Pechlivanova, D.; Denev, R.; Nikolova, M.; Georgieva, L.; Sidjimova, B.; Bakalov, D.; Tafradjiiska, R.; Stoynev, A.; Momekov, G.; et al. GC-MS analysis of *Amaryllidaceae* and *Sceletium*-type alkaloids in bioactive fractions from *Narcissus* cv. Hawera. *Rapid Commun. Mass. Spectrom.* **2021**, *35*, e9116. [CrossRef]
46. Hummel, J.; Strehmel, N.; Selbig, J.; Walther, D.; Kopka, J. Decision tree supported substructure prediction of metabolites from GC-MS profiles. *Metabolomics* **2010**, *6*, 322–333. [CrossRef] [PubMed]
47. Balouiri, M.; Sadiki, M.; Ibnsouda, S.K. Methods for in Vitro Evaluating Antimicrobial Activity: A Review. *J. Pharm. Anal.* **2016**, *6*, 71–79. [CrossRef] [PubMed]
48. Nikolova, M.; Lyubenova, A.; Yankova-Tsvetkova, E.; Georgiev, B.; Berkov, S.; Aneva, I.; Trendafilova, A. *Artemisia Santonicum* L. and *Artemisia Lerchiana* Web. Essential Oils and Exudates as Sources of Compounds with Pesticidal Action. *Plants* **2023**, *12*, 3491. [CrossRef] [PubMed]
49. Schneider, C.A.; Rasband, W.S.; Eliceiri, K.W. NIH Image to ImageJ: 25 years of image analysis. *Nat. Methods* **2012**, *9*, 671–675. [CrossRef]
50. Zygadlo, J.A.; Guzman, C.A.; Grosso, N.R. Antifungal Properties of the Leaf Oils of *Tagetes minuta* L. and *T. filifolia* Lag. *J. Essent. Oil Res.* **1994**, *6*, 617–621. [CrossRef]
51. Ellman, G.L.; Courtney, K.D.; Andres, V.; Featherstone, R.M. A New and Rapid Colorimetric Determination of Acetylcholinesterase Activity. *Biochem. Pharmacol.* **1961**, *7*, 88–95. [CrossRef]
52. López, S.; Bastida, J.; Viladomat, F.; Codina, C. Acetylcholinesterase Inhibitory Activity of Some Amaryllidaceae Alkaloids and Narcissus Extracts. *Life Sci.* **2002**, *71*, 2521–2529. [CrossRef]
53. Atak, M.; Mavi, K.; Uremis, I. Bio-herbicidal effects of oregano and rosemary essential oils on germination and seedling growth of bread wheat cultivars and weeds. *Rom. Biotechnol. Lett.* **2016**, *21*, 11149–11159.
54. Chalchat, J.-C.; Gorunovic, M.S.; Maksimovic, Z.A. Essential Oil of *Satureja kitaibelii* Wierzb. F. Aristata (Vand.) Hayek, *Lamiaceae* from Eastern Serbia. *J. Essent. Oil Res.* **1999**, *11*, 691–692. [CrossRef]
55. Konakchiev, A.; Tsankova, E. The Essential Oils of *Satureja montana* ssp. kitaibelii Wierzb. and Satureja pilosa var. pilosa Velen from Bulgaria. *J. Essent. Oil Res.* **2002**, *14*, 120–121. [CrossRef]

56. Marchese, A.; Arciola, C.R.; Barbieri, R.; Silva, A.S.; Nabavi, S.F.; Tsetegho Sokeng, A.J.; Izadi, M.; Jafari, N.J.; Suntar, I.; Daglia, M.; et al. Update on Monoterpenes as Antimicrobial Agents: A Particular Focus on p-Cymene. *Materials* **2017**, *10*, 947. [CrossRef] [PubMed]
57. Balahbib, A.; El Omari, N.; Hachlafi, N.E.; Lakhdar, F.; El Menyiy, N.; Salhi, N.; Mrabti, H.N.; Bakrim, S.; Zengin, G.; Bouyahya, A. Health beneficial and pharmacological properties of p-cymene. *Food Chem. Toxicol.* **2021**, *153*, 112259. [CrossRef] [PubMed]
58. de Cássia da Silveira e Sá, R.; Andrade, L.N.; de Sousa, D.P. A review on anti-inflammatory activity of monoterpenes. *Molecules* **2013**, *18*, 1227–1254. [CrossRef] [PubMed]
59. Hu, X.; Yan, Y.; Liu, W.; Liu, J.; Fan, T.; Deng, H.; Cai, Y. Advances and perspectives on pharmacological activities and mechanisms of the monoterpene borneol. *Phytomedicine* **2024**, *132*, 155848. [CrossRef]
60. Bendre, R.; Bagul, S.; Rajput, J. Carvacrol: An excellent natural pest control agent. *Nat. Prod. Chem. Res.* **2018**, *6*, 349. [CrossRef]
61. Muñoz, M.; Torres-Pagán, N.; Peiró, R.; Guijarro, R.; Sánchez-Moreiras, A.M.; Verdeguer, M. Phytotoxic effects of three natural compounds: Pelargonic acid, carvacrol, and cinnamic aldehyde, against problematic weeds in Mediterranean crops. *Agronomy* **2020**, *10*, 791. [CrossRef]
62. Jesus, J.A.; Lago, J.H.; Laurenti, M.D.; Yamamoto, E.S.; Passero, L.F. Antimicrobial activity of oleanolic and ursolic acids: An update. *Evid. Based Complement. Alternat Med.* **2015**, *2015*, 620472. [CrossRef]
63. Chung, Y.K.; Heo, H.J.; Kim, E.K.; Kim, H.K.; Huh, T.L.; Lim, Y.; Kim, S.K.; Shin, D.H. Inhibitory effect of ursolic acid purified from *Origanum majorana* L on the acetylcholinesterase. *Mol. Cells* **2001**, *11*, 137–143. [CrossRef]
64. Stępnik, K.; Kukula-Koch, W.; Plazinski, W.; Rybicka, M.; Gawel, K. Neuroprotective Properties of Oleanolic Acid—Computational-Driven Molecular Research Combined with In Vitro and In Vivo Experiments. *Pharmaceuticals* **2023**, *16*, 1234. [CrossRef]
65. López-Cobo, A.; Gómez-Caravaca, A.M.; Švarc-Gajić, J.; Segura-Carretero, A.; Fernández-Gutiérrez, A. Determination of phenolic compounds and antioxidant activity of a Mediterranean plant: The case of *Satureja montana* subsp. kitaibelii. *J. Funct. Foods* **2015**, *18*, 1167–1178. [CrossRef]
66. Pan, L.; He, F.; Liang, Q.; Bo, Y.; Lin, X.; Javed, Q.; Ullah, M.S.; Sun, J. Allelopathic Effects of Caffeic Acid and Its Derivatives on Seed Germination and Growth Competitiveness of Native Plants (Lantana indica) and Invasive Plants (*Solidago canadensis*). *Agriculture* **2023**, *13*, 1719. [CrossRef]
67. Widmer, T.L.; Laurent, N. Plant extracts containing caffeic acid and rosmarinic acid inhibit zoospore germination of *Phytophthora* spp. pathogenic to *Theobroma cacao. Eur. J. Plant Pathol.* **2006**, *115*, 377–388. [CrossRef]
68. Jian, Y.; Chen, X.; Ma, H.; Zhang, C.; Luo, Y.; Jiang, J.; Yin, Y. Limonene formulation exhibited potential application in the control of mycelial growth and deoxynivalenol production in *Fusarium graminearum*. *Front. Microbiol.* **2023**, *14*, 1161244. [CrossRef] [PubMed]
69. Li, H.; Liao, H.; Li, Y.; Qi, Y.; Ni, H.; Zou, Z.; Liu, Z. Chemical composition and antifungal activity of *Cinnamomum camphora* chvar. Borneol essential oil obtained using solvent-free microwave-assisted method. *Arab. J. Chem.* **2023**, *16*, 104996. [CrossRef]
70. Lopez, M.D.; Pascual-Villalobos, M.J. Mode of inhibition of acetylcholinesterase by monoterpenoids and implications for pest control. *Ind. Crops Prod.* **2010**, *31*, 284–288. [CrossRef]
71. Caputo, L.; Amato, G.; De Martino, L.; De Feo, V.; Nazzaro, F. Anti-Cholinesterase and Anti--Amylase Activities and Neuroprotective Effects of Carvacrol and p-Cymene and Their Effects on Hydrogen Peroxide Induced Stress in SH-SY5Y Cells. *Int. J. Mol. Sci.* **2023**, *24*, 6073. [CrossRef]
72. Gülçin, İ.; Scozzafava, A.; Supuran, C.T.; Koksal, Z.; Turkan, F.; Çetinkaya, S.; Bingöl, Z.; Huyut, Z.; Alwasel, S.H. Ros-marinic acid inhibits some metabolic enzymes including glutathione S-transferase, lactoperoxidase, acetylcholin-esterase, butyrylcholinesterase and carbonic anhydrase isoenzymes. *J. Enzyme Inhib. Med. Chem.* **2016**, *31*, 1698–1702. [CrossRef]
73. Kamli, M.R.; Sharaf, A.A.M.; Sabir, J.S.M.; Rather, I.A. Phytochemical Screening of *Rosmarinus officinalis* L. as a Potential Anticholinesterase and Antioxidant–Medicinal Plant for Cognitive Decline Disorders. *Plants* **2022**, *11*, 514. [CrossRef]
74. Khan, S.; Taning, C.N.T.; Bonneure, E.; Mangelinckx, S.; Smagghe, G.; Ahmad, R.; Fatima, N.; Asif, M.; Shah, M.M. Bioactivity-guided isolation of rosmarinic acid as the principle bioactive compound from the butanol extract of *Isodon rugosus* against the pea aphid, *Acyrthosiphon pisum*. *PLoS ONE* **2019**, *14*, e0215048. [CrossRef]
75. Silva, F.V.M.; Martins, A.; Salta, J.; Neng, N.R.; Nogueira, J.M.F.; Mira, D.; Gaspar, N.; Justino, J.; Grosso, C.; Urieta, J.S.; et al. Phytochemical Profile and Anticholinesterase and Antimicrobial Activities of Supercritical versus Conventional Extracts of *Satureja montana*. *J. Agric. Food Chem.* **2009**, *57*, 11557–11563. [CrossRef]
76. Cabana, R.; Silva, L.R.; Valentão, P.; Viturro, C.I.; Andrade, P.B. Effect of different extraction methodologies on the recovery of bioactive metabolites from *Satureja parvifolia* (Phil.) Epling (*Lamiaceae*). *Ind. Crop Prod.* **2013**, *48*, 49–56. [CrossRef]

77. Les, F.; Galiffa, V.; Cásedas, G.; Moliner, C.; Maggi, F.; López, V.; Gómez-Rincón, C. Essential Oils of Two Subspecies of *Satureja montana* L. against Gastrointestinal Parasite Anisakis simplex and Acetylcholinesterase Inhibition. *Molecules* **2024**, *29*, 19. [CrossRef] [PubMed]
78. Uslu, Ö.; Gedik, O.; Kaya, A.; Khan, M.; Tassever, M.; Turkkaya, E. Allelopathic Effects of Flower Extract of Oleander (Nerium Oleander) on the Germination of Seed and Seedling Growth of *Lolium Multiflorum*. *J. Inst. Sci. Technol.* **2018**, *8*, 309–3017. [CrossRef]
79. El-Mergawi, R.A.; Al-Humaid, A.I. Searching for natural herbicides in methanol extracts of eight plant species. *Bull. Nat. Res. Cent.* **2019**, *43*, 22. [CrossRef]
80. Ben Kaab, S.; Lins, L.; Hanafi, M.; Bettaieb Rebey, I.; Deleu, M.; Fauconnier, M.-L.; Ksouri, R.; Jijakli, M.H.; De Clerck, C. Cynara cardunculus Crude Extract as a Powerful Natural Herbicide and Insight into the Mode of Action of Its Bioactive Molecules. *Biomolecules* **2020**, *10*, 209. [CrossRef]
81. Wahyuni, D.S.C.; Van der Kooy, F.; Klinkhamer, P.G.L.; Verpoorte, R.; Leiss, K. The Use of Bio-Guided Fractionation to Explore the Use of Leftover Biomass in Dutch Flower Bulb Production as Allelochemicals against Weeds. *Molecules* **2013**, *18*, 4510–4525. [CrossRef]
82. Hernández, I.; Munné-Bosch, S. Naringenin inhibits seed germination and seedling root growth through a salicylic acid-independent mechanism in *Arabidopsis thaliana*. *Plant Physiol. Biochem.* **2012**, *61*, 24–28. [CrossRef]

*agronomy*

MDPI

*Article*

# Faunistic Study of Auchenorrhyncha in Olive Orchards in Greece, Including First Records of Species

**Zoi Thanou [1], Myrto Stamouli [2], Anastasia Magklara [1], David Theodorou [1], Georgia Stamatakou [1], Georgios Konidis [1], Panagiota Koufopoulou [1], Christos Lyberopoulos [1], Sofia Tribonia [1], Petros Vetsos [1], Andreas Katribouzas [1], Argyro Kalaitzaki [3], Georgios Papadoulis [1] and Antonios Tsagkarakis [2,*]**

[1] Laboratory of Agricultural Zoology and Entomology, Agricultural University of Athens, Iera Odos 75, 118 55 Athens, Greece; zoethaua@gmail.com (Z.T.); anastasiamag108@gmail.com (A.M.); theodoroudth@gmail.com (D.T.); gstamatak@yahoo.gr (G.S.); giorgoskonidis@hotmail.com (G.K.); giotakouf612@gmail.com (P.K.); xristos.libe@gmail.com (C.L.); sofiatribo@gmail.com (S.T.); pet.veccio@gmail.com (P.V.); andreaskatrib132@gmail.com (A.K.); gpapadoulis@aua.gr (G.P.)

[2] Laboratory of Sericulture and Apiculture, Agricultural University of Athens, Iera Odos 75, 118 55 Athens, Greece; myrtostamouli@aua.gr

[3] Institute of Olive Tree, Subtropical Plants and Viticulture, Hellenic Agricultural Organization 'DEMETER', 731 00 Chania, Greece; kalaitzaki@elgo.gr

* Correspondence: atsagarakis@aua.gr; Tel.: +30-2105294578

**Abstract:** The study of Auchenorrhyncha species composition in Greek olive orchards is crucial due to the potential threat of *Xylella fastidiosa* invading the region. Recent studies have begun exploring agricultural landscapes, particularly olive and citrus orchards. From 2016 to 2022, biodiversity surveys were conducted in thirteen olive orchards across three regions of Greece: Peloponnese, Sterea Ellada, and the Northeast Aegean. Malaise traps were installed in each orchard and monitored monthly, supplemented by sweep net sampling in two orchards to capture less mobile species and assess their association with host plants. A total of 14,771 specimens were collected, representing 125 species predominantly feeding on weeds. The dominant species were the Typhlocybinae *Hebata decipiens* and *Zyginidia pullula*, while *Euscelis lineolata* was the most common Deltocephalinae. Aphrophoridae, including *Philaenus spumarius* and *Neophilaenus campestris*, were more effectively collected with sweep nets, primarily from *Avena sterilis* L. This study offers a detailed overview of the Auchenorrhyncha fauna in Greek olive orchards, providing essential insights for developing strategies to prevent the invasion of *Xylella fastidiosa*.

**Keywords:** leafhoppers; olive; *Xylella fastidiosa*; potential vector; Aphrophoridae; new records

## 1. Introduction

*Xylella fastidiosa* Wells et al. is a Gram-negative bacterial plant pathogen that causes a variety of economically devastating diseases in numerous crops, including Pierce's disease (PD) of grapes, citrus variegated chlorosis (CVC), phony peach disease, and olive quick decline syndrome (OQDS). This bacterium is of significant concern due to its extensive genetic and phenotypic diversity, which has led to its classification into four subspecies: *X. fastidiosa fastidiosa*, *X. fastidiosa pauca*, *X. fastidiosa multiplex*, and *X. fastidiosa sandyi*. In North America, subspecies like *X. fastidiosa multiplex* and *X. fastidiosa fastidiosa* affect almond, peach, oak, and grapevine, while *X. fastidiosa pauca* has caused significant damage in South America, particularly in coffee, citrus, and olive crops [1–3].

In recent years, *X. fastidiosa* has been increasingly detected across Europe, threatening a wide range of crops. The first major outbreak occurred in southern Italy, where *X. fastidiosa pauca* was detected in olive orchards in the Apulia region, causing olive quick decline syndrome [4,5]. This outbreak affected over 10,000 hectares of olive trees, leading to severe economic and agricultural impacts. Since then, the pathogen has been detected in several other European countries.

In France, *X. fastidiosa* was first identified in Corsica in 2015 on ornamental plants like *Polygala myrtifolia* L. and later spread to mainland France, affecting regions such as Provence-Alpes-Côte d'Azur [6]. The subspecies *X. fastidiosa multiplex* was responsible for these infections. In Spain, *X. fastidiosa* was detected in the Balearic Islands in 2016, where both *X. fastidiosa fastidiosa* and *X. fastidiosa multiplex* were reported, with infections extending to almond trees, sweet cherries, and other ornamental plants [7,8]. In mainland Spain, outbreaks occurred in Alicante, infecting almond trees, and in other areas of Valencia, further escalating concerns about the pathogen's spread in the Iberian Peninsula [9]. Additionally, in 2019, *X. fastidiosa* was detected in Portugal, affecting olive trees in the Vila Nova de Gaia region [10]. This detection marked another critical spread of the pathogen in Mediterranean Europe, which has a high density of host plants susceptible to *X. fastidiosa*.

The rapid spread of *X. fastidiosa* across Europe has led to significant quarantine measures and concerted efforts to monitor and control its vectors. The pathogen's ability to infect a wide variety of plant species, combined with the similarities in climatic conditions across much of the Mediterranean region, raises concerns that other countries, such as Greece, may soon face similar outbreaks. Given Greece's extensive olive cultivation, with approximately 900,000 hectares of olive trees representing a significant portion of the agricultural landscape [11], the introduction of *X. fastidiosa* would have severe consequences for the country's economy and agriculture.

*X. fastidiosa* is primarily transmitted by xylem-feeding Auchenorrhyncha (Hemiptera), including species from the families Cicadellidae (leafhoppers and sharpshooters), Aphrophoridae (spittlebugs), and Cercopidae (froghoppers). These insects acquire the bacterium when they feed on the xylem of infected plants and can subsequently transmit it to healthy plants, facilitating the spread of the disease [12–15]. According to Purcell [16], every xylem fluid-feeding hemipteran should be considered a potential vector of the bacterium, highlighting the need for comprehensive faunistic studies to identify and manage these vectors.

In Europe, two primary xylem-feeding species have been identified as the main vectors of *X. fastidiosa*: *Philaenus spumarius* (L.) and *Neophilaenus campestris* (Fallén), both of which have been found to carry the pathogen in olive orchards in southern Italy [17]. However, recent studies have expanded the scope of potential vectors. For example, Elbeaino et al. [18] discovered that *Euscelis lineolata* (Brullé), a phloem-feeding leafhopper, was infected with *X. fastidiosa*, suggesting that specialized phloem feeders could also become infected by probing xylem vessels. Similarly, Chuche et al. [19] demonstrated that *Scaphoideus titanus* Ball, another phloem feeder, can reach xylem tissues and feed on them for extended periods, further complicating the understanding of vector dynamics.

Given the expanding knowledge of *X. fastidiosa* vectors and the potential for new species to act as transmitters of the bacterium, it is crucial to investigate the entire Auchenorrhyncha fauna present in regions at risk of pathogen introduction, not only to identify current vectors but also to assess the potential for other species to become vectors under certain conditions. In Italy, while considerable data on Auchenorrhyncha fauna exists [20–22], the lack of specific studies in olive orchards before the arrival of *X. fastidiosa* delayed the implementation of effective vector management strategies. However, following the pathogen's introduction, a series of studies quickly prioritized the monitoring of Auchenorrhyncha populations in olive orchards [17,18,23].

Although recent studies have begun to explore the Auchenorrhyncha fauna in Greek olive orchards [24,25], there is still a significant knowledge gap, particularly regarding the role of these insects as potential vectors of *X. fastidiosa*. These studies have provided valuable insights into the species composition, diversity, and seasonal abundance of Auchenorrhyncha under different management systems. However, more extensive research is required to fully understand the seasonal dynamics, regional variation, and the potential role of these insects in the transmission of *X. fastidiosa* in Greece. Developing effective monitoring and control strategies is crucial to mitigate the risk of *X. fastidiosa* becoming established in this key olive producing region. This study aims to build on previous research by providing a comprehensive survey of Auchenorrhyncha species in Greek olive orchards, with a focus on documenting the full spectrum of Auchenorrhyncha fauna rather than solely potential vector species. The research was conducted across three of the country's most important olive-growing regions: Central (Sterea Ellada), Southern (Peloponnese), and Northern Greece (Northern Aegean, Lesvos Island). In addition to documenting the seasonal appearance and abundance of Auchenorrhyncha species, this study contributes new data on the regional distribution and composition of these insects in different agroecological zones collected by sweep net sampling of specific ground vegetation cover. By examining both xylem and phloem feeders, this research provides crucial insights into the Auchenorrhyncha fauna associated with olive cultivation in Greece, expanding the knowledge necessary for effective pest and vector management strategies in areas at risk for *X. fastidiosa* introduction.

## 2. Materials and Methods

### 2.1. Sampling Areas

This study was conducted in 13 olive orchards in Greece, between 2016 and 2022. The orchards were distributed across three geographical regions of Greece: Peloponnese (Achaia, Argolida, Ileia, Messinia), Sterea Ellada (Attica and the island of Euboea), and Lesvos Island (North Aegean) (Figure 1). The orchards included a range of management systems: some complied with organic standards according to European Union (EU) legislation (Council Regulation (EC) 834/2007), others adhered to the EU Common Agricultural Policy (CAP) framework for conventional farming, and a few received no treatments. The surrounding vegetation mainly consisted of other orchards, including olives and citrus, with occasional vineyards. Differences in climatic conditions between regions—ranging from Mediterranean coastal climates to more temperate inland conditions—could influence insect population dynamics and are relevant for interpreting the results.

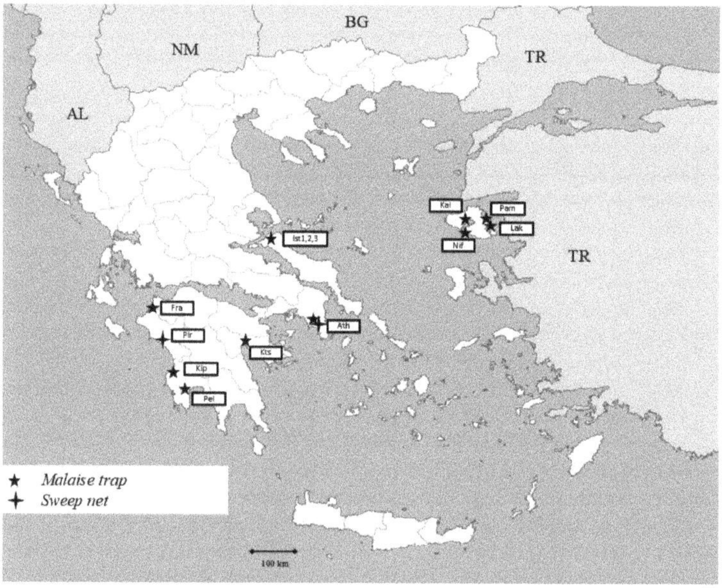


**Figure 1.** Geographical distribution of sampling sites across Greece. The different symbols indicate the different sampling methods. Pam = Pamfila, Lak = Lakerda, Kal = Kalloni, Nif = Nifida, Ist1,2,3 = Istiea 1,2,3, Ath = Athens, Kts = Koutsopodi, Fra = Fragka, Pir = Pirgos, Kip = Kiparissia, and Pel = Pelekanada.

#### 2.1.1. Sterea Ellada

Athens (Ath), Agricultural University of Athens campus (Attica): (37°98′18.57″ N, 23°70′68.59″ E). Olive tree varieties: more than 30 varieties, including Greek, Spanish, Italian, and other European varieties. Surrounding orchards: olive yards, citrus, stone fruits, apples, pears, etc. Also, there are pistachio trees, a vineyard, and a botanical garden with many ornamental trees. Weed control was performed using a mechanical tiller attached to a tractor. No pesticides were applied. Sampling period: 2016–2019 (with Malaise trap), 2020–2021 (with sweep net).

Istiea–North Euboea conventional (Ist1), Euboea Island: (38°57′47.7″ N, 23°09′03.0″ E). Olive tree varieties: Kalamon and Amfissis. Surrounding orchards: olive yards. Herbicides were used for weed control. Sampling period: 2018–2019 (with Malaise trap).

Istiea–North Euboea organic (Ist2), Euboea Island: (38°58′06.8″ N, 23°07′36.3″ E). Olive tree variety: Kalamon. Surrounding orchards: olive yards. Weed control was achieved using a mechanical tiller attached to a tractor. Sampling period: 2018–2019 (with Malaise trap).

Istiea–North Euboea–no treatment (Ist3), Euboea Island: (38°57′46.7″ N, 23°08′37.7″ E). Olive tree variety: Amfissis. Surrounding orchards: olive yards. The wild vegetation remained intact. Sampling period: 2018–2019 (with Malaise trap).

#### 2.1.2. Northeast Aegean

Lakerda (Lak), Lesvos Island: (39°04′46.1″ N, 26°31′57.5″ E). Olive tree variety: Kolovi (Valanolia). Surrounding orchards: olive yards. This organic orchard used a lawnmower for weed control. Sampling period: 2018–2022 (with Malaise trap).

Kalloni (Kal), Lesvos Island: (39°14′08.2″ N, 26°13′01.8″ E). Olive tree variety: Kolovi (Valanolia). Surrounding orchards: olive yards and nearby river and the saltpans of Kalloni gulf. Conventional orchard, with herbicides and mechanical tiller attached to a tractor used for weed control. Sampling period: 2018–2020 (with Malaise trap).

Nifida (Nif), Lesvos Island: (39°08′66.5″ N, 26°13′33.2″ E). Olive tree variety: Kolovi (Valanolia). Surrounding orchards: olive yards, seaside region. Conventional orchard, with herbicides and mechanical tiller attached to a tractor used for weed control. Sampling period: 2018–2020 (with Malaise trap).

Pamfila (Pam), Lesvos Island: (39°09′10.3″ N, 26°31′40.8″ E). Olive tree variety: Kolovi (Valanolia). Surrounding orchards: olive yards. Conventional orchard, with herbicides and mechanical tiller attached to a tractor used for weed control. Sampling period: 2017–2018 (with Malaise trap).

#### 2.1.3. Peloponnese

Koutsopodi–Argolida (Kts): (37°67′36.9″ N, 22°68′97.2″ E). Olive tree variety: Koroneiki. Surrounding orchards: olive yards and some citrus trees. Organic orchard, with mechanical tiller attached to a tractor used for weed control. Sampling period: 2018–2019 (with Malaise trap).

Pelekanada–Messinia (Pel): (37°05′09.28″ N, 21°83′68.94″ E). Olive tree varieties: Mavrolia Messinias and Koroneiki. Surrounding orchards: olive yards, vineyards, and wild shrubs and trees, such as *Erica manipuliflora* Salisb., *Pistacia* spp., *Vitex agnus-castus* L., *Sarcopoterium spinosum* (L.) Spach., *Quercus* spp., *Pyrus* spp. Conventional orchard, with herbicides used for weed control. Sampling period: 2016–2017 (with Malaise trap).

Kiparissia–Messinia (Kip): (37°15′35.94″ N, 21°40′42.85″ E). Olive trees variety: Koroneiki. Surrounding orchards: olive yards and some citrus trees. Conventional orchard, use of mechanical tiller attached to a tractor for weed control. Sampling period: 2019–2020 (with Malaise trap).

Fragka–Achaia (Fra): (38°04′20.5″ N, 21°29′08.1″ E). Olive tree variety: Koroneiki. Surrounding orchards: olive yards. A mechanical tiller attached to a tractor was used for weed control and no pesticides were applied. Sampling period 2019–2020 (with Malaise trap).

Pirgos–Ileia (Pir): (37°42′39.4″ N, 21°27′33.5″ E). Olive tree variety: Kalamon and Koroneiki. Surrounding orchards: olive yards, some citrus trees, and almonds. Conventional orchard. Mechanical tiller attached to a tractor was used for weed control. Sampling period 2021–2022 (with sweep net).

In each orchard, a white-colored custom-made Malaise trap was installed at the center to study the presence, seasonal appearance, and abundance of Auchenorrhyncha species. Each trap measured 170 cm in height at the top end and 110 cm at the lower end, with a length of 160 cm and a width of 180 cm, providing a total interception area of 165 × 110 cm$^2$. A 600-mL plastic container was attached in each trap with 98% ethanol as a preservation fluid. Samples were collected monthly and were then sent to the Laboratory of Zoology and Entomology of the Agricultural University of Athens for species identification. Samplings from each orchard lasted for at least one year, with some orchards (Athens and Lakerda) sampled for 3 consecutive years.

Additionally, sweep net samplings were conducted in two orchards, Pirgos' and Athens', for one year each to target key vector species like aphrophorids, which are less likely to be captured by Malaise traps. These samplings were conducted from dominant weeds (recorded number of Auchenorrhyncha species per weed species), woody hosts and from the canopy of olives with a custom-made entomological sweeping net (39 cm diameter). Every fortnight, 10 consecutive sweeps were undertaken from the dominant plants in the field in 5 different areas in the orchard with that plant species.

### 2.2. Classification

The taxonomic classification of the collected Auchenorrhyncha species followed the taxonomic keys of Ribaut [26,27], Ossiannilsson [28–30], Guglielmino and Bückle [31], and Dmitriev [32]. A very useful supporting tool was the Auchenorrhyncha collection of the late Prof. Dr. Sakis Drosopoulos, housed at the laboratory of Agricultural Zoology and Entomology. Male genitalia were dissected, kept in KOH (10%) (Merck KGaA, Darmstadt, Germany) for 2 h (30 min. for Typhlocybinae), mounted on microscope slides in glycerol, and observed under stereoscopic and microscopic microscope. Females were identified in genus level, and, subsequently, if all the specimens of a genus in a specific sampling belonged to one species, the female specimens were added to that species. The nomenclature that followed was according to Dmitriev [32], the website TaxonPages.

### 2.3. Data Analysis

Insect dominance for each orchard was determined using the classification system proposed by Curry [33], Cusack et al. [34], and Emmanuel [35]. Species were categorized as 'dominant' (>10% of the total individuals), 'influent' (5–10%), or 'recedent' (<5%).

Diversity was estimated by Simpson's Diversity Index (1-D) [36], where 0 represents low diversity and 1 represents infinite diversity. The value of D was calculated using the following formula:

$$D = \sum_{i=1}^{s} \frac{n_i(n_i - 1)}{N(N - 1)}$$

$n_i$ = the number of individuals of a particular auchenorrhynchan species; $N$ = the total number of individuals of all auchenorrhynchan species.

Simpson's Diversity Index represents the probability that two individuals randomly selected from a sample belong to different species. This index is particularly useful for assessing the evenness (E) and richness (S) of species within each orchard ecosystem. Richness (S) is the total number of species in every orchard, and evenness or equitability (E) can be calculated by taking Simpson's Reciprocal Index (1/D) and expressing it as a proportion of S. Equitability takes a value between 0 to 1, with 1 being complete evenness (i.e., where there are exactly equal numbers of individuals per species). The formula used was

$$E = \frac{1}{D} \times \frac{1}{S}$$

## 3. Results

### *3.1. Species Composition*

In total, 14,771 Auchenorrhyncha belonging to 109 species were collected using a Malaise trap (Table 1). Of these, 385 individuals belonged to the suborder Fulgoromorpha, while the remaining 14,386 belonged to Cicadomorpha. The collected species were classified into seven families: Cicadellidae (95 species), Aphrophoridae (4 species), Issidae (4 species), Delphacidae (2 species), Flatidae (2 species), Dictyopharidae (1 species), and Cercopidae (1 species). Across all regions, Cicadellidae was the most abundant and diverse family, represented mainly by Typhlocybinae (35 species), which also accounted for 52.05% of the total population collected with Malaise traps. The second most abundant but the most diverse subfamily was Deltocephalinae (50 species), consisting of 40.42% of the total population. Fewer species were found in the subfamilies Aphrodinae (two species), Eurymelinae (two species), Iassinae (one species), and Megopthalminae (five species), which contributed smaller percentages to the total population.

**Table 1.** Percentage of every species collected in each orchard with Malaise trap (+ indicates the species collected in less than 5%. – indicates absence from that orchard) (Ath = Athens, Lak = Lakerda, Lesvos, Kal = Kalloni, Lesvos, Nif = Nifida, Lesvos Pam = Pamfila, Lesvos, Ist1 = Istiea conventional, Euboea, Ist2 = organic, Euboea, Ist3 = abandoned, Euboea, Kts = Koutsopodi, Peloponnese, Pel = Pelekanada, Peloponnese, Kip = Kiparissia, Peloponnese, Fra = Fragka, Peloponnese).

| Species | Ath (%) | Lak (%) | Kal (%) | Nif (%) | Pam (%) | Ist1 (%) | Ist2 (%) | Ist3 (%) | Kts (%) | Pel (%) | Kip (%) | Fra (%) |
|---|---|---|---|---|---|---|---|---|---|---|---|---|
| Total Number | 2566 | 314 | 1442 | 311 | 320 | 2761 | 1090 | 2652 | 1231 | 366 | 1533 | 185 |
| Aphrophoridae, Aphrophorinae | | | | | | | | | | | | |
| *Lepyronia coleoptrata* (L.) | – | – | – | – | – | + | – | + | – | – | – | – |
| *Philaenus spumarius* (L.) | + | + | + | + | + | + | + | + | + | + | + | + |
| *Neophilaenus (Neophilaenulus) campestris* (Fallén) | – | 5.79 | + | + | + | + | + | + | – | – | – | + |
| *Neophilaenus (Neophilaenus) lineatus* (L.) | – | – | – | – | – | + | – | – | – | – | – | – |
| Cercopidae, Cercopinae | | | | | | | | | | | | |
| *Cercopis sanguinolenta* (Scopoli) | – | – | – | – | – | + | + | + | + | – | – | – |
| Cicadellidae, Aphrodinae | | | | | | | | | | | | |
| *Anoscopus albifrons* (L.) | – | + | – | – | – | – | – | + | + | – | – | – |
| *Aphrodes bicincta* (Schrank) | – | – | – | – | – | – | + | + | – | + | – | – |
| Cicadellidae, Deltocephalinae | | | | | | | | | | | | |
| *Allygus modestus* (Scott) | + | – | – | – | – | + | + | + | + | 8.71 | – | – |
| *Anaconura acuticeps* (Ribaut) | – | – | – | – | – | – | + | – | – | – | – | – |
| *Anoplotettix putoni* (Ribaut) | + | – | – | – | – | – | – | – | + | 8.43 | + | – |
| *Anoplotettix fuscovenosus* (Ferrari) | – | + | – | – | – | + | + | + | – | – | – | – |
| *Arocephalus (Arocephalus) longiceps* (Kirschbaum) | + | – | – | – | – | – | – | – | – | – | – | – |
| *Balclutha frontalis* (Ferrari) | + | + | + | + | – | 8.84 | 10.46 | + | – | – | 5.94 | + |
| *Balclutha punctata* (Fabricius) | 10.02 | – | – | – | – | – | – | – | – | – | 5.68 | + |
| *Balclutha saltuella* (Kirschbaum) | + | – | + | – | – | + | + | + | – | – | – | + |
| *Cicadula (Cicadula) lineatopunctata* (Matsumura) | – | – | – | – | – | + | + | + | – | – | – | – |
| *Cicadulina bipunctata* (Melichar) | + | + | 8.20 | – | – | 17.17 | + | + | – | – | 6.52 | – |
| *Docotettix cornutus* (Ribaut) | – | + | + | + | – | – | – | – | – | – | – | – |
| *Eohardya fraudulenta* (Horváth) | + | – | + | – | – | – | – | + | – | – | – | – |
| *Epistagma (Epistagma) guttulinervis* (Kirschbaum) | – | – | – | – | – | – | – | – | – | – | + | + |
| *Eupelix cuspidata* (Fabricius) | – | – | – | – | – | – | – | + | – | – | – | – |
| *Euscelidius mundus* (Haupt) | – | + | + | – | – | – | – | – | – | – | – | – |
| *Euscelidius variegatus* (Kirschbaum) | + | – | – | – | – | – | – | – | – | – | – | + |
| *Euscelis alsia* (Ribaut) | + | + | + | 5.47 | – | + | + | – | – | – | – | + |
| *Euscelis lineolata* (Brullé) | + | 7.40 | 9.92 | + | + | + | 14.50 | + | – | 12.64 | + | + |
| *Exitianus capicola* (Stål) | + | + | + | + | – | + | + | + | + | + | + | + |
| *Fieberiella florii* (Stål) | – | – | – | – | – | – | – | + | – | – | – | – |
| *Fieberiella septentrionalis* (Wagner) | – | – | – | – | – | – | – | – | + | + | – | – |
| *Goniagnathus (Goniozygotes) bolivari* (Melichar) | – | – | – | + | – | – | – | – | – | – | – | + |
| *Goniagnathus (Goniagnathus) brevis* (Herrich-Schäffer) | – | + | + | + | – | – | – | – | – | – | – | – |
| *Grypotellus staurus* (Ivanoff) | + | – | – | – | – | – | + | + | – | – | – | – |

**Table 1.** *Cont.*

| Species | Ath (%) | Lak (%) | Kal (%) | Nif (%) | Pam (%) | Ist1 (%) | Ist2 (%) | Ist3 (%) | Kts (%) | Pel (%) | Kip (%) | Fra (%) |
|---|---|---|---|---|---|---|---|---|---|---|---|---|
| *Hecalus glaucescens* (Fieber) | – | – | – | – | – | + | + | – | – | – | – | – |
| *Jassargus (Obtujargus) obtusivalvis* (Kirschbaum) | – | – | – | – | – | – | + | – | – | – | – | – |
| *Macrosteles quadripunctulatus* (Kirschbaum) | – | – | + | – | + |  | – | – | – | – | + | – |
| *Macrosteles ramosus* (Ribaut) | – | – | + | – | – | – | – | – | – | – | – | – |
| *Macrosteles sexnotatus* (Fallén) | – | – | + | – | – | – | – | – | – | – | – | – |
| *Maiestas schmidtgeni* (Wagner) | + | + | + | + | + | + | + | + | + | + | + | – |
| *Melillaia desbrochersi* (Lethierry) | – | + | – | – | – | – | – | – | – | – | – | – |
| *Mocydia crocea* (Herrich-Schäffer) | – | – | – | – | – | – | – | – | – | + | – | – |
| *Neoaliturus (Circulifer) haematoceps* (Mulsant & Rey) | + | + | + | – | – | + | – | – | + | – | – | – |
| *Neoaliturus (Neoaliturus) fenestratus* (Herrich-Schäffer) | – | – | + | – | – | + | + | + | + | – | – | – |
| *Nesoclutha erythrocephala* (Ferrari) | – | – | – | – | – | + | – | – | – | – | – | – |
| *Opsius stactogalus* (Fieber) | – | – | – | + | – | – | – | – | – | – | – | – |
| *Orosius orientalis* (Matsumura) | + | – | + | – | – | – | – | – | – | – | – | – |
| *Paralimnus (Paralimnus) zachvatkini* (Emeljanov) | – | – | – | – | – | + | + | – | – | – | – | – |
| *Paramesodes lucaniae* (Dlabola) | – | – | – | – | – | + | – | – | – | – | – | – |
| *Phlepsius intricatus* (Herrich-Schäffer) | + | – | + | + | – | + | + | + | + | + | – | + |
| *Phlogotettix cyclops* (Herrich-Schäffer) | – | – | – | – | – | – | – | + | – | – | – | – |
| *Proceps acicularis* (Mulsant & Rey) | – | + | – | – | – | – | – | – | – | – | – | – |
| *Psammotettix alienus* (Dahlbom) | 5.56 | + | 12.54 | + | – | + | + | + | – | + | + | + |
| *Psammotettix confinis* (Dahlbom) | – | – | + | – | – | – | – | – | – | – | – | – |
| *Psammotettix notatus* (Melichar) | + | – | – | – | + | – | – | – | + | + | – | – |
| *Selenocephalus pallidus* (Kirschbaum) | – | + | + | 6.11 | – | – | + | – | – | – | – | – |
| *Streptanus (Streptanulus) albanicus* (Horváth) | – | + | – | – | – | – | – | – | – | + | – | – |
| *Synophropsis lauri* (Horváth) | + | – | + | – | – | + | + | + | – | + | – | + |
| *Thamnotettix zelleri* (Kirschbaum) | + | 25.08 | + | 49.19 | + | + | 15.41 | – | + | + | + | – |
| *Varta rubrostriata* (Horváth) | – | – | – | – | – | + | + | + | – | – | – | – |
| Cicadellidae, Eurymelinae | | | | | | | | | | | | |
| *Acericerus vittifrons* (Kirschbaum) | + | – | – | – | – | – | – | – | – | – | – | – |
| *Sulamicerus stali* (Fieber) | + | – | – | – | – | – | – | – | – | – | – | – |
| Cicadellidae, Iassinae | | | | | | | | | | | | |
| *Batracomorphus (Batracomorphus) irroratus* (Lewis) | – | – | – | – | – | + | – | + | – | – | – | – |
| Cicadellidae, Megopthalminae | | | | | | | | | | | | |
| *Agallia consobrina* (Curtis) | + | – | – | – | – | + | + | + | + | + | – | – |
| *Anaceratagallia (Anaceratagallia) glabra* (Dmitriev) | – | + | + | + | – | + | + | + | + | + | + | + |
| *Anaceratagallia (Anaceratagallia) ribauti* (Ossiannilsson) | – | – | + | + | + | – | + | + | – | – | + | – |
| *Austroagallia sinuata* (Mulsant & Rey) | – | + | + | + | – | + | – | – | – | – | – | – |
| *Megopthalmus scabripennis* (Edwards) | + | + | + | + | – | + | + | + | + | – | – | – |
| Cicadellidae, Typhlocybinae | | | | | | | | | | | | |
| *Anzygina honiloa* (Kirkaldy) | + | + | – | – | – | – | – | – | – | – | – | – |
| *Arboridia parvula* (Boheman) | – | – | – | – | – | – | – | – | – | + | – | – |
| *Arboridia (Arboridia) versuta* (Melichar) | – | – | – | – | – | – | – | – | – | + | – | – |
| *Assymetrasca decedens* (Paoli) | + |  | + | + |  | + | + | + | – | – | + | – |
| *Edwardsiana platanicola* (Vidano) | – | – | – | – | – | – | – | + | – | – | – | – |
| *Eupteryx (Eupteryx) collina* (Flor) | – | – | – | – | – | – | – | – | – | 7.02 | – | – |
| *Eupteryx (Eupteryx) curtisii* (Flor) | – | – | – | – | – | – | – | + | – | – | – | – |
| *Eupteryx (Eupteryx) decemnotata* (Rey) | + | – | – | – | – | – | – | – | – | – | – | – |
| *Eupteryx (Eupteryx) filicum* (Newman) | + | – | – | – | – | – | – | – | – | – | – | – |
| *Eupteryx (Eupteryx) gyaurdagica* (Dlabola) | – | 14.47 | + | + | – | – | – | – | – | – | – | – |
| *Eupteryx (Eupteryx) insulana* (Ribaut) | – | – | + | – | – | + | + | + | – | – | – | – |
| *Eupteryx (Eupteryx) melissae* (Curtis) | + | – | – | – | + | + | + | + | + | – | – | – |
| *Eupteryx (Eupteryx) rostrata* (Ribaut) | – | – | – | – | – | – | – | – | – | + | – | – |
| *Eupteryx (Eupteryx) urticae* (Fabricius) | + | – | – | – | – | – | – | – | – | – | – | – |
| *Eupteryx (Eupteryx) zelleri* (Kirschbaum) | – | – | – | – | – | + | + | 39.52 | – | – | 7.44 | – |
| *Ficocyba ficaria* (Horváth) | + | – | – | – | – | + | – | + | – | – | – | – |
| *Frutioidia (Frutioidia) bisignata* (Mulsant & Rey) | – | – | – | – | – | – | – | + | – | – | – | – |
| *Hauptidia (Hauptidia) provincialis* (Ribaut) | + | + | + | – | – | + | + | + | + | 9.83 | 20.74 | – |
| *Hebata (Alboneurasca) decipiens* (Paoli) | + | – | 14.96 | + | 32.19 | 12.57 | 5.69 | + | 6.91 | + | 27.46 | 6.49 |
| *Hebata (Signatasca) vitis* (Göthe) | + | – | – | – | – | – | – | – | – | – | + | + |
| *Liguropia juniperi* (Lethierry) | + | – | – | – | – | – | – | – | – | – | – | – |
| *Lindbergina cretica* (Asche) | – | + | – | – | – | – | – | – | – | – | – | – |
| *Ribautiana cruciata* (Ribaut) | – | – | + | + | – | – | + | – | – | – | – | – |
| *Ribautiana tenerrima* (Herrich-Schäffer) | – | – | – | – | – | + | + | 18.89 | – | – | – | – |
| *Zygina (Hypericiella) hyperici* (Herrich-Schäffer) | – | – | – | – | – | – | – | – | + | – | – | – |

**Table 1.** *Cont.*

| Species | Ath (%) | Lak (%) | Kal (%) | Nif (%) | Pam (%) | Ist1 (%) | Ist2 (%) | Ist3 (%) | Kts (%) | Pel (%) | Kip (%) | Fra (%) |
|---|---|---|---|---|---|---|---|---|---|---|---|---|
| *Zygina (Zygina) angusta* (Lethierry) | + | – | – | – | – | – | – | – | – | – | – | – |
| *Zygina (Zygina) nivea* (Mulsant & Rey) | + | – | – | – | – | – | – | – | – | – | – | – |
| *Zygina (Zygina) rhamni* (Ferrari) | – | – | – | – | – | + | + | + | – | – | – | – |
| *Zygina (Zygina) roseipennis* (Tollin) | – | – | – | + | – | – | – | – | – | – | – | – |
| *Zygina (Zygina) suavis* (Rey) | – | – | – | – | – | – | – | – | + | + | – | – |
| *Zygina (Zygina) tiliae* (Fallén) | – | – | – | – | – | – | – | – | – | + | – | – |
| *Zyginella pulchra* (Löw) | + | – | – | – | – | + | – | – | – | – | + | – |
| *Zyginidia adamczewskii* (Dworakowska) | + | – | – | + | – | – | – | – | – | – | – | – |
| *Zyginidia pullula* (Boheman) | 16.62 | + | 10.06 | + | 40.00 | 36.22 | 18.53 | + | 72.38 | 10.11 | + | – |
| *Wagneriala sinuata* (Then) | – | – | – | – | – | – | – | + | – | – | – | – |
| Delphacidae, Stenocraninae | | | | | | | | | | | | |
| *Stenocranus fuscovittatus* (Stål) | – | – | – | – | – | + | – | + | – | – | – | – |
| Delphacidae, Delphacinae | | | | | | | | | | | | |
| *Laodelphax striatellus* (Fallén) | – | – | + | – | – | + | – | + | – | – | – | – |
| *Toya (Metadelphax) propinqua* (Fieber) | – | + | + | – | – | + | + | + | + | + | – | – |
| Dictyopharidae, Dityopharinae | | | | | | | | | | | | |
| *Dictyophara (Dictyophara) europaea* (L.) | – | – | – | – | + | – | + | + | – | – | – | – |
| Flatidae, Flatinae | | | | | | | | | | | | |
| *Metcalfa pruinosa* (Say) | – | – | – | – | – | – | + | + | – | – | – | – |
| *Phantia subquadrata* (Herrich-Schäffer) | – | + | + | + | – | – | + | + | – | – | – | – |
| Issidae, Hysteropterinae | | | | | | | | | | | | |
| *Agalmatium bilobum* (Fieber) | – | – | – | – | – | + | + | + | – | – | – | – |
| *Agalmatium flavescens* (Olivier) | – | + | – | – | – | – | – | – | – | + | – | – |
| *Latilica antalyica* (Dlabola) | + | + | – | – | – | – | – | – | – | – | – | – |
| *Latilica maculipes* (Melichar) | 5.73 | – | – | – | – | – | – | – | – | – | + | – |

The dominance ranking of species was calculated for each orchard individually. The results showed that two Typhlocybinae species were dominant in most orchards: *Zyginidia pullula* (with a population ranging from 10.06% to 72.38% of the total population) and *Hebata decipiens* (ranging from 12.57% to 32.19% of the total population). Both species were present even in small numbers in most of the orchards. Other common Typhlocybinae included *Assymetrasca decedens* and species of the genus *Eupteryx*, where they were recorded in every olive grove, with *Eupteryx zelleri* being dominant in the untreated Istiea's grove (Euboea Island) with a relative abundance of 39.52%, while *Eupteryx gyardagica* was dominant in Lakerda's grove (Lesvos Island) at 14.47%. Moreover, *Hauptidia provincialis* was present in most orchards and was influent in some, reaching 20.74% of the total population in Kiparissia's orchard (Messinia).

Moreover, several Deltocephalinae species were either dominant or influent in the orchards. *Balclutha* spp. and *Euscelis* spp. were recorded in every orchard. *Balclutha frontalis* was dominant in Istiea's organic orchard (10.46%), while *Balclutha punctata* was influent in the Athens's orchard (10.02%). *Euscelis lineolata* was dominant in both the Pelekanada orchard (Messinia) (12.64%) and the organic Istiea orchard (14.5%) and was also influent in Kalloni (Lesvos Island) (9.92%). *Cicadulina bipunctata* was dominant in Istiea's conventional orchard (17.17%) and influent in Kalloni (8.20%). Other widespread species, present in almost all orchards, included *Synophropsis lauri*, *Phlepsius intricate*, *Thamnotettix zelleri* (with relative abundances of 15.41% in organic Istiea orchard, 49.19% in Nifida, and 25.08% in Lakerda on Lesvos Island), *Maiestas schmidtgeni*, and *Exitianus capicola*, and several species of the genus *Psammotettix*, with *Psammotettix alienus* being the most common representative (12.54% in Kalloni's orchard on Lesvos Island).

Megopthalminae were represented by a small number of species in each orchard, with *Anaceratagallia glabra* (formerly known as *Anaceratagallia laevis*) being the most common.

Aphrophorids were collected in small numbers using Malaise traps, with *Philaenus spumarius* present in almost every orchard. *Neophilaenus campestris* was found in low numbers in a few orchards. Other species from the family Aphrophoridae, such as *Lepyronia coleoptrata*, *Neophilaenus lineatus*, and *Cercopis sanguinolenta*, were collected sporadically.

Fulgoromorpha were generally found more sporadically. The delphacid species *Toya (Metadelphax) propinqua* was the most common and was found in most orchards. The family Issidae was represented in relatively high numbers in some orchards, such as in Athens, with *Latilica maculipes* accounting for 5.73% of the total population.

### 3.2. Seasonal Fluctuations

The seasonal fluctuation for the Auchenorrhyncha population was calculated for the Athens's orchard and Lakerda's orchard on Lesvos Island, where data were available for three consecutive years. The population showed clear peaks in the warmer months in both locations.

In the Athens orchard, population growth began in late April each year, with the peak occurring during the summer months. The peaks of the population were n = 306 in August 2016, n = 714 in June 2017, n = 189, and in July 2018 (Figure 2). In the Lakerda orchard, the population increase started earlier, at the beginning of April, with the peak occurring in May, showing population sizes of n = 42 in May 2018, n = 67 in May 2020, and n = 26 in May 2021 (Figure 3). In both regions, the lowest numbers were recorded during the winter, when temperatures are lower and adult Auchenorrhyncha are less mobile.

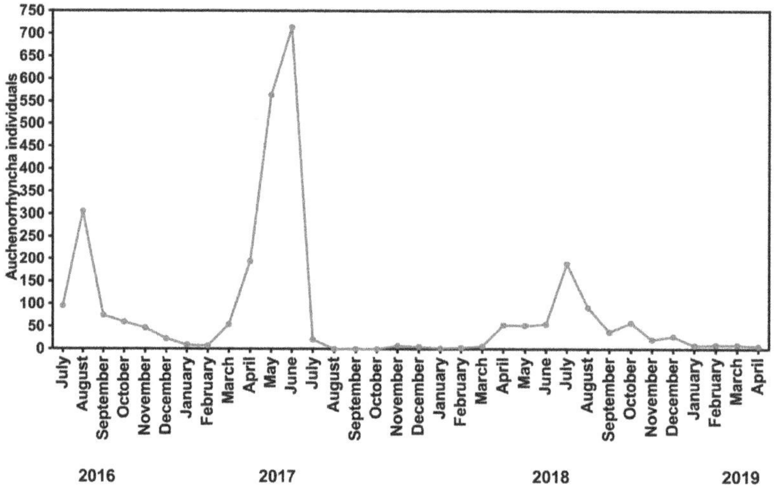


**Figure 2.** Seasonal fluctuation of Auchenorrhyncha adults captured during a 3-year period in Malaise trap placed in olive orchard in Athens.

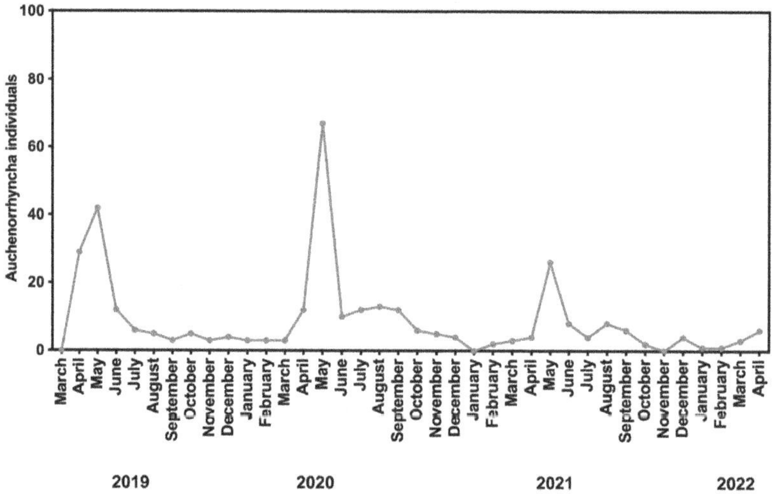


**Figure 3.** Seasonal fluctuation of Auchenorrhyncha adults captured during a 3-year period in Malaise trap placed in olive orchard in Lakerda, Lesvos.

The Simpson diversity index was calculated for each orchard (Table 2), with the highest values observed in the organic orchards of Athens, Kalloni, and Lakerda, ranging from 0.901 to 0.932. Interestingly, high diversity was also found in the conventional orchard at Pelekanada, where we had the greatest evenness in comparison with other populations collected with a Malaise trap. In contrast, the lowest diversity and evenness were recorded in the organic orchard of Koutsopodi.

**Table 2.** Species richness, Simpson's Diversity and Evenness Indices for each locality.

| Management System | Locality | Richness (S) | Simpson's Index of Diversity (1-D) | Evenness (E) |
|---|---|---|---|---|
| Organic | Athens | 62 | 0.932 | 0.240 |
| Organic | Lakerda | 52 | 0.901 | 0.194 |
| Conventional | Kalloni | 64 | 0.924 | 0.203 |
| Conventional | Nifida | 43 | 0.746 | 0.092 |
| Conventional | Pamfila | 17 | 0.735 | 0.222 |
| Conventional | Istiea | 48 | 0.824 | 0.118 |
| Organic | Istiea | 45 | 0.899 | 0.220 |
| No-treatments | Istiea | 57 | 0.815 | 0.095 |
| Organic | Koutsopodi | 55 | 0.468 | 0.034 |
| Conventional | Pelekanada | 47 | 0.928 | 0.296 |
| Conventional | Kiparissia | 35 | 0.858 | 0.201 |
| Organic | Fragka | 44 | 0.881 | 0.191 |
| Conventional | Pirgos (sweep net) | 32 | 0.897 | 0.303 |
| Organic | Athens (sweep net) | 18 | 0.637 | 0.153 |

### *3.3. Sweep Net Results*

Sweep net sampling was performed in Pirgos and Athens orchards. A total of 1533 Auchenorrhyncha were collected, belonging to 43 species. Six families were recorded: Aphrophoridae (2 species), Cicadellidae (32 species), Delphacidae (4 species), Flatidae (1 species), Dictyopharidae (1 species), and Issidae (3 species). Most of these species were collected from plants of the family Poaceae, as shown in Table 3. Seven species were collected from the canopy of olives: *P. spumarius* during spring and autumn, *Al. modestus*, *A. putoni*, *B. punctata*, *S. lauri*, *T. zelleri*, and the issid *Latilica antalyica*.

**Table 3.** Percentage of Auchenorrhyncha species collected with sweep net from Pirgos (Peloponnese) and Athens (Attica) (Pir = Pirgos, Ath = Athens). + indicates percentage less than 5%, – indicates absence in that field.

| Species | Pir (%) | Pirgos' Plant Species | Ath (%) | Athens' Plant Species |
|---|---|---|---|---|
| *Philaenus spumarius* (L.) | 14.10 | AS | 15.85 | PL, OE, AS, HM, CD, ASV, PA, SE, MS |
| *Neophilaenus (Neophilaenulus) campestris* (Fallén) | 16.98 | AS | + | PA, SE, AS |
| *Aphrodes bicincta* (Schrank) | – | – | + | HM, SO |
| *Allygidius (Dicrallygus) mayri* (Kirschbaum) | + | PO | – | – |
| *Allygus modestus* (Scott) | – | – | + | OE, PL, MS |
| *Anaconura acuticeps* (Ribaut) | – | – | + | CD |
| *Anoplotettix putoni* (Ribaut) | + | PO | + | OE, HM, ASV |
| *Balclutha frontalis* (Ferrari) | 14.24 | AC | – | – |
| *Balclutha punctata* (Fabricius) | + | AC | + | OE, ASV |
| *Balclutha saltuella* (Kirschbaum) | + | AC | – | – |
| *Cicadulina bipunctata* (Melichar) | + | PO | – | |
| *Doratura stylata* (Boheman) | + | AC, CD | – | – |

**Table 3.** *Cont.*

| Species | Pir (%) | Pirgos' Plant Species | Ath (%) | Athens' Plant Species |
|---|---|---|---|---|
| *Epistagma (Epistagma) guttulinervis* (Kirschbaum) | + | CD | – | – |
| *Euscelidius variegatus* (Kirschbaum) | – | – | + | AS, HM, PHA |
| *Euscelis lineolata* (Brullé) | + | AS, CD | – | – |
| *Euscelis ohausi* (Wagner) | + | PO | – | – |
| *Exitianus capicola* (Stål) | 6.19 | CD | + | CD, AS |
| *Exitianus nanus* (Distant) | + | PO | – | – |
| *Hecalus storai* (Lindberg) | + | CD | – | – |
| *Maiestas schmidtgeni* (Wagner) | + | CD | – | – |
| *Mocydiopsis longicauda* (Remane) | + | PO | – | – |
| *Mocydiopsis monticola* (Remane) | + | PO | – | – |
| *Nesoclutha erythrocephala* (Ferrari) | + | AS, HM, CD, ASV, PA | – | – |
| *Phlepsius intricatus* (Herrich-Schäffer) | + | PO | – | – |
| *Psammotettix alienus* (Dahlbom) | 14.96 | AS, HM, CD, ASV, PA | + | CD |
| *Streptanus (Streptanulus) albanicus* (Horváth) | – | – | + | AS, HM |
| *Synophropsis lauri* (Horváth) | – | – | 8.17 | OE, ASV |
| *Thamnotettix zelleri* (Kirschbaum) | + | PO | + | OE, AS, MS, SA, HM, ASV |
| *Anaceratagallia (Anaceratagallia) glabra* (Dmitriev) | + | AS | – | – |
| *Anaceratagallia (Anaceratagallia) ribauti* (Ossiannilsson) | – | – | + | SO |
| *Sulamicerus stali* (Fieber) | – | – | 57.20 | PL |
| *Hauptidia (Hauptidia) provincialis* (Ribaut) | + | MA, TR | – | – |
| *Hebata (Alboneurasca) decipiens* (Paoli) | + | MA, TR | – | – |
| *Zyginidia pullula* (Boheman) | + | MA, TR | – | – |
| *Asiraca clavicornis* (Fabricius) | + | PO | – | – |
| *Euidopsis truncata* (Ribaut) | + | PO | – | – |
| *Eurysella brunnea* (Melichar) | + | PO | – | – |
| *Toya (Metadelphax) propinqua* (Fieber) | + | AC, CD | + | CD |
| *Dictyophara (Dictyophara) europaea* (Linnaeus) | – | – | + | AV |
| *Phantia subquadrata* (Herrich-Schäffer) | + | PO | – | – |
| *Clybeccus declivum* (Dlabola) | + | AS | – | – |
| *Latilica antalyica* (Dlabola) | – | – | + | OE, ASV |
| *Latilica maculipes* (Melichar) | + | AS | – | – |

N$^{\alpha}$: number of individuals captured. Plant species and abbreviations: *Amaranthus viridis* L. = AV; *Apera spica-venti* (L.) P. Beauv. = ASV; *Agrostis capillaris* L. = AC; *Avena sterilis* L. = AS; *Cynodon dactylon* (L.) Pers. = CD; *Hordeum murinum* L. = HM; *Malva silvestris* L. = MS; *Malva* sp. = MA; *Olea europaea* L. = OE; *Phalaris arundinacea* L. = PHA; *Pistacia lentiscus* L. = PL; *Poa annua* L. = PA; Poaceae mix = PO; *Sinapis alba* L. = SA; *Solanum elaeagnifolium* Cav. = SE; *Sonchus oleraceus* L. = SO; *Trifolium* sp. = TR.

The dominance ranking from the sweep net sampling was quite different from that of the Malaise trap. In the Pirgos orchard, the most collected species with a sweep net was *Neophilaenus campestris* (16.98%), followed by two deltocephalins, *Psammotettix alienus* (14.96%), and *Balclutha frontalis* (14.25%). *Philaenus spumarius* (14.10%) was the fourth most collected species. In the Athens orchard, *P. spumarius* was the most collected species with a sweep net (15.85%), while *Su. stali* was collected in significant numbers (57.2%) but only above lentisk shrubs present in the field.

Two population peaks were recorded in the Athens orchard during autumn (early October to early December) and during spring (mid-May to August) (Figure 4a). Similar results were recorded in the Pirgos orchard, with two peaks a little earlier (early September to late November and late March to early July) (Figure 4b). The spittlebug population showed two peaks throughout the year. The first peak occurred in spring, with adults collected from mid-April to early June, and a population peak occurred at the end of April in both orchards. The second peak occurred in autumn, with adults collected from late October to late December, peaking in early November in the Athens orchard, and late September to early December in the Pirgos orchard (Figure 4a,b).

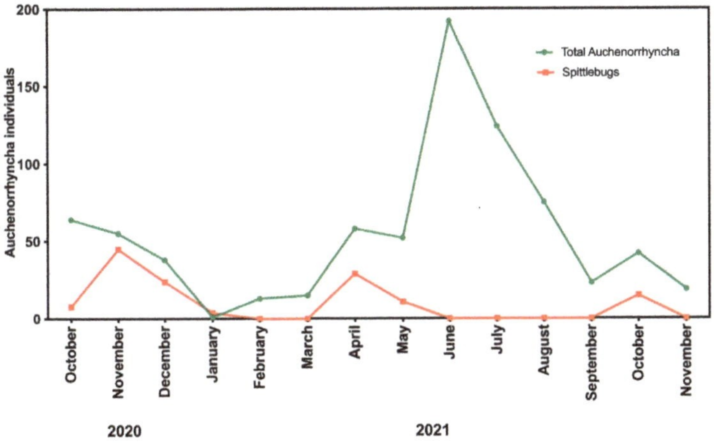


(a)

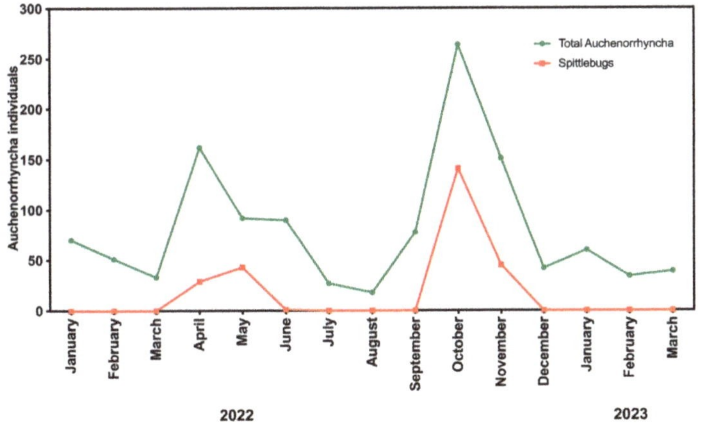


(b)

**Figure 4.** Seasonal fluctuation of total Auchenorrhyncha (green line) and spittlebugs (red line) adults, collected with sweep net in olive orchards in (**a**) Athens from October 2020 to November 2021 and (**b**) in Pirgos from January 2022 to March 2023.

Most adult spittlebugs were collected above Poaceae weeds. In Pirgos, all the spittlebug specimens of both *P. spumarius* and *N. campestris* were found exclusively on *Avena sterilis* (Figure 4b), while in Athens they were collected from various hosts (Figure 5a,b): *P. spumarius* primarily was collected above *A. sterilis* from mid-October to late December of 2020 and early October to late December of 2021 peaking in both years in early November. Another important host in spring was *Hordeum murinum* where most adults were collected in April of 2021 and significant numbers were collected on *Apera apica-venti* in November of 2020 and April of 2021 (Figure 5a). *Philaenus spumarius* was also collected from the olive canopy in both spring and autumn, albeit in small numbers. Additionally, 1–2 adults were collected from *Solanum elaeagnifolium*, *Malva silvestris*, and *Pistacia lentiscus*, but these were not included in the diagram due to their low numbers. *N. campestris* was collected in November and December of 2021 from *Poa annua* and from *A. sterilis* during winter in both years, and one adult was collected also from *So. elaeagnifolium* (Figure 5b).

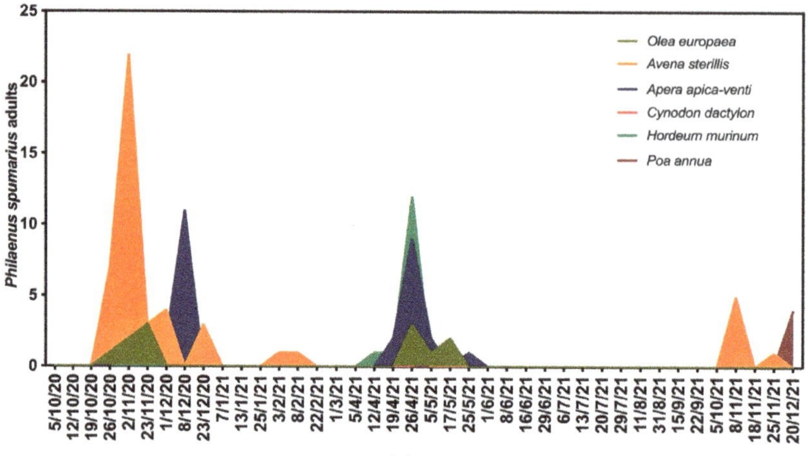


(a)

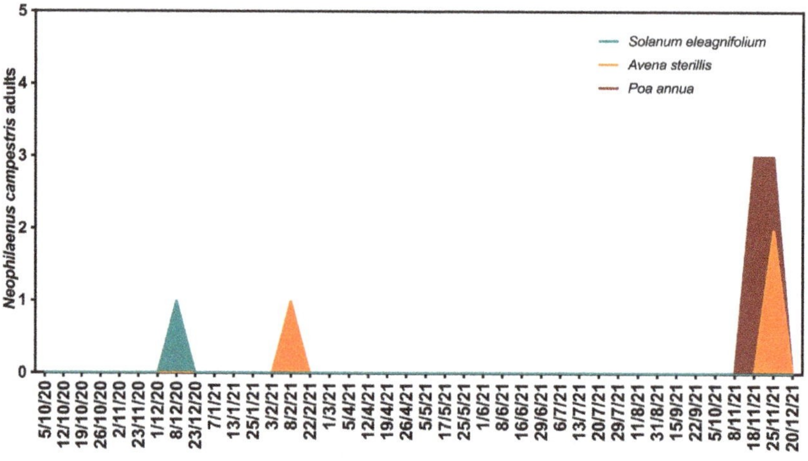


(b)

**Figure 5.** Adults of (**a**) *Philaenus spumarius* and (**b**) *Neophilaenus campestris*, collected with sweep net from different plant species in the Athens orchard.

## 4. Discussion

The species composition of Auchenorrhyncha in olive orchards in Greece is heavily influenced by geographical locality, neighboring plant species, and the weeds present within the orchards. The most common and abundant species observed during this study were primarily oligophagous or polyphagous, inhabiting dry meadows and shrubs. The most dominant species, *Zyginidia pullula*, feeds primarily on grasses of the family Poaceae [37–39], which are typical ground vegetation in Greek orchards. Other prevalent Typhlocybinae species such as *Hebata decipiens*, *Eupteryx* spp., and *Hauptidia provincialis* are also herbivorous, feeding on ground vegetation and herbaceous plants [31,40].

Similarly, the dominant Deltocephalinae species observed were grass-feeding Auchenorrhyncha. *Balclutha frontalis* and *Balclutha punctata*, both feeding on grasses, especially those of the family Poaceae [38,41], were very common, with *B. frontalis* found in 10 out of 13 orchards. These species were collected in large numbers from *Agrostis capillaris* using a sweep net.

*Euscelis lineolata*, a species preferring both Poaceae and Fabaceae, was another common Auchenorrhyncha species in olive and citrus orchards in previous studies undertaken in Greece [24,42,43], and also in orchards in Italy and France [44,45]. In this study, it was found in all orchards in large numbers, characterized as dominant or frequent in most of them. With a sweep net, it was collected above *Avena sterillis* and *Cynodon dactylon*, both grasses of the family Poaceae.

Other widespread frequently observed Deltocephalinae species included *Psammotettix alienus, Exitianus capicola, Allygus modestus, Cicadulina bipunctata, Maiestas schmidtgeni, Phlespsius intricatus* and *Thamnotettix zelleri. Maiestas schmidtgeni, Ph. intricatus, Psammotettix* spp., *Al. modestus,* and *T. zelleri* are polyphagous and commonly found contributing to the biodiversity in vineyard and olive agroecosystems [46–48], while *T. zelleri* also feeds on woody plants [22,38]. *Synophropsis lauri,* which was collected in 8 out of 13 fields and in high numbers above the canopy of olives, as in previous studies in Greece [42], feeds on woody plants, including olives [21,22,49]. Some species, such as *S. lauri, T. zelleri, Ph. intricatus, and Anoplotettix putoni,* were likely underrepresented in the Malaise trap collections due to their canopy-dwelling habits, whereas the Typhlocybinae species were collected in high numbers like *Z. pullula, Eupteryx* spp., *Hebata* spp., and *Hauptidia* sp. since they are more mobile and inhabit ground vegetation.

Most of the species found in the olive orchards are common and widespread throughout Greece, with their presence largely dependent on the composition of the undergrowth vegetation. The specific location and the neighboring plant species, whether they are other cultivated fields or rural gardens, play a crucial role in shaping the composition of Auchenorrhyncha species in the orchard. As a result, many species may occur in olive orchards incidentally. For instance, *Zyginella pulchra* and *Sulamicerus stali,* are species that feed on *Pistacia* spp. *Zyginella pulchra* is polyphagous, feeding on trees of the family Sapindaceae, while *Su. stali* is monophagous, exclusively feeding on *Pistacia* spp. [37,40,50]. Both species were found in the Athens orchard, with *Su. stali* being particularly abundant during spring, as it was also collected using a sweep net. Their presence can be attributed to the neighboring pistachio field and the scattered lentisk shrubs, which were heavily infested with *Su. stali.* A similar situation was observed with two species found in the Athens orchard: *Liguropia juniperi* and *Acericerus vittifrons,* which feed on cypress and acer trees, respectively [40,49,50]. These species were likely present due to the proximity of the botanical garden adjacent to the olive orchard. In Nifida's orchard, which is located in a seaside region, the species *Opsius stactogalus* was recorded. This species feeds monophagously on *Tamarix* spp. [40], and was likely present due to the nearby *Tamarix* trees along the coastline near the road. It is possible that a larger population of *O. stactogalus* was feeding on these trees and subsequently moved into the neighboring olive orchard. *Macrosteles* spp. generally prefer moist, wet habitats, with their nymphs feeding on sedges (*Carex*), woodrushes (*Luzula*), rushes (*Juncus*), and grasses (Poaceae) [38]. Both *Macrosteles ramosus* and *Macrosteles sexnotatus* were found in Kalloni's orchard, which is located near saltpans and a large wetland. The third species collected, *Macrosteles quadripunctulatus*, is the only species in the genus that also inhibits dry regions and it was the only one found in other orchards [40]. Although the majority of species collected during this study were not directly associated with olive trees, non-crop hosts such as weeds, neighboring cultivations, and even gardens can contribute to the pest pressure. This is because Auchenorrhyncha species are highly mobile and capable of dispersing and alternating between hosts [51].

Differences in the auchenorrhynchan fauna between olive orchards also exist due to different localities since some species are native to a specific region or have restricted distribution in Greece. *Docotettix cornutus*, the olive leafhopper, is a species distributed in Cyprus, Turkey, and Greece, mainly in the islands of the northeast Aegean [52]. Here, it was collected only from the olive orchards of Lesvos. In Turkey, this species was collected in the canopy of olives in high numbers, so it could have a larger population in olives of Lesvos, occurring in the canopy of the trees and could be a potential pest [53].

During this study, new distribution data were obtained for several species. *Clybeccus declivum* was found in abundance in the Pirgos orchard in the Peloponnese, collected with a sweep net from *Avena sterilis* L. This is the first record of the species in the Peloponnese, as it was previously reported only on Rhodes island in the Dodecanese [54]. *Eupteryx zelleri* was previously known only from the Rodopi, Olympus, and Pindos mountains in northern and northwestern Greece, according to the collection of Sakis Drosopoulos. However, our study revealed that this species also occurs in Sterea Ellada and Peloponnese, significantly broadening its known range. Similarly, *Balclutha frontalis* was previously recorded in Sterea Ellada and Crete Island, but we now confirm its presence in the Peloponnese and Lesvos Island as well. The Australian-native species *Anzygina honiloa*, first recorded in Greece in 2018 in Athens, [55], was also recorded in Lesvos Island during this study. This suggests that the species has spread to other localities in Greece within just a few years, indicating a rapid expansion of its distribution.

Moreover, two new records for Greece were made during this study: *Hecalus storai* and *Exitianus nanus*, both collected using a sweep net in the Pirgos olive orchard (Figures 6 and 7). *H. storai* is distributed in France and Bulgaria [32], and in this study, it was collected from *Cynodon dactylon* in the olive orchard. On the other hand, *E. nanus* has a broader distribution across Asia, Africa, Australia, and southwest America. In Europe, it has been recorded only in Italy [32].

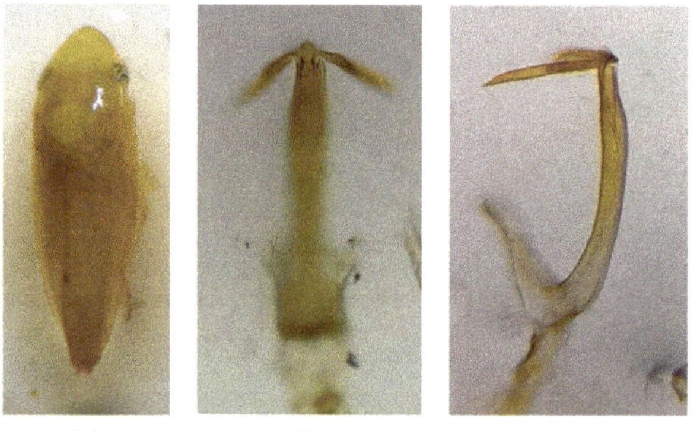

(a) (b) (c)

**Figure 6.** *Hecalus storai* (**a**) Adult, (**b**) Aedeagus (front side), (**c**) Aedeagus (lateral side).

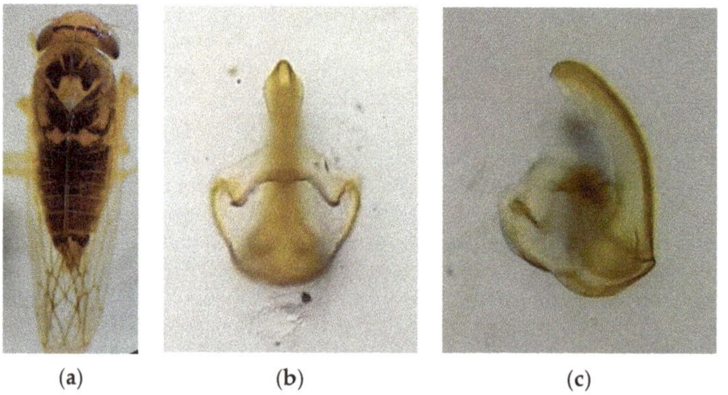

(a) (b) (c)

**Figure 7.** *Exitianus nanus* (**a**) Adult, (**b**) Aedeagus (front side), (**c**) Aedeagus (lateral side).

### 4.1. Potential Vectors

In this study, the most common spittlebug collected was *Philaenus spumarius*, found in almost every field, albeit in low numbers, using Malaise traps. The second most common species was *Neophilaenus campestris*, which was found in 7 out of 13 fields. Higher numbers of both species were collected using a sweep net in comparison with Malaise, highlighting the efficiency of sweep nets in capturing aphrophorids. Also, according to Dongiovanni et al. [56], yellow sticky traps are more efficient, even from a sweep net, for monitoring spittlebugs from the olive canopy, especially during summer, while the combination with a sweep net helps to accurately estimate the spittlebug population by finding the right period for insecticide application. This suggests that these species may exist in higher numbers in other fields where only Malaise traps were used. In Athens, where we used both trapping methods, we can clearly see the huge difference in collecting spittlebugs. With a sweep net, both spittlebugs were collected, and *P. spumarius* was dominant, while with a Malaise trap, only *P. spumarius* was collected, at 0.61% of the total population. The most collected species with a Malaise trap was *Z. pullulan*, with a percentage of 16.62%, while it was not collected with a sweep net. Previous studies in Greece [25,42] embrace these results, with *P. spumarius* and *N. campestris* effectively collected using a sweep net from both the canopy of olives and surrounding weeds across the country. Weeds from the Poaceae family, particularly *Avena sterilis*, appear to support a significant proportion of the adult population of spittlebugs, probably serving as a preferred host for both *P. spumarius* and *N. campestris*. Thompson et al. [57] assert that records of nymph feeding provide the most reliable evidence on species' host plants. We cannot, therefore, be sure that adult spittlebugs were feeding on the plants from which they were swept. They may simply have been resting on these plants or associated with other plants within the vegetation. Nonetheless, nymphs of *P. spumarius* have been recorded feeding on *A. sterilis* in the Mediterranean, suggesting that it may also be a valid host for adults [57,58].

The spittlebug fauna observed in this study is similar to that of other Mediterranean European countries, such as Spain and Italy, where the bacterium *X. fastidiosa* has been established [17,23,59]. However, unlike the studies of Ben Moussa et al. [17] and Cornara et al. [23], our results align more closely with those of Tsagkarakis et al. [24], Antonatos et al. [42], and Theodorou et al. [25], where *Philaenus spumarius* was captured during both spring and autumn in higher numbers. This pattern suggests that this species may be bivoltine, or as Antonatos et al. [42] proposed, it may migrate away from the olive orchards during summer and return in autumn after the first rains. The observations of Drosopoulos and Asche [60] also support the idea of *P. spumarius* being bivoltine or partly bivoltine at elevations below 1000 m, where the majority of olive orchards are located. This bivoltine cycle could make olive orchards more vulnerable to the transmission of *X. fastidiosa* if the bacterium were to enter Greece.

Other spittlebug species collected during the survey were found sporadically in low numbers. These species are included in the EFSA [61] list of potential vectors for Europe, although they were not as prominent in this study.

### 4.2. Biodiversity Indices

When examining species diversity, it appears that the surrounding environment plays a more significant role than the cultivation system itself. The highest biodiversity indices were recorded in the conventional orchards of Kalloni and Pelekanada and the organic orchards of Athens. In Kalloni, the orchard is located next to the saltpans of Kalloni, which is considered a hotspot of biodiversity, providing a moist environment that supports a rich variety of common wild plants. The Athens orchard is located near a botanical garden, a vineyard, and an area with a variety of fruit trees. Similarly, in Pelekanada, the presence of numerous forest trees and shrubs contributed to the higher biodiversity. There, the greatest evenness was observed, even though the number of species collected was lower than the other two ($S_{Pel} = 47 < S_{kal} = 64$, $S_{ath} = 62$). In contrast, the lowest biodiversity index and evenness were recorded in the organic orchard of Koutsopodi, which resembled a monoculture of olives. The field, covering 2.5 acres, was surrounded by more olive trees, limiting the diversity of other plant species and likely reducing overall insect biodiversity. The species *Z. pullula* consisted of 72.37% of the total population, which decreased the biodiversity dramatically. In the Istiea region, where both organic and conventional cultivation systems were present, a clear difference in biodiversity was observed. The organic orchard exhibited higher biodiversity and greater evenness than its conventional counterpart, suggesting that while the surrounding environment is the primary driver of species diversity, the cultivation system still plays a secondary role in shaping Auchenorrhyncha populations [25].

### 4.3. Seasonal Fluctuations

The population peaks of Auchenorrhyncha were observed in the spring and summer months, particularly in samples collected with Malaise traps. This could be attributed to the fact that many species alternate hosts during these seasons, and most adults have emerged from the nymphal stage. The higher temperatures during spring and summer likely increase insect activity, while during rainy days, Auchenorrhyncha species are less active and more likely to hide, reducing their presence in the traps. Additionally, the emergence of many wild plant species during spring and summer provides a more diverse food and habitat source for Auchenorrhyncha, further contributing to the increased species diversity during these months.

On the other hand, sweep net catches were higher during autumn. The cooler temperatures during this period may slow down the insects' metabolic rate, reducing their rapid 'jump' reaction and making them easier to catch [62]. Moreover, the wide variety of wild plants available in the fields during autumn likely supports a higher Auchenorrhyncha population, despite the lower overall temperature.

## 5. Conclusions

This study focused on the Auchenorrhyncha fauna in olive orchards across three geographical divisions in Greece: Sterea Ellada, Peloponnese, and Lesvos Island. The results demonstrate that species composition is influenced by the ground vegetation, the surrounding environment, geographical locality, and the trapping method employed. Most of the species collected were weed feeding, emphasizing the importance of weeds as a food source, alternative host, and site for oviposition. *P. spumarius*, *B. frontalis*, *C. bipunctata*, *E. alsia*, *E. lineolata*, *Ex. capicola*, *M. schmidtgeni*, *Ph. intricatus*, *Ps. alienus*, *S. lauri*, *T. zelleri*, *Ha. provincialis*, *H. decipiens*, *Z. pullula*, *An. Glabra*, and the delphacid *T. propinqua* were ubiquitous, even occurring in small numbers. Malaise traps proved more efficient in capturing smaller, more active species, like Typhlocybinae, which predominantly inhabit ground vegetation. The dominant species of orchards of Lesvos Island were *T. zelleri* for the Lakerda and Nufida orchards, *H. decipiens* and *Z. pullula* for the Kalloni and Pamfila orchards, *P. alienus* for the Kalloni orchard, and *Eu. gyardagica* for the Lakerda orchard. All orchards of Sterea Ellada (Euboea and Athens) had *Z. pullula* as the dominant species. *Eu. zelleri* was dominant in the Istiea orchard with no treatments; *B. frontalis*, *T. zelleri*, and *E. lineolatus* in the Istiea organic field; *H. decipiens* and *C. bipunctata* in the Istiea conventional field; and *Al. modestus* and *A. putoni* in the Athens field. For Peloponnese, *H. decipiens* were dominant in the Fragka, Koutsopodi and Kiparissia orchards; *Z. pullula* in the Pelekanada orchard; and Koutsopodi and *Ha. provincialis* in the Kiparissia field. In contrast, a sweep net is more efficient for capturing larger, less active species, including potential vectors of *X. fastidiosa*. *P. spumarius* was collected in low numbers in almost every field with Malaise traps, while capturing with a sweep net was dominant in both fields in which this trapping method was used. *N. campestris* was collected in small numbers in the Lesvos and Peloponnese fields with a Malaise trap and it was dominant with a sweep net in the Pirgos field. Additionally, this study contributed new distribution data for several species, including the first records of *Hecalus storai* and *Exitianus nanus* in Greece. These findings expand the known distribution of Auchenorrhyncha in Greece and highlight the importance of continued biodiversity monitoring.

Further studies should investigate the preferred plant species for *P. spumarius* oviposition and nymphal development, which could be used as a management tool to suppress its population, if necessary. The potential bivoltine cycle of *P. spumarius* could increase the effectiveness of *X. fastidiosa* transmission if the bacterium is established in Greece. This, in turn, may complicate efforts to control the population of these vectors.

**Author Contributions:** Conceptualization, A.T.; methodology, Z.T. and A.T.; validation, Z.T. and A.T.; investigation, Z.T., M.S., A.M., D.T., G.S., G.K., P.K., C.L., S.T., P.V. and A.K. (Andreas Katribouzas); data curation, Z.T.; writing—original draft preparation, Z.T. and A.T.; writing—review and editing, Z.T., A.K. (Argyro Kalaitzaki), G.P. and A.T.; visualization, Z.T. and M.S.; supervision, A.T.; project administration, A.T. and Z.T. All authors have read and agreed to the published version of the manuscript.

**Funding:** This research received no external funding.

**Data Availability Statement:** The data presented in this study are available upon request from the corresponding author. The data are not publicly available due to also forming part of an ongoing study.

**Conflicts of Interest:** The authors declare no conflicts of interest.

## References

1. Coletta-Filho, H.D.; Francisco, C.S.; Lopes, J.R.S.; De Oliveira, A.F.; Da Silva, L.F.D.O. First Report of Olive Leaf Scorch in Brazil, Associated with *Xylella fastidiosa* Subsp. *pauca*. *Phytopathol. Mediterr.* **2016**, *55*, 130–135. [CrossRef]
2. Li, W.-B.; Pria, W.D., Jr.; Teixeira, D.C.; Miranda, V.S.; Ayres, A.J.; Franco, C.F.; Costa, M.G.; He, C.-X.; Costa, P.I.; Hartung, J.S. Coffee Leaf Scorch Caused by a Strain of *Xylella fastidiosa* from Citrus. *Plant Dis.* **2001**, *85*, 501–505. [CrossRef] [PubMed]
3. Rapicavoli, J.; Ingel, B.; Blanco-Ulate, B.; Cantu, D.; Roper, C. Xylella fastidiosa: An Examination of a Re-Emerging Plant Pathogen: *Xylella fastidiosa*. *Mol. Plant Pathol.* **2018**, *19*, 786–800. [CrossRef] [PubMed]
4. Loconsole, G.; Potere, O.; Boscia, D.; Altamura, G.; Djelouah, K.; Elbeaino, T.; Frasheri, D.; Lorusso, D.; Palmisano, F.; Pollastro, P.; et al. Detection of *Xylella fastidiosa* in Olive Trees by Molecular and Serological Methods. *J. Plant Pathol.* **2014**, *96*, 7–14. [CrossRef]
5. Saponari, M.; Boscia, D.; Nigro, F.; Martelli, G.P. Identification of Dna Sequences Related to *Xylella fastidiosa* in Oleander, Almond and Olive Trees Exhibiting Leaf Scorch Symptoms in Apulia (Southern Italy). *J. Plant Pathol.* **2013**, *95*, 668. [CrossRef]
6. EPPO. *Situation of Xylella fastidiosa in France*; EPPO Reporting Service no. 10—2016 Num. article: 2016/193 2016; Bayer Corporation: Leverkusen, Germany, 2016.
7. EPPO. *First Report of Xylella fastidiosa in Spain*; EPPO Reporting Service no. 11—2016 Num. article: 2016/213 2016; Bayer Corporation: Leverkusen, Germany, 2016.
8. Olmo, D.; Nieto, A.; Adrover, F.; Urbano, A.; Beidas, O.; Juan, A.; Marco-Noales, E.; López, M.M.; Navarro, I.; Monterde, A.; et al. First Detection of *Xylella fastidiosa* Infecting Cherry (*Prunus avium*) and *Polygala myrtifolia* Plants, in Mallorca Island, Spain. *Plant Dis.* **2017**, *101*, 1820. [CrossRef]
9. Landa, B.B.; Castillo, A.I.; Giampetruzzi, A.; Kahn, A.; Román-Écija, M.; Velasco-Amo, M.P.; Navas-Cortés, J.A.; Marco-Noales, E.; Barbé, S.; Moralejo, E.; et al. Emergence of a Plant Pathogen in Europe Associated with Multiple Intercontinental Introductions. *Appl. Environ. Microbiol.* **2020**, *86*, 3. [CrossRef]
10. EPPO. *First Report of Xylella fastidiosa subsp. multiplex in Portugal*; EPPO Reporting Service no. 01/2019. Num. article 2019/017 2019; Bayer Corporation: Leverkusen, Germany, 2019.
11. El-Kholy, M. *Following Olive Footprints (Olea europaea L.). Cultivation and Culture, Folklore and History, Traditions and Uses*; International Society for Horticultural Science: Leuven, Belgium, 2012.
12. Frazier, N.W.; Freitag, J.H. Ten Additional Leafhopper Vectors of the Virus Causing Pierce's Disease of Grapes. *Phytopathology* **1946**, *36*, 634–637.
13. Hewitt, W.B.; Houston, B.R.; Frazier, N.W.; Freitag, J.H. Leaf-Hopper Transmission of the Virus Causing Pierce's Disease of Grape and Dwarf of Alfalfa. *Phytopathology* **1946**, *36*, 117–128.
14. Redak, R.A.; Purcell, A.H.; Lopes, J.R.S.; Blua, M.J.; Mizell, R.F., III; Andersen, P.C. The Biology of Xylem Fluid-Feeding Insect Vectors of *Xylella fastidiosa* and Their Relation to Disease Epidemiology. *Annu. Rev. Entomol.* **2004**, *49*, 243–270. [CrossRef]
15. Severin, H.H.P. Spittle-Insect Vectors of Pierce's Disease Virus. II. Life History and Virus Transmission. *Hilgardia* **1950**, *19*, 357–376.
16. Purcell, A.H. *The Ecology of Bacterial and Mycoplasma Plant Diseases Spread by Leafhoppers and Planthoppers*; John Wiley & Sons: Hoboken, NJ, USA, 1985; pp. 351–380.
17. Ben Moussa, I.E.B.; Mazzoni, V.; Valentini, F.; Yaseen, T.; Lorusso, D., Speranza, S.; Digiaro, M.; Varvaro, L.; Krugner, R.; D'Onghia, A.M. Seasonal Fluctuations of Sap-Feeding Insect Species Infected by *Xylella fastidiosa* in Apulian Olive Groves of Southern Italy. *J. Econ. Entomol.* **2016**, *109*, 1512–1518. [CrossRef] [PubMed]

18. Elbeaino, T.; Yaseen, T.; Valentini, F.; Ben Moussa, I.E.; Mazzoni Valerio, V.; D'Onghia, A.M. Identification of Three Potential Insect Vectors of *Xylella fastidiosa* in Southern Italy. *Phytopathol. Mediterr.* **2014**, *53*, 328–332. [CrossRef]
19. Chuche, J.; Sauvion, N.; Thiéry, D. Mixed Xylem and Phloem Sap Ingestion in Sheath-Feeders as Normal Dietary Behavior: Evidence from the Leafhopper *Scaphoideus titanus*. *J. Insect Physiol.* **2017**, *102*, 62–72. [CrossRef] [PubMed]
20. Guglielmino, A.; D'Urso, V.; Alma, A. Auchenorrhyncha (Insecta, Homoptera) from Sardinia (Italy): A Faunistic, Ecological and Zoogeographical Contribution. *Mitt. Mus. Nat. Kd. Berl. Dtsch. Entomol.* **2000**, *47*, 161–172. [CrossRef]
21. Guglielmino, A.; Bückle, C.; Remane, R. Contribution to the Knowledge of the Auchenorrhyncha Fauna of Central Italy (Hemiptera, Fulgoromorpha et Cicadomorpha). *Marbg. Entomol. Publ.* **2005**, *3*, 13–98.
22. Guglielmino, A.; Modola, F.; Scarici, E.; Speranza, S.; Bückle, C. The Auchenorrhyncha Fauna (Insecta, Hemiptera) of Villa Lante, Bagnaia (Italy): A Study of an Urban Ecosystem. *Bull. Insectology* **2015**, *68*, 239–253.
23. Cornara, D.; Cavalieri, V.; Dongiovanni, C.; Altamura, G.; Palmisano, F.; Bosco, D.; Porcelli, F.; Almeida, R.P.P.; Saponari, M. Transmission of *Xylella fastidiosa* by Naturally Infected *Philaenus spumarius* (Hemiptera, Aphrophoridae) to Different Host Plants. *J. Appl. Entomol.* **2017**, *141*, 80–87. [CrossRef]
24. Tsagkarakis, A.E.; Afentoulis, D.G.; Matared, M.; Thanou, Z.N.; Stamatakou, G.D.; Kalaitzaki, A.P.; Tzobanoglou, D.K.; Goumas, D.; Trantas, E.; Zarboutis, I.; et al. Identification and Seasonal Abundance of Auchenorrhyncha with a Focus on Potential Insect Vectors of *Xylella fastidiosa* in Olive Orchards in Three Regions of Greece. *J. Econ. Entomol.* **2018**, *111*, 2536–2545. [CrossRef]
25. Theodorou, D.; Koufakis, I.; Thanou, Z.N.; Kalaitzaki, A.P.; Chaldeou, E.; Afentoulis, D.G.; Tsagkarakis, A.E. Management System Affects the Occurrence, Diversity and Seasonal Fluctuation of Auchenorrhyncha, Potential Vectors of *Xylella fastidiosa*, in the Olive Agroecosystem. *Bull. Insectology* **2021**, *74*, 27–40.
26. Ribaut, H. Homoptères Auchénorhynches I. (Typhlocybidae). In *Faune de France*; Lechevalier, P., Ed.; Paul Lechevalier et fils: Paris, France, 1936; Volume 31.
27. Ribaut, H. Homoptères Auchénorhynches II. (Jassidae). In *Faune de France*; Lechevalier, P., Ed.; Paul Lechevalier et fils: Paris, France, 1952; Volume 57.
28. Ossiannilsson, F. Nagra for Sverige Nya Stritar (Hom. Auchenorrh.). *Med. En. Synonymisk Anmarkning. Entomol. Tidskr.* **1955**, *76*, 131–133.
29. Ossiannilsson, F. The Auchenorrhyncha (Homoptera) of Fennoscandia and Denmark. Part 2: The Families Cicadidae, Cercopidae, Membracidae, and Cicadellidae (Excl. Deltocephalinae). *Entomol. Scand.* **1981**, *7*, 223–593.
30. Ossiannilsson, F. The Auchenorrhyncha (Homoptera) of Fennoscandia and Denmark. Part 3: The Family Deltocephalinae, Catalogue, Literature and Index. *Fauna Entomol. Scand.* **1983**, *7*, 594–979.
31. Guglielmino, A.; Bückle, C. Revision of Errhomeninae and Aphrodinae (Hemiptera, Cicadomorpha) in Italy with Remarks on Their Variability and Distribution in Adjacent Regions and Description of Three New Taxa. *Zootaxa* **2015**, *3906*, 1–66. [CrossRef] [PubMed]
32. Dmitriev, D.A.; Anufriev, G.A.; Bartlett, C.R.; Blanco-Rodríguez, E.; Borodin, O.I.; Cao, Y.-H.; Deitz, L.L.; Dietrich, C.H.; Dmitrieva, M.O.; El-Sonbati, S.A.; et al. World Auchenorrhyncha Database. TaxonPages. Available online: https://Hoppers.Speciesfile.Org (accessed on 25 September 2024).
33. Curry, J.P. The Arthropods Associated with the Decomposition of Some Common Grass and Weed Species in the Soil. *Soil Biol. Biochem.* **1973**, *5*, 645–657. [CrossRef]
34. Cusack, P.D.; Evans, G.O.; Brennan, P.A. A Survey of the Mites of Stored Grain and Grain Products in the Republic of Ireland. *Sci. Proc. R. Dublin Soc.* **1975**, *3*, 273–329.
35. Emmanuel, N. Aspects of the Biology of Mites Associated with Cereals During Growth and Storage. Ph.D. Thesis, University College Dublin, Dublin, Ireland, 1977.
36. Simpson, E.H. Measurement of Diversity. *Nature* **1949**, *163*, 688. [CrossRef]
37. Malenovský, I.; Baňař, P.; Kment, P. A Contribution to the Faunistics of the Hemiptera (Cicadomorpha, Fulgoromorpha, Heteroptera, and Psylloidea) Associated with Dry Grassland Sites in Southern Moravia (Czech Republic). *Acta Musei Morav. Sci. Biol.* **2011**, *96*, 41–187.
38. Nickel, H.J. The Leafhoppers and Planthoppers of Germany (Hemiptera, Auchenorrhyncha): Patterns and Strategies in a Highly Diverse Group of Phytophagous Insects. In *Pensoft Series Faunistica*; Pensoft Publ: Sofia, Bulgaria, 2003; ISBN 978-954-642-169-2.
39. Nickel, H.J.; Remane, R. Check List of the Planthoppers and Leafhoppers of Germany, with Notes on Food Plants, Diet Width, Life Cycles, Geographic Range and Conservation Status (Hemiptera, Fulgoromorpha and Cicadomorpha). *Beiträge Zur. Zikadenkunde* **2002**, *5*, 27–64.
40. Mazzoni, V. Contribution to the Knowledge of the Auchenorrhyncha (Hemiptera Fulgoromorpha and Cicadomorpha) of Tuscany (Italy). *Redia* **2005**, *88*, 85–102.
41. Söderman, G. Taxonomy, Distribution, Biology and Conservation Status of Finnish Auchenorrhyncha. In *The Finnish Environment*; Finnish Environment Institute: Helsinki, Finland, 2007.
42. Antonatos, S.; Papachristos, D.P.; Kapantaidaki, D.E.; Lytra, I.C.; Varikou, K.; Evangelou, V.I.; Milonas, P. Presence of Cicadomorpha in Olive Orchards of Greece with Special Reference to *Xylella fastidiosa* Vectors. *J. Appl. Entomol.* **2020**, *144*, 1–11. [CrossRef]
43. Thanou, Z.N.; Kontogiannis, E.G.; Tsagkarakis, A.E. Impact of Weeds on Auchenorrhyncha Incidence and Species Richness in Citrus Orchards. *Phytoparasitica* **2021**, *49*, 333–347. [CrossRef]

44. Albre, J.; García, C.; José, M.; Gibernau, M. Ecology of the Meadow Spittlebug *Philaenus Spumarius* in the Ajaccio Region (Corsica)—I: Spring. *Bull. Entomol. Res.* **2021**, *111*, 246–256. [CrossRef]
45. Albre, J.; Gibernau, M. Diversity and Temporal Variations of the Hemiptera Auchenorrhyncha Fauna in the Ajaccio Region (France, Corsica). *Ann. Soc. Entomol. Fr.* **2019**, *55*, 497–508. [CrossRef]
46. Bosco, D.; Alma, A.; Arzone, A. Studies on Population Dynamics and Spatial Distribution of Leafhoppers in Vineyards (Homoptera: Cicadellidae). *Ann. Appl. Biol.* **1997**, *130*, 1–11. [CrossRef]
47. Karavin, M.; Çalişkan, E.; Dede, O. Determination of Auchenorrhyncha Species Distributed in Apple Orchards in Amasya, Turkey with a New Record for Turkish Fauna. *Int. J. Sci. Lett.* **2021**, *3*, 32–51. [CrossRef]
48. Tholt, G.; Kiss, B. Host Range of Psammotettix Alienus (Dahlbom). *Növényvédelem* **2011**, *47*, 229–235.
49. Holzinger, W.; Kammerlander, I.; Nickel, H. *The Auchenorrhyncha of Central Europe. Die Zikaden Mitteleuropas, Volume 1: Fulgoromorpha, Cicadomorpha Excl. Cicadellidae*; Brill: Leiden, The Netherlands, 2003; 674p.
50. Malenovský, I.; Lauterer, P. Leafhoppers and Planthoppers (Hemiptera: Auchenorrhyncha) of the Bílé Karpaty Protected Landscape Area and Biosphere Reserve (Czech Republic). *Acta Musei Morav. Sci. Biol.* **2012**, *96*, 155–322.
51. Bianchi, F.J.J.A.; Booij, C.J.H.; Tscharntke, T. Sustainable Pest Regulation in Agricultural Landscapes: A Review on Landscape Composition, Biodiversity and Natural Pest Control. *Proc. R. Soc. B* **2006**, *273*, 1715–1727. [CrossRef]
52. Ribaut, H. On the Insect Fauna of Cyprus. Results of the Expedition of 1939 by Harald, Hakan and P.H. Lindberg. *Soc. Sci. Fenn.* **1948**, *1–2*, 2–14.
53. Başpinar, H.; Yıldırım, E.M.; Xing JiChun, X.J. Determination and Population Fluctuations of Cicadellidae (Hemiptera: Cicadomorpha) Species in Pomegranate Orchards in Aydin Province, Turkey. *Türkiye Entomoloji Derg.* **2013**, *37*, 3–11.
54. Gnezdilov, V.M.; Drosopoulos, S.; Wilson, M.R. New Data on Taxonomy and Distribution of Some Fulgoroidea (Homoptera, Cicadina). *Zool. Inst.* **2004**, *12*, 217–223.
55. Thanou, Z.N.; Afentoulis, D.G.; Koufopoulou, P.; Ampatzi, A.P.; Lekkou, S.D.; Koutsogiannopoulou, A.; Bravou, A.A.; Stamatakou, G.D.; Voulgaraki, K.N.; Piperkas, A.; et al. New Records and Updated Checklist of Cicadomorpha (Hemiptera: Auchenorrhyncha) Species from Greece. *Zootaxa* **2018**, *4413*, 133. [CrossRef]
56. Dongiovanni, C.; Di Carolo, M.; Fumarola, G.; Tauro, D.; Tedone, B.; Ancona, S.; Palmisano, V.; Carrieri, M.; Cavalieri, V. Comparing Different Sticky Traps to Monitor the Occurrence of *Philaenus spumarius* and *Neophilaenus campestris*, Vectors of Xylella fastidiosa, in Different Crops. *Insects* **2023**, *14*, 777. [CrossRef]
57. Thompson, V.; Harkin, C.; Stewart, A.J.A. The Most Polyphagous Insect Herbivore? Host Plant Associations of the Meadow Spittlebug, *Philaenus spumarius* (L.). *PLoS ONE* **2023**, *18*, e0291734. [CrossRef]
58. Antonatos, S.; Papachristos, D.P.; Varikou, K.; Vahamidis, P.; Kapranas, A.; Milonas, P. Seasonal Appearance, Abundance, and Host Preference of *Philaenus Spumarius* and *Neophilaenus Campestris* (Hemiptera: Aphrophoridae) in Olive Groves in Greece. *Environ. Entomol.* **2021**, *50*, 1474–1482. [CrossRef]
59. Morente, M.; Cornara, D.; Plaza, M.; Durán, J.M.; Capiscol, C.; Trillo, R.; Ruiz, M.; Ruz, C.; Sanjuan, S.; Pereira, J.A.; et al. Distribution and Relative Abundance of Insect Vectors of *Xylella fastidiosa* in Olive Groves of the Iberian Peninsula. *Insects* **2018**, *9*, 175. [CrossRef] [PubMed]
60. Drosopoulos, S.; Asche, M. Biosystematic Studies on the Spittlebug Genus Philaenus with the Description of a New Species. *Zool. J. Linn. Soc.* **1991**, *101*, 169–177. [CrossRef]
61. European Food Safety Authority. Statement of EFSA on Host Plants, Entry and Spread Pathways and Risk Reduction Options for Xylella fastidiosa Wells et al. *EFSA J.* **2013**, *11*, 3468. [CrossRef]
62. Delong, D.M. Some Problems Encountered in the Estimation of Insect Populations. *Ann. Ent. Soc. Amer Columb. Ohio* **1932**, *25*, 13–17.

*agronomy*

MDPI

*Article*

# Longitudinal Analysis of Honey Bee Colony Health as a Function of Pesticide Exposure

**Susan E. Kegley [1,*], Rosemarie Radford [1], Timothy J. Brown [1], Jeff Anderson [2], Darren Cox [3], Steve Ellis [4] and Geoffrey W. Marcy [5]**

1 Pesticide Research Institute, Santa Rosa, CA 95404, USA; tjbrown17@gmail.com (T.J.B.)
2 California-Minnesota Honey Farms, Oakdale, CA 95361, USA; jfand57@gmail.com
3 Cox Honey of Utah, Petersboro, UT 84325, USA; coxhoney@gmail.com
4 Old Mill Honey Farm, Barrett, MN 56311, USA; nhbabsellis@gmail.com
5 Space Laser Awareness, Santa Rosa, CA 95404, USA; geoff@spacelaserawareness.com
* Correspondence: skegley@pesticideresearch.com

**Abstract:** Sixty commercial honey bee colonies were monitored over the course of one year with the goal of assessing potential correlations between measured colony strength and environmental stressors, including exposures to pesticides and pathogens. We developed a new method for assessing colony health by determining the fractional change in population of the four colonies on each pallet between peak population on 1 June and the last population assessment prior to winter on 1 October. This fractional change in population was evaluated as a function of pesticide load per pallet for each of the 37 pesticide chemicals detected, grouping pallets by beekeeper. The analysis of individual chemicals showed that both imidacloprid and cyprodinil were negatively correlated with colony health, while tau-fluvalinate and dinotefuran (at very low concentrations) were positively correlated, possibly because of effects on *Varroa* control. Exposure to groups of chemicals was also evaluated. Normalization of each chemical concentration to the maximum observed for that chemical provided an equal weighting for each chemical, even though their relative occurrence in the environment and their effective toxicities might differ. A total of 24 chemical groups whose members share a structural commonality, a functional commonality, or suspected synergistic actions were considered, demonstrating negative correlations between colony health and exposures to neonicotinoids as a group and neonicotinoids in combination with (1) methoxyfenozide (2) organophosphates, and (3) diflubenzuron. Analysis of several groups of fungicides applied to almonds during pollination also showed negative correlations with colony health.

**Keywords:** honey bees; pollinators; pesticides; colony collapse; pollination

## 1. Introduction

Since 2006, commercial beekeepers have experienced extensive colony losses, both during the overwintering period and, recently, during the summer as well [1]. Beekeepers are currently mitigating these colony losses at significant cost in resources and staff time [2], which strains the sustainability of commercial beekeeping operations in the U.S. The stresses experienced by honey bees affect not only managed bees and the beekeepers who rely on them for their livelihood but also wild pollinators, farmers that depend on pollinators, and ultimately our civilization that depends on a bountiful and diverse food supply.

Many potential causes of colony decline among the honey bees in North America, Europe, Asia, and the Middle East have been proposed [3–6]. These include exposure to agricultural chemicals, climate change, lack of diverse forage, hive management practices, and a proliferation of pests and pathogens. No single factor has been definitively identified as the primary cause of the recent increase in annual honey bee colony losses. Indeed, the interdependence of these factors cannot be dismissed. For example, certain pesticides have been shown to down-regulate genes that control the function of the immune system [7], which is essential for protecting the honey bee from viruses transmitted by the *Varroa* mite. Determining the contribution of the different factors to the problem of colony losses is paramount to solving it and requires longitudinal studies to evaluate multiple colony stressors over time, as well as their potential interactions.

### 1.1. Relevant Prior Work

Previous work has demonstrated the necessity of longitudinal tracking of multiple parameters that may affect colony health. Van Englesdorp et al. conducted a comparison of multiple stressors (pesticides, pathogens, and parasites) on collapsed colonies vs. healthy colonies, but only at a single point in time [8]. Stoner and Eitzer evaluated pesticide residues in Connecticut colonies once per year over a two- to five-year period, documenting the presence of 60 different pesticides or metabolites in pollen [9]. More recently, Traynor et al. tracked the presence of pesticides in pollen over a 7-year period, with samples taken from a subset of colonies randomly surveyed for the National Honey Bee Disease Survey (NHBDS) to determine a baseline of pesticide exposure [10]. However, because the loads of pesticides and pathogens change over the course of a year, and their effects on colony health may not appear until weeks or even months after exposure, single-point analyses of a colony are insufficient for determining cause and effect.

Runckel, Flenniken, et al. tracked pathogen loads for a migratory bee operation, assessing their changes over time and correlations to colony health [11,12]. This study was the first of its kind and showed that pathogen types and loads in commercial honey bee colonies vary over time, even in visibly healthy colonies. The U.S. National Honey Bee Disease Survey conducted a longitudinal study of pathogens in both migratory and stationary honey bee colonies between 2009 and 2014, which established a baseline of honey bee diseases and their seasonality in the U.S. [13]. Further work by Faurot-Daniels et al. [14] also shows annual variation in pathogen populations. Measurement of pesticide residues and *Varroa* mite loads was not within the scope of these studies, and thus no conclusions could be drawn regarding potential for interactions with other stressors.

Using samples from the study presented here, Glenny et al. explored potential correlations of colony health as a function of the prevalence of seven common pathogens detected in the colonies, finding no statistically significant correlations between colony strength and pathogen prevalence for the primary pathogens known to infect honey bees [15]. Much work remains to be done to understand the interactions between honey bee pathogens and other colony stressors [16].

Pesticides, including insecticides, herbicides, and fungicides, have garnered particular attention as potential stressors because of their known toxicity and widespread use in crop production. Multiple studies show that honey bee exposures to pesticides are common and that exposures change depending on the season of application, the availability of contaminated pollens and nectars, and the season in which they are consumed [17,18].

The neonicotinoid insecticides have received the most intensive study because of their high acute toxicity to honey bees, persistence in the environment, and their high water solubility, which facilitates systemic uptake and distribution into pollen and nectar, thereby increasing exposure potential for pollinators. Sublethal effects of neonicotinoids have been observed at environmentally relevant concentrations and include impairment of immune function [7,19], reduced queen lifespan and fertility [20,21], decreased sperm viability in drones [22,23] interference with foraging ability [24–26] and navigation [27], and reduced brood survival [28]. Recent work indicates that sublethal exposure can affect reproduction in native bee species over months to years [29,30].

Certain fungicides, while not acutely toxic to bees, may also impair colony health [31,32]. Mussen documented beekeeper reports of time-delayed brood kills after fungicide applications to almonds [33]. McArt et al. found a landscape-scale correlation between use of fungicides and increased Nosema prevalence in bumblebees [34], a result consistent with the observation of Pettis et al. of increased probability of Nosema infection in honey bees consuming fungicide-contaminated pollen [35]. Yoder et al. indicate that fungicide contaminants reduce the diversity and number of beneficial fungi essential for the conversion of pollen into bee bread within the hive, with potential to impact the nutritional status of the colony [36].

Fungicides have been observed to interfere with the production of cytochrome P450 enzymes responsible for detoxification of both naturally occurring and synthetic chemicals in the honey bees' environment [37–39]. Mao et al. found that azole fungicides bind to a cytochrome P450 (CYP) enzyme (CYP9Q1) responsible for metabolism of quercetin and other naturally occurring components of pollen, reducing the ability of the bee to process these components of its diet [40]. The result was impaired mitochondrial regeneration, resulting in production of less ATP as energy for flight muscles. Traynor et al. found that the presence of fungicides that inhibit sterol biosynthesis (azoles) and those with multi-site contact activity were correlated with queen failures [5]. Ricke et al. found that fungicides used during almond bloom reduced queen longevity [41].

Insect growth regulators impair honey bee colony growth and longevity, with effects ranging from mortality of larvae and pupae to physical abnormalities in exposed adult bees [42], as well as reduced hypopharyngeal gland development [43], inhibition of vitellogenin synthesis [44], and reduced forager survival [45].

Reductions in CYP enzyme function can also enhance the toxicity of environmental chemicals by slowing metabolic detoxification pathways. Indeed, the addition of a "synergist" that disables insect detoxification systems has long been used as a resistance management tool, with the role of CYP enzymes in this process elucidated long after the effect was known [46]. Johnson et al. measured up to 100-fold increases in the toxicity of insecticides in the presence of azole fungicides [47]. Similar results have been noted by Han et al. for Asian honey bees [48]. Sgolastra et al. noted a synergistic toxic effect between clothianidin and propiconazole in both honey bees and native bees [49]. Synergies have been noted by Tosi et al. between the butenolide neonicotinoid insecticide flupyradifurone and propiconazole as well [50]. Indeed, a substantial number of fungicide patents document synergistic enhancement of insecticidal toxicity as a characteristic [51].

### 1.2. Study Background and Goals

In this paper, we describe a study in which we conducted longitudinal monitoring of 20 colonies in each of three commercial beekeeping operations (60 total colonies) from January 2014 to January 2015. Commercial beekeepers offer a statistically large population of honey bees that are exposed to a variety of environments during each year, enabling temporal measurements of colony stressors in the hive along with colony health to search for correlations that may illuminate fundamental causes of colony failures. The goals of the study were to assess the dependence of colony health and survival on the occurrence of specific pests, pathogens, and pesticides or combinations of pesticides found in the 60 colonies over time. The results of the dependence of colony health on a range of pathogens have already been published, with no statistically significant correlations found between colony health and any specific pathogen [15]. This current paper focuses on assessment of colony health as a function of pesticide concentrations and mite loads.

In the spring of 2014, significant colony losses were experienced by many commercial beekeepers after pollinating almonds, with numerous reports of dead or malformed brood during and immediately after almond bloom [52]. There was speculation that application of a fungicide/insect growth regulator (IGR) tank mix was responsible for the spring losses. Wade et al. conducted a follow-up in vitro bioassay to test this hypothesis, evaluating the effects on larval survival of two insect growth regulators and an insecticide (diflubenzuron, methoxyfenozide, and chlorantraniliprole) and four fungicides (propiconazole, iprodione, and a combination of boscalid and pyraclostrobin), both singly and in combination [53]. The results demonstrated increased larval mortality after exposure to the insect growth regulator diflubenzuron alone and to combinations of propiconazole or iprodione with chlorantraniliprole. Neither methoxyfenozide nor any methoxyfenozide-fungicide combination increased mortality. Fisher et al. studied the effects of combined exposure to methoxyfenozide, pyriproxyfen, and bifenazate in a laboratory setting, finding adverse effects on foragers [45]. These results are consistent with the pattern of 2014 losses, but the studies did not evaluate the full range of pesticides used on almonds. The results presented here indicate that other fungicides and combinations of fungicides and other pesticides may be equally or more problematic for colony health than those tested previously.

## 2. Materials and Methods

Sixty honey bee colonies from three different commercial migratory beekeeping operations were included in this study and monitored for one year, from January 2014 to January 2015. Colony strength, chemical concentrations in the pollen, pathogen loads, and *Varroa* mite counts in the hives were measured at various times throughout the year. Beekeeper-specific details on colony management are provided in Appendix A.

### *2.1. Colony Selection and Study Design*

Three commercial beekeeping operations were included in this study, identified as "JA", "SE", and "DC", associated with co-authors of this paper. Each beekeeper contributed twenty honey bee colonies for the study, with four colonies on each of five pallets, thus incorporating 15 pallets supporting 60 colonies. Pallets were coded by beekeeper, as JA-1, JA-2, etc. At the start of the study, all colonies were queen-right and had successfully survived the 2013–2014 overwintering period. Colonies selected for the study contained on average 9 $\pm$ 2.6 frames of bees, based on the conventional frame-count method used in contracts for commercial pollination to assess hive strength—the number of frames within a hive having a minimum number of bees [54,55].

### *2.2. Sampling Hive Materials for Pesticides*

Pollen as beebread was sampled for pesticide analysis four times during the year: (1) Before almond pollination (February 2014); (2) Immediately after almond pollination (March 2014); (3) Between April and June 2014; and (4) In early October 2014. For each hive, 3-inch core samples of a comb near the brood nest containing wax, pollen, and honey were extracted for pesticide residue analysis. Composite pollen samples were prepared for each pallet from the ensemble of individual hive cores by mixing equal masses of pollen from each of the available hives to create a single homogeneous 3 g sample for residue analysis. Sampling details, colony locations, and colony characteristics and management are described in Appendix A.

### *2.3. Laboratory Pesticide Analysis*

Hive matrix samples (pollen/bee bread, wax, honey) were analyzed by the USDA lab in Gastonia, NC using the QuEChERS-NSL (National Standard Lab) extraction procedure for a multi-residue screen for approximately 180 different chemicals, including pesticides and common metabolites [56]. Analytes, their detection limits (LODs), current US EPA registration status, use type, and chemical classifications can be found in the Supplementary Material, Table S1. Over the course of the project, spiked samples were also prepared and sent to the USDA lab to validate laboratory accuracy. Standards of azoxystrobin, chlorpyrifos, coumaphos, λ-cyhalothrin, cyprodinil, fluvalinate, imidacloprid, methoxyfenozide, pendimethalin, and thiamethoxam at 100 μg/mL were purchased from AccuStandard and diluted to 20 μg/mL stock solutions. Three pollen samples of known residue composition were spiked by adding the appropriate amount of stock solution for several different pesticides typically found in the samples to produce concentrations in the range of the analytical method. All spike results were within $\pm$30% of the spiked value, indicating that the analyses were reliable within the limitations of the method. Recoveries of the spiked amounts averaged 110% (SD = 27%).

A subset of the samples (early summer and late summer samples) was submitted to the Harvard Center for Environmental Health laboratory for a higher-sensitivity analysis of neonicotinoids in pollen, with limits of detection (LOD) 2–5-fold lower than those of the USDA laboratory. A modified QuEChERS method was used for the extraction, as described in Chen et al. [57]. One pollen sample spiked with known amounts of neonicotinoid pesticides was sent to the Harvard lab. Recoveries ranged from 74% to 79% of the spiked amount, averaging 76% (SD = 4%). Project samples found to have imidacloprid and clothianidin concentrations between 1 ppb and 7 ppb by the Harvard lab were systematically identified as <LOD by the USDA lab, even though the USDA LODs were listed as 1 ppb, indicating systematic inaccuracies in the USDA methods at lower concentrations.

### *2.4. Measuring Changes in Colony Strength over Time*

Colony strength was measured at 7 to 10 different times, depending on beekeeper, during the period from January 2014 to January 2015. The bee population was measured by using "frame counts" as the metric of colony strength, as described and listed in Appendix B within Tables A1–A3.

A total of 35 of the 60 colonies were split in the spring of 2014. For colonies that were split, the old queen was left with the original study hive, so there was no break in the brood cycle. The split was moved away from the study hive pallets and was no longer considered part of the study, except as described here. Frame counts from splits were accounted for by calculating a split factor (SF), defined as the number of frames remaining in the original hive after the split divided by the total number of frames in the original hive, e.g., for an 18-frame colony from which a 4-frame split was made, SF = 14/18 = 0.78. Corrected frame counts were calculated by dividing the measured frame counts by SF for all frame count measurements made after the split. This correction assumes that the population of the split changes at the same rate as the original colony. Many of the SE hives were split twice, necessitating the use of a second split factor, $SF_2$, defined as the number of frames remaining in the split hive after the second split divided by the total number of frames in the original hive. The corrected frame count was calculated as the measured frame count divided by the product of SF and $SF_2$.

In order to match results with composite pesticide samples across hives on a pallet, frame counts for all four hives residing on each pallet were summed to arrive at a pallet population. Summing frame counts over a pallet improves the statistical integrity of the measured bee population, reducing inter-colony variability due to queen performance and solving the mathematical problem of zero frame counts for individual colonies that died during the study.

The "bee population" (frame count) for each pallet was plotted as a function of time. A spline interpolation was computed between the (approximately) eight points of pallet population measurement to estimate the population in between actual measurements to produce a continuous curve of frame counts vs. time during the year.

Using these population curves, a metric of population change was computed: the percent change in the honey bee population between 1 June and 1 October. This percent change represents the fractional decline in the bee population between the early summer peak population and the fall season when a reduced population of a colony is preparing for winter. This metric of population change was defined as the total pallet frame count on 1 October minus the total pallet frame count on 1 June, divided by the pallet frame count on 1 June, i.e., the fractional change in the number of frames stated as a percentage.

### *2.5. Construction of a Metric of Chemical Exposure*

A straightforward metric of exposure to each of the 37 chemicals was developed by summing the four measurements of concentration in the pollen from each pallet. The table of these 37 exposure metrics for each pallet is given in Appendix C, as Table A4.

In order to search for correlations between the fractional change in bee population and the combined group concentrations of several chemicals with similar modes of action, a simple metric based on concentrations normalized to the maximum observed concentration was constructed. Each pesticide concentration was divided by the maximum observed concentration ever measured in this study for that chemical. Normalization gives each chemical equal weight, even though their relative occurrence in the environment and their effective toxicities might differ. For each chemical group and pallet, the max-normalized concentrations of each chemical were summed over the entire study period to obtain an estimate of exposure. A total of 24 chemical groups whose members share a structural commonality, a functional commonality, or suspected synergistic actions were considered (see Sections 3 and 4).

### 2.6. Assessment of Potential Correlations

Potential correlations of colony health to chemical exposure were assessed by plotting the fractional change in the bee population vs. the sum of concentrations (in ppb) for individual chemicals and the sums of the max-normalized concentrations (unitless, see above) for groups of chemicals. There is a clear grouping of data points by beekeeper, likely reflecting beekeeper-specific locations/exposures and management practices. Thus, we averaged the fractional change in bee population and the sum of the concentrations for each beekeeper and plotted these averages. Thus, each plot contains three points, with each point representing a beekeeper. The error bars represent the uncertainty in the mean of the two measured quantities from the five pallets. An unweighted linear least-squares fit was computed to provide a trendline for visual inspection. The limited number of beekeepers precluded a formal statistical assessment of significance.

### 2.7. Measurement of Varroa and Pathogen Prevalence

*Varroa* mites per 100 bees were counted in each hive 5–6 times during the year, depending on beekeeper, using either the ethanol wash [58] or sugar roll [59] method. Ultimately, it was not possible to use these data in statistical analysis due to the variability in *Varroa* management timing and methods used by the different beekeepers.

Adult hive bees were sampled five times during the year from selected colonies to assess the occurrence of pathogens, specifically. A single frame from the center of the brood nest covered with bees was selected and shaken into a bin. One-third of a cup of bees (approximately 150 bees) were placed in a pre-labeled quart-sized Ziploc® bag, immediately packed in dry ice for transport, then placed in long-term storage at −20 °C. The analytical methods used to assess pathogen presence and the results of this analysis are described in reference [15].

## 3. Results

### 3.1. Colony Population Changes with Time

Figure 1 shows plots of the bee population over time during the study, as experienced by the 15 pallets from all three beekeepers. The points show the measured bee populations for a specific date, and the dashed line represents the spline interpolation between the points.

The percentage change in bee population between 1 June and 1 October for the 15 pallets is shown in Figure 1. The populations exhibit the well-known seasonal pattern of an increase from February to May, a plateau during early summer, and a decline in late summer and early autumn. We use this percentage population change for each of the 15 pallets for all upcoming analyses as a simple and transparent metric related to the change in the health of the bees from early summer to fall. The percentage changes in bee populations are negative except for two, DC-2 and DC-5, which nonetheless suffered their declines a few weeks after 1 October. On average, the percent population change for the 15 pallets is −43% and the median is −54%. A total of 28 of the 60 colonies (47%) died before the end of the study, seven from the DC operation, eleven from the JA operation, and ten from the SE operation. The full data set of colony population, over time, is available in Appendix B (as previously noted).

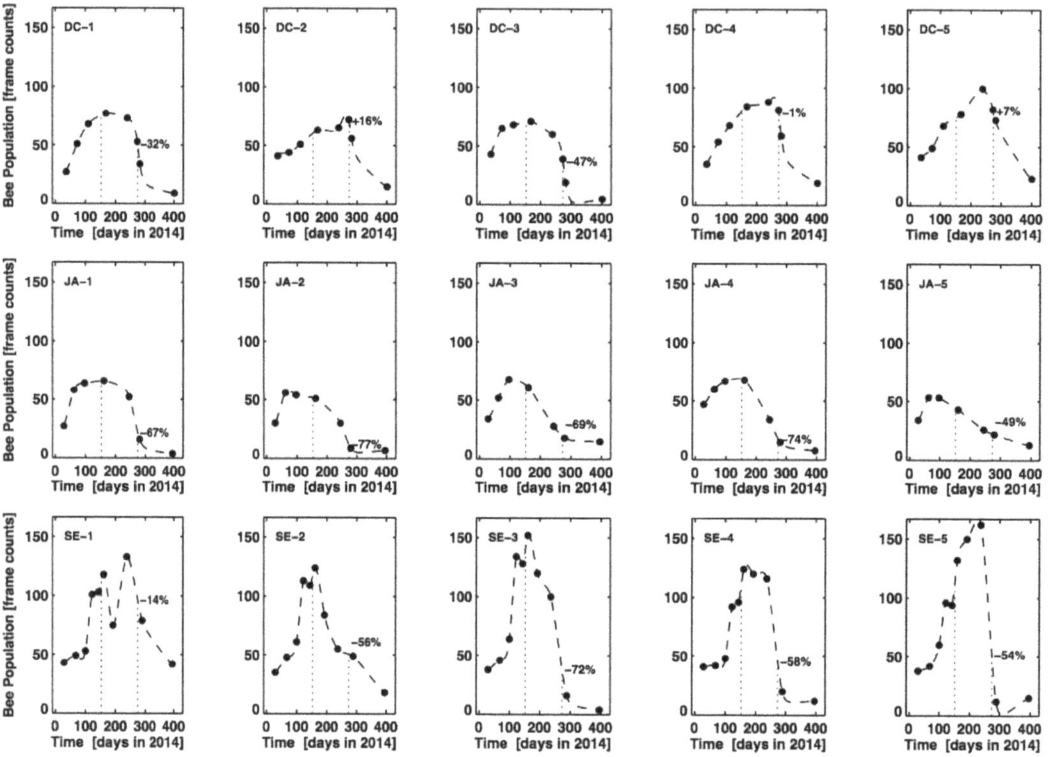

**Figure 1.** Bee population (in counts of healthy frames) vs. time during 2014 for all 15 hives in the study. The points represent the measured bee population in units of frame counts on each pallet. The dashed line shows a spline interpolation. Two vertical dotted lines mark the dates 1 June and 1 October, during which the percentage change in the bee population (in frame counts) is noted at the right.

### 3.2. Pesticide Concentrations

Analytical results revealed the presence of 37 different pesticide chemicals in hive pollen over the course of the study that exhibited at least one measurement significantly above the LOD. Ten of these chemicals had either three or fewer significant detections across all samples, namely captan, cypermethrin, flubendiamide, myclobutanil, pyrimethanil, pyriproxyfen, spirodiclofen, tebufenozide, tetrahydrophthalimde, trifloxystrobin, and trifluralin. Thirteen additional pesticides or degradation products were detected in trace quantities at or near the LOD, including aldrin, bifenthrin, chlorferon, DDE, endosulfan II (beta), fenpropathrin, fenpyroximate, metalaxyl, norflurazon, phorate, quinoxyfen, sethoxydim, and tebuconazole. The data were insufficient to conduct a statistical analysis for these 23 chemicals and they are not included in the analyses described below in Figures 2–4. A comprehensive table of results by pallet is given in Appendix C, Table A4.

Of the 37 pesticides detected, 33 are still in current use, one (flubendiamide) was canceled by US EPA in 2016 for excessive risk to aquatic ecosystems, and the remaining three chemicals (coumaphos-oxon, THPI, and 2,4-DMPF) are degradation products of registered pesticides (see Supplementary Material, Table S1).

At the time of the first measurement in February 2014, before the hives went to almond orchards, the pollen analyzed was largely free of pesticide residues (see Table 1). The miticide thymol and the amitraz metabolite 2,4-DMPF were most frequently detected, along with the herbicides pendimethalin and oxyfluorfen at modest levels.

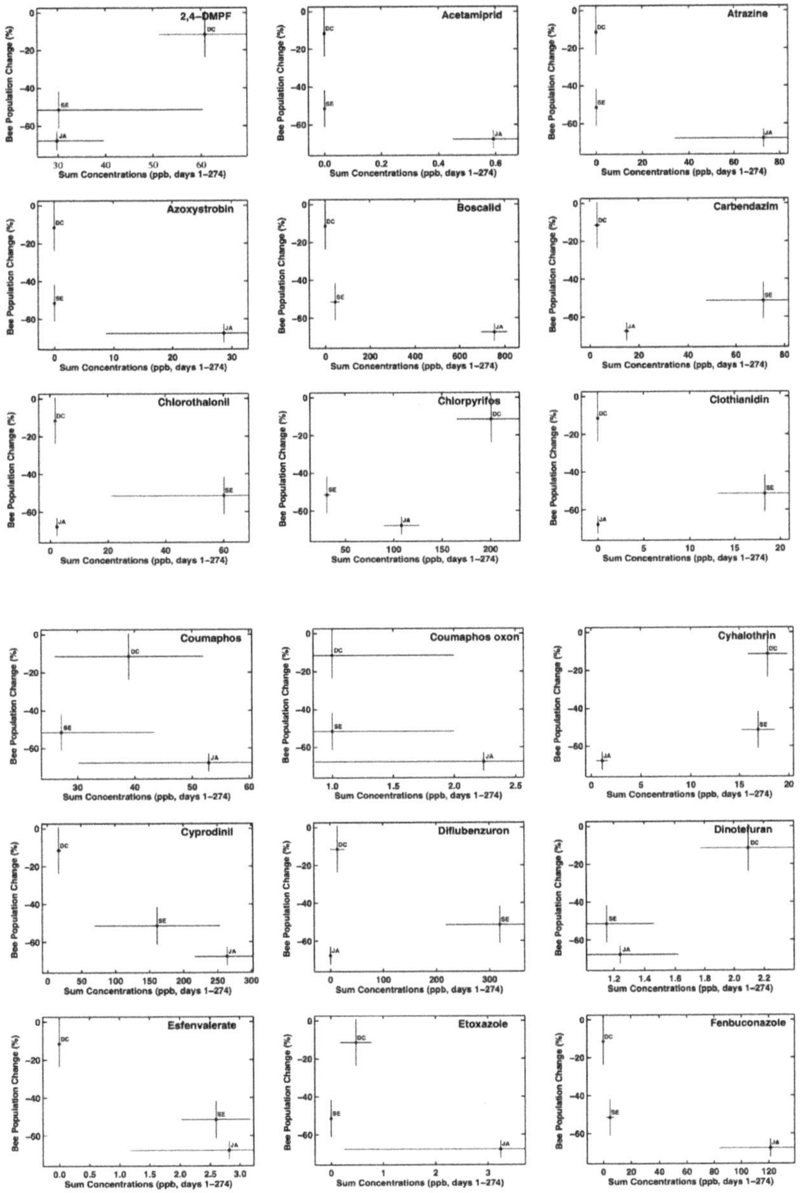


**Figure 2.** *Cont.*

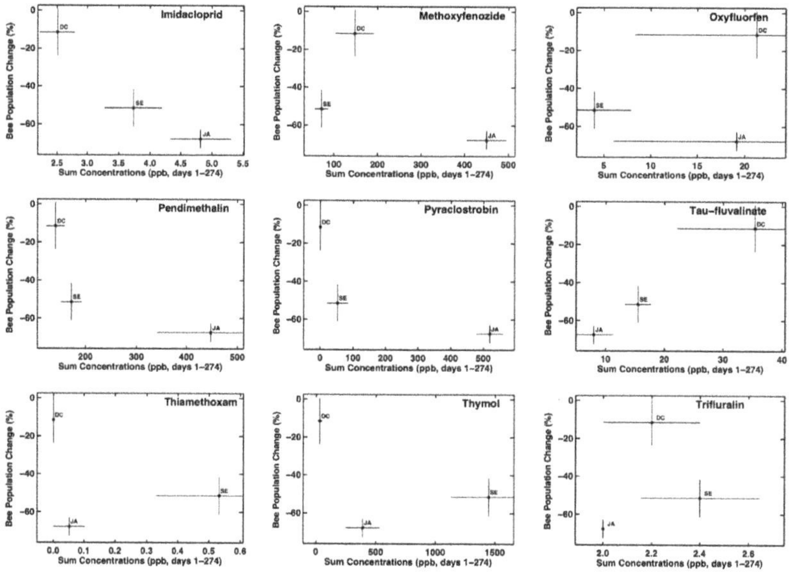


**Figure 2.** Bee population change from 1 June to 1 October vs. the sum of chemical concentration measurements. Each panel represents a specific chemical. Each point represents one beekeeper and the average concentration of the chemical over five pallets containing four hives. The error bars show the uncertainty in the mean.

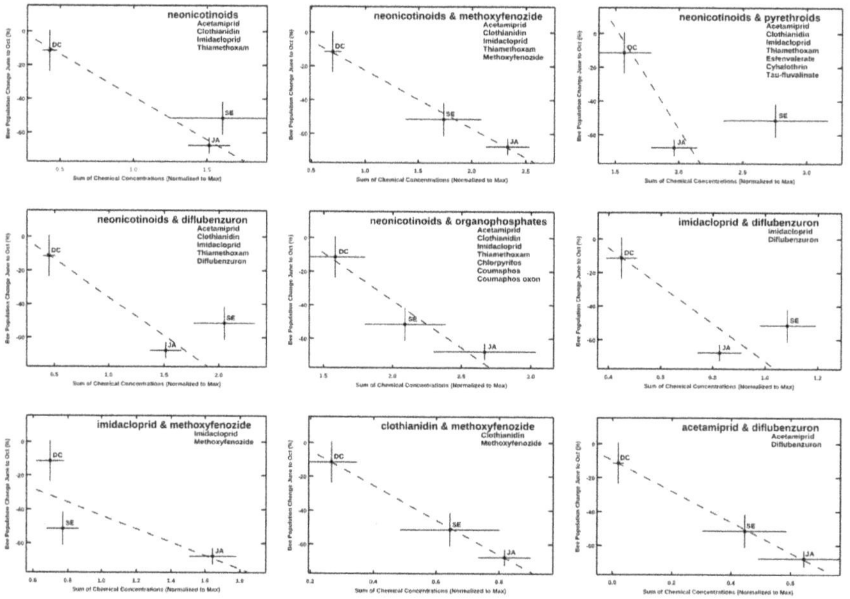


**Figure 3.** Neonicotinoid groups. The plots show bee population change vs. concentrations of combinations of neonicotinoids and other chemicals. Each plot shows the percent change in the frame count from 1 June to 1 October vs. the sum of chemical concentration measurements (max-normalized, see text) for the group of chemicals listed at upper right. Each of the three points represents one beekeeper. The dashed line is the best straight-line fit to the data including errors in both quantities.

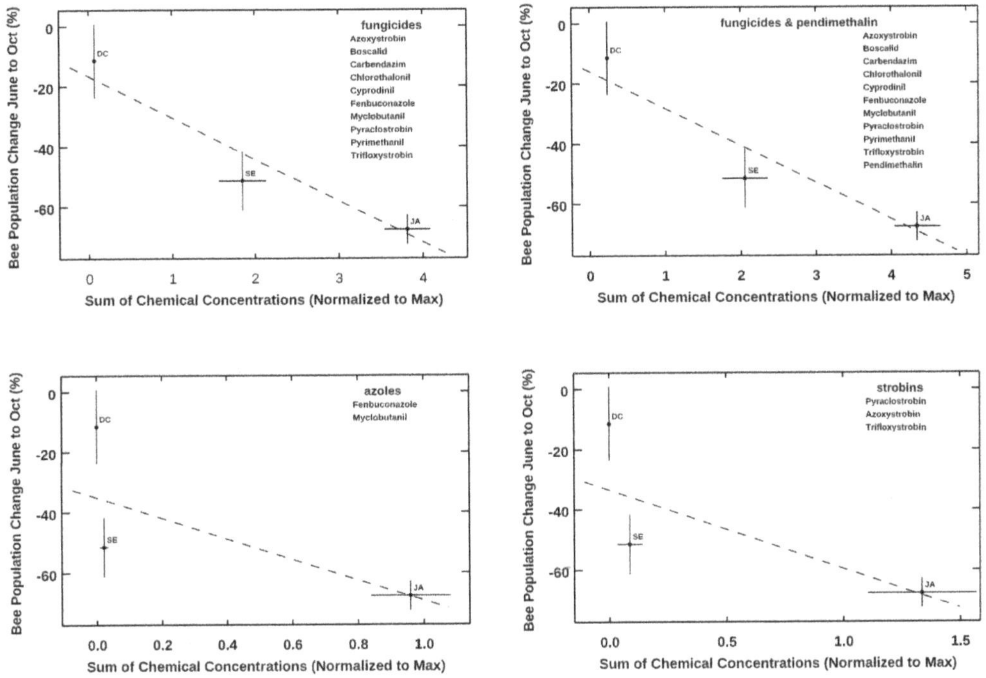


**Figure 4.** Fungicide groups. The four plots show bee population change vs. concentrations of fungicide groups, as with Figure 3 but with fungicides. The sum of concentrations of all four groups of fungicides correlates with the fractional change in bee population from June to October. Thus the fractional decline in honey bee population is negatively correlated with increasing fungicide concentration.

The measurements made after the colonies were removed from the almond orchards revealed 17 different chemicals in the pollen samples from the 15 pallets, including insecticides, fungicides, and insect growth regulators.

Although pallets from each beekeeper were placed in different almond orchards in February (with the exception that SE-1 and SE-4 were placed with the JA pallets), the mix of pesticides found in the post-almond pollination samples from each beekeeping operation was similar, dominated by pesticides applied during almond bloom. Most of the pesticides detected were permitted by law to be applied to almonds during bloom; others are likely a result of bees foraging in nearby fields. Once the beekeepers left California for their different summer locations, the types and amounts of pesticides detected varied substantially across operations. Some pesticides were found only in the hives of a single beekeeper.

Data collection for one of the beekeepers (JA) allowed the assessment of pesticide exposure over several months. The JA hives were moved to cherries immediately after pollinating almonds, and samples were taken both after almonds (in March) and after cherries (in April). An additional set of samples was taken in June, when all JA hives appeared to be in crisis (dead bees in front of the hives and queen failures). These samples contained several of the same pesticide residues detected both after almond pollination and after cherry pollination (methoxyfenozide, boscalid, pyraclostrobin, pendimethalin, chlorpyrifos, and thymol), indicating continuous exposure to these chemicals over the course of several months. The highest absolute concentrations over time were measured for boscalid, pyraclostrobin, methoxyfenozide, and pendimethalin.

**Table 1.** Pesticide detection before and after almond pollination.

| | | Before Almond Pollination | | | After Almond Pollination | | |
|---|---|---|---|---|---|---|---|
| | LOD (ppb) | Avg Conc. (ppb) | Conc. Range (ppb) | Detection Frequency (%) [a] | Avg Conc. (ppb) | Conc. Range (ppb) | Detection Frequency (%) [a] |
| **Fungicides** | | | | | | | |
| Boscalid | 4 | ND | ND | 0% | 283 | 19–821 | 67% |
| Carbendazim | 5 | ND | ND | 0% | 33 | 5–124 | 87% |
| Chlorothalonil | 1 | 1 | 1 | 73% | 22 | 1–195 | 93% |
| Cyprodinil | 1 | ND | ND | 0% | 129 | 10–476 | 100% |
| Etoxazole | 1 | 1 | 1 | 13% | 8 | 1–15 | 13% |
| Fenbuconazole | 10 | ND | ND | 0% | 76 | 10–214 | 47% |
| Pyraclostrobin | 2 | ND | ND | 0% | 243 | 15–554 | 67% |
| Pyrimethanil | 3 | ND | ND | 0% | 4 | 3–5 | 20% |
| Tetrahydrophthalimide [b] | 50 | ND | ND | 0% | 50 | 50 | 7% |
| **Herbicides** | | | | | | | |
| Oxyfluorfen | 1 | 19 | 4–37 | 20% | 42 | 20–65 | 13% |
| Pendimethalin | 6 | 50 | 6–81 | 27% | 153 | 85–233 | 100% |
| Trifluralin | 1 | 1 | 1 | 7% | 1 | 1 | 100% |
| **Insecticides** | | | | | | | |
| Acetamiprid | 0.05 | ND | ND | | 1 | ND–0.92 | 33% |
| Chlorpyrifos | 1 | 3 | 1–8 | 33% | 48 | 12–123 | 100% |
| Cyhalothrin, lambda | 1 | 1 | 1 | 13% | 4 | 1–6 | 53% |
| Cypermethrin | 4 | ND | ND | 0% | 4 | 4 | 7% |
| Dinotefuran | 1.00 | ND | ND | 0% | 1 | 1.00–1.97 | 60% |
| Esfenvalerate | 2 | 7 | 7 | 7% | 2 | 2 | 20% |
| Imidacloprid | 1.00 | ND | ND | 0% | 4 | 1.67–5.83 | 100% |
| Spirodiclofen | 1 | ND | ND | 0% | 5 | 5 | 13% |
| **Insect Growth Regulators** | | | | | | | |
| Diflubenzuron | 10 | ND | ND | 0% | 277 | 69–717 | 40% |
| Flubendiamide | 25 | ND | ND | 0% | 558 | 182–863 | 20% |
| Methoxyfenozide | 2 | ND | ND | 0% | 133 | 41–350 | 100% |
| Pyriproxyfen | 1 | ND | ND | 0% | 6 | 6–7 | 13% |
| Tebufenozide | 5 | ND | ND | 0% | 30 | 30 | 7% |
| **Miticides and their Degradates** | | | | | | | |
| 2,4-DMPF [c] | 4 | 41 | 20–65 | 53% | 24 | 15–34 | 20% |
| Coumaphos | 1 | 5 | 3–14 | 40% | 18 | 3–92 | 93% |
| Coumaphos oxon | 5 | 5 | 5 | 7% | ND | ND | 0% |
| Tau-fluvalinate | 1 | 16 | 7–36 | 27% | 12 | 4–43 | 87% |
| Thymol | 50 | 175 | 50–709 | 67% | 461 | 50–2310 | 80% |

ND = Not detected. [a] Detection frequency reported as a percentage of 15 pallets tested. [b] Chlorothalonil degradation product. [c] Amitraz degradation product.

Analysis of the individual hive samples allowed for an assessment of the variability of pesticide measurements in pollen across hives in the same location and in different locations. The data indicate that while concentrations may vary by a factor of up to 11 across individual hives, pallet concentrations only vary by up to a factor of 3.6 (see Appendix C).

### 3.3. Reproductive Failures

Observations of reproductive failures were noted throughout the study, including queen supersedure, poor brood pattern or no brood, drone-laying queens, or laying worker bees. All of these observations are indicative of colony exposure to stressors that impair the queen's ability to reproduce or the larval ability to grow and mature into adult bees. Over the course of the study, 42 colonies exhibited symptoms of reproductive failures at some point during the study, and 18 did not. Of the colonies that did not experience reproductive failures, 82% survived to the end of the study. In contrast, only 64% of the 42 colonies experiencing reproductive failures survived to the end of the study.

### 3.4. Correlation of Fractional Change in Bee Population with Individual Chemicals

Figure 2 contains 27 plots, each corresponding to an individual chemical, showing the percentage change in the bee population from 1 June to 1 October vs. the measured metric of chemical exposure, as specified above. Each plot contains three points, each point representing a beekeeper. The error bars are the uncertainty in the mean of the quantities from the five pallets for each beekeeper.

Figure 2 indicates that several individual chemicals exhibit a correlation between the specific chemical concentration in the hives and the fractional change in bee population. The plots for imidacloprid and cyprodinil show the strongest negative correlations between concentration and fractional declines in bee populations. Tau-fluvalinate and dinotefuran exhibit a positive correlation, with a lesser decline in bee population with a greater exposure to the chemical. Robust $p$-values cannot be computed from three data points, but the fit to a linear least-squares trendline is tight.

### 3.5. Correlation of Fractional Change in Bee Population with Groups of Chemicals

In Figures 3 and 4 we plot the average percent change in population from 1 June to 1 October vs. the average sum of max-normalized concentrations of chemicals for the five pallets of each of three beekeepers, with the included substances listed at the upper right of each panel.

Figure 3 shows combination plots for different neonicotinoids, as well as neonicotinoids combined with other chemicals to search for potential synergies. All plots exhibit a downward trend, indicating the fractional decline in the bee population was greater with increasing exposure to neonicotinoids or neonicotinoids plus other chemicals.

The dashed lines in Figure 3 show the best-fit straight line that includes the uncertainties in both dimensions, i.e., in the sum of the max-normalized concentrations of the neonicotinoids and in the percent bee population change from the five pallets of each beekeeper. The uncertainty in the mean of the five pallets captures the fluctuations for a given beekeeper, caused by colonies bringing in or consuming different amounts of pesticides and/or sampling inhomogeneity.

The observed downward correlation between the change in bee population and the increasing amount of neonicotinoids appears significant, as the population decline exceeds the size of the error bars and the straight line fit goes through the error bars. A robust $p$-value is not possible due to the paucity of beekeepers, i.e., only three.

Figure 3 also contains a similar diagnostic plot for eight other groups of chemicals that include neonicotinoids, five of which exhibit a downward trend. The five groups are neonicotinoids plus methoxyfenozide, neonicotinoids plus diflubenzuron, imidacloprid plus methoxyfenozide, clothianidin plus methoxyfenozide, and acetamiprid plus diflubenzuron. The persistent downward trends suggests that neonicotinoids in combination with the other chemicals are associated with greater end-of-season bee population decline (but this is not necessarily causal; see Discussion). We note that dinotefuran is not included with the set of neonicotinoids in this analysis because of its previously noted positive correlation with population change that may be due to its deleterious effects on *Varroa* at concentrations that appear to have minimal effects on honeybees.

Figure 4 shows plots for chemical groups containing fungicides and one herbicide. The first panel of Figure 4 at upper left shows the percent bee population change vs. the sum of max-normalized concentrations of ten fungicides, with the other three plots providing data on groups of strobin and azole fungicides, and fungicides plus the herbicide pendimethalin. The persistently greater fractional declines in bee population with increased fungicide concentrations in all four combinations supports the authenticity of the correlation. Again, the presence of only three points and their error bars are insufficient to perform a robust statistical test of the significance of the correlation.

## 4. Discussion

The goals of the study were to assess the dependence of colony health on the occurrence of specific pests, pathogens, and pesticides or combinations of pesticides found in the 60 colonies tracked over one year. We were not able to determine dependence of colony health on *Varroa* populations because of the substantial differences in methods of mite management used by the different beekeepers. The results of the dependence of colony health on a range of pathogens have already been published [15], with no statistically significant correlations found between colony health and any specific pathogen. Below, we discuss the implications of the pesticide results obtained in this study. It is worth noting that because most of the commercial bee population in the U.S. is transported to California for almond pollination, and because similar pesticides are used across different almond orchards, our results apply more broadly to the population of honey bees used to pollinate almonds.

### 4.1. Exposures over Time

As pesticide concentrations within a hive change substantially over time, frequent measurements are required to determine exposures as hive resources are consumed or stored. There can be a time lag between when chemicals enter the hive and when they are consumed and/or exhibit observable effects. For example, if exposure affects the queen and her egg-laying ability, the 21-day brood maturation period must pass before colony population changes significantly. During times when uncontaminated pollen is abundant (such as for the SE colonies in Angels Camp), previously stored pesticide-laden pollen may not be consumed immediately, further delaying any observable effects. These factors complicate determination of a direct cause and effect from pesticide-contaminated pollen.

The duration of exposure to a chemical also affects toxicity. A short-term high-dose exposure may be tolerable by an organism, while a longer-term exposure of a lower dose may be toxic [60,61]. The JA colonies that pollinated almonds and cherries contained pollen with substantial levels (tens to hundreds of ppb) of methoxyfenozide, boscalid, pyraclostrobin, and pendimethalin over several months. Assuming the bees consumed the available pollen in the hive, the "dose" of these chemicals received by the colony lasted not a single day or week but instead lasted at least several months between March 3 and 1 June. For worker bees in the spring, with an average life span of 6–8 weeks, the presence of these chemicals in the hive for three months represents a lifetime of exposure for several generations of honey bees and a substantial fraction of a queen bee life span. The fact that these four pesticides (and others as well) were present in the colonies simultaneously over the same time period also provided ample opportunity for any synergistic toxic effects to occur. Our simple sum of concentrations during the year has the advantage of capturing some measure, however sub-optimal, of the two temporal and dosage regimes of exposure: brief, high-dose exposures causing acute effects and chronic exposures to sublethal doses associated with a variety of adverse effects which impair colony health and reproduction.

### 4.2. Variability in Pesticide Measurements

There are several sources of variability in the pesticide measurements from the different study hives, including the inhomogeneity from sampling only a single core sample from each hive, the inhomogeneity of the pollens inside the cells selected for analysis, the different foraging patterns of the colonies, the date the contaminated pollen was brought into the hive, the date on which the pollen was consumed, and the different consumption rates of the pollen within the colony. None of these variables can be controlled, hence our decision to average the observed pesticide residue concentrations over the four hives on a pallet, reducing variability from a factor of 11 across the 20 hives placed in the same location to a factor of 3.6 across the five pallets (see Appendix C).

### 4.3. Correlation of Fractional Change in Bee Population with Pesticide Exposure

We constructed simple metrics of both the fractional change in the bee population from 1 June to 1 October and the year-long exposure to various pesticide active ingredients, allowing us to perform transparent searches for any correlation between colony health and exposure to pesticides. The diagnostic plots of the fractional change in the bee population versus exposure to individual chemicals (Figure 2) suggested negative correlations of colony health with exposure to cyprodinil and imidacloprid. Only two chemicals showed a possible positive correlation, tau-fluvalinate and dinotefuran. Tau-fluvalinate is used as a miticide, which likely explains this result, as it may have limited the population of *Varroa*. The same may be true of the low levels of dinotefuran found (1–2 ppb). None of these results can be considered statistically significant due to the limited number of beekeepers involved in the study (only three). Future longitudinal studies including more beekeepers at different locations with four to eight hives per beekeeper would enhance the statistical power of these types of studies.

The consideration of groups of chemicals that share common molecular structure, mode of action, or known biological synergies allowed for a more statistically powerful search for correlations, as shown in Figures 3 and 4. The inclusion of many chemicals within a group helped both illuminate any additive or synergistic effects of multiple sub-lethal doses and provided a sum of chemical concentrations having less fractional uncertainty than did any one chemical alone. The results for groups of neonicotinoids, fungicides, and insect growth regulators (Figures 3 and 4) revealed certain groups of chemicals that persistently correlated with a greater fractional decline in the bee population between the peak population in early June and early October. Specifically, the negative correlations observed for combinations of neonicotinoids and methoxyfenozide and neonicotinoids and diflubenzuron are consistent with laboratory experiments of synergies between these chemicals, e.g., [37,62]. The calculation of a statistically robust $p$-value was not possible due to the limited number of beekeepers, three, in the study.

## 5. Conclusions

### 5.1. Pesticide Exposures Negatively Affect Honey Bee Colony Health

This study provided extensive data on pesticide concentrations in commercial honey bee colonies over the course of a year, as beekeepers conducted their normal pollination and honey-making activities. These data provide evidence of pesticide effects on honey bees at field-realistic doses and with typical commercial management practices. Our findings indicate that the fractional decline in honey bee populations from peak population in early June to a low in early October was systematically greater with increasing amounts of certain groups of chemicals examined. Groups with visible correlations include neonicotinoids as a group, neonicotinoids with methoxyfenozide, neonicotinoids with organophosphates, and fungicides as a group. While we cannot calculate a precise $p$ value for these effects, the predominant negative change in fractional decline of the bee population with many of the chemical groups indicates the correlation is not governed by random chance. If there were no correlation, one would expect as many positive as negative correlations. Such was not the case. The fractional change in honey bee populations between 1 June and 1 October was systematically greater with increasing amounts of the groups of chemicals examined. The indications of correlations observed with fungicides and insect growth regulators merit further scrutiny of the sublethal effects of these pesticides.

### 5.2. Improving Methods for Assessment of Mixtures of Pesticides

Our methods for combining the concentrations of multiple chemicals brought value to this study in multiple ways. Assessing combinations provides a preliminary estimate of potential additive or synergistic effects of different pesticides. Bees may survive a sub-lethal dose of one chemical, but they may not survive the additive effects of sub-lethal doses of several chemicals. In addition, pairs of chemicals may act synergistically to impact the health of bees more than either chemical alone. Such synergies often involve disabling of the cytochrome P450 detoxification mechanisms and have been found in other studies [46,47,49,51]. Indeed, many of these synergistic combinations of pesticides have been patented specifically for these synergies [51]. Combining the concentrations of several chemicals reduces the fractional noise of the sum of measured concentrations compared to the fractional noise of an individual chemical, thereby improving the detectability and statistical integrity of any correlations.

This study faced several challenges. One was constructing a useful metric or "formula" to compute a sum of concentrations of multiple chemicals. It was impossible to know what weight (i.e., coefficient) to assign to the concentrations of each chemical within a group when "adding" them together. Because only acute toxicity ($LD_{50}$) data are available for most chemicals, we also did not know the functional dependence of each concentration on bee health exclusive of morbidity, with nonlinearities being likely. We also did not know the range of sublethal effects of each chemical within a group, nor did we have comprehensive knowledge about synergies between different chemicals.

Still, by simply adding together the max-normalized concentrations of different chemicals, any correlation of this metric of chemical exposure with the fractional change in the bee population provides a clue about causal dependencies or stressors associated with those chemicals. While other factors besides chemical exposure could be the reason for population declines, the observed correlations are consistent with extensive laboratory work demonstrating that bees exposed to certain individual chemicals experience adverse effects on their health [5,6,17,19,47,49]. The correlations found in this longitudinal field study suggest that the deleterious health effects of these chemicals found in the lab do indeed occur in the field.

### 5.3. Protection of the Honey Bee Population

In order to protect honey bees from adverse effects due to pesticide exposure, a reduction in pesticide use on bee-attractive crops during pollination season would need to occur. In 2018, the Almond Board of California published a "Best Practices" guide for almond growers that describes potential problems related to synergistic effects from applying tank mixtures of fungicides and/or insect growth regulators (IGRs) with other pesticides and recommends eliminating problematic tank-mix applications and reducing fungicide and IGR applications during bloom [63]. However, an assessment of the California pesticide use data for San Joaquin county as a representative almond-growing region shows that progress in protecting honey bees from pesticide exposure is mixed.

Between 2014 when this study was conducted and 2022 (the latest year for which pesticide use data are available), overall fungicide use per bearing acre (BA) of almonds increased from 3.05 lbs/BA in 2014 to 4.16 lbs/BA in 2022 (see Supplementary Materials, Table S2).

For example, for the azole fungicides, use per bearing acre increased from 0.16 lbs/BA in 2014 to 0.20 lbs/BA in 2022. Applications during February (the month when almond pollination occurs) decreased slightly, from 2122 lbs over 12,411 acres to 1679 lbs over 11,320 acres, a decrease in application rate from 0.17 lbs/acre treated to 0.15 lbs/acre treated.

Similarly, application of strobin fungicides almost tripled, from 0.10 lbs/BA in 2014 to 0.27 lbs/BA in 2022, with the application rate during February remaining constant at 0.10 lbs/treated acre, but with the absolute amount of strobin fungicides used in February more than tripling, from 502 lbs over 5280 acres to 1604 lbs over 16,050 acres.

Use of the insect growth regulator methoxyfenozide increased from 6445 lbs total in 2014 to 21,599 lbs in 2022; however, use during the month of February decreased substantially, with the absolute number of acres treated in February decreasing dramatically from 5092 acres to 100 acres. Application rates were lower as well, from 0.18 lbs/treated acre in 2014 to 0.14 lbs/treated acre in 2022.

These mixed results indicate that grower education and outreach have the potential to mitigate pollinator exposures to problematic pesticides, but more needs to be done to raise awareness of the potential hazards of certain pest management practices to growers and beekeepers. More effective would be for US EPA to require pesticide label changes prohibiting applications of problematic pesticides during bloom and to limit the practice of tank-mixing and co-application of pesticide chemicals. Further laboratory work to more clearly characterize problematic pesticide combinations is urgently needed.

**Supplementary Materials:** The following supporting information can be downloaded at: https://www.mdpi.com/article/10.3390/agronomy14112505/s1, Table S1: USDA Lab Analytes for the Study, including LODs, Current US EPA Registration Status, Use Type, and Chemical Classification; Table S2: Comparison of Pesticide Use on Almonds in San Joaquin County between 2014 and 2022[a].

**Author Contributions:** Conceptualization, S.E.K. and G.W.M.; Data curation, S.E.K., R.R., T.J.B., J.A., D.C. and S.E.; Formal analysis, S.E.K., R.R., T.J.B. and G.W.M.; Funding acquisition, S.E.K. and S.E.; Investigation, S.E.K., R.R., T.J.B., J.A., D.C. and S.E.; Methodology, S.E.K., T.J.B., J.A., D.C. and S.E.; Project administration, S.E.K.; Resources, S.E.K., J.A., D.C., S.E. and G.W.M.; Software, S.E.K., R.R. and G.W.M.; Supervision, S.E.K., J.A., S.E. and D.C.; Validation, R.R., J.A., D.C. and S.E.; Writing—original draft, S.E.K. and G.W.M.; Writing—review and editing, S.E.K. and G.W.M. All authors have read and agreed to the published version of the manuscript.

**Funding:** We thank the National Honey Board and the Pollinator Stewardship Council and its donors for providing partial funding for this study. The funders played no role in the study design, data collection, analysis, and interpretations.

**Data Availability Statement:** The original contributions presented in the study are included in the article/Supplementary Material, further inquiries can be directed to the corresponding author/s.

**Acknowledgments:** We thank James Cook, Samantha Jones, Alyssa Anderson, and Moroni Roundy for their work in the field, Chensheng (Alex) Lu for neonicotinoid analysis of pollen samples, Michelle Flenniken for helpful comments on the manuscript, and Brian Hill for preliminary assistance with statistical analysis.

**Conflicts of Interest:** Author Jeff Anderson was employed by the company California-Minnesota Honey Farm. Author Darren Cox was employed by the company Cox Honey of Utah. Author Steve Ellis was employed by the company Old Mill Honey Farm. Author Geoffrey Marcy was employed by the company Space Laser Awareness. All authors declare that the research was conducted in the absence of any commercial or financial relationships that could be construed as a potential conflict of interest.

## Appendix A. Colony Selection and Measurements

### *Appendix A.1. Colony Selection*

Colonies selected for the study contained on average nine frames of bees with a standard deviation of 2.6 frames, based on the conventional frame-count method used in contracts for commercial pollination to assess hive strength (see below). Colony strength was assessed by counting the number of frames in each colony approximately 7–10 times (depending on beekeeper) during the period from January 2014 to January 2015, as described in the following section. Pollen, wax, and honey were sampled for pesticide analysis four times during the year. Pesticide analyses were focused primarily on pollen samples, with representative honey and wax samples analyzed at each sampling date.

The initial assessment of colony strength and chemical concentrations was made at the end of January 2014, prior to the placement of the hives in almond orchards for almond pollination. These measurements established the initial hive conditions, including the concentration of pesticide residues, the presence of pathogens and *Varroa* mites, and colony strength. The second sampling was conducted in late February and early March, immediately after the hives were removed from almond orchards. Two additional samplings of hive matrices to determine pesticide concentrations in the hives were conducted for each beekeeping operation. For the JA operation, the third sampling occurred in early April after cherry pollination and the fourth in late August. After almond pollination, the SE colonies were moved to the California foothills, and the DC colonies were move directly to

Utah. For both DC and SE operations, the third and fourth sampling dates were in early June and late August.

Throughout the year, beekeeping practices that may have affected colony health were monitored, including colony location, colony manipulations performed by beekeeper (i.e., splitting colonies, requeening), colony inputs (i.e., feeding sugar and protein patties, application of miticides or antibiotics), and queen performance and/or events (i.e., death, supersedure, fecundity). In addition, blooming crops near each apiary were noted. Lastly, observations such as unusual bee behavior, symptoms of disease, failure to thrive as expected for the season, the presence of dead bees at the hive entrance, and colony death were also recorded.

*Appendix A.2. Colony Characteristics and Maintenance*

Beekeepers DC, JA, and SE followed generally standard practices for migratory commercial beekeeping operations, including feeding sugar syrup and pollen patties as needed and administering treatments for *Varroa* mites and foulbrood throughout the course of the year. The DC and JA colonies were requeened when the absence of a queen or a failing queen was noted; the SE colonies were allowed to requeen themselves. Separate apiary locations were used by each beekeeper, with each pollinating a different almond orchard in Central California (with one exception—see below) and each then traveling to a unique summer location in Minnesota (JA and SE) or Utah (DC). Specifics for each beekeeping operation are described below.

*Appendix A.3. DC Operation*

The DC operation is based in northern Utah, in an agricultural area where alfalfa, small grains, safflower, mustard, fruit trees, pumpkins and melons are grown. Bees also forage on native plants, as well as dandelions, thistles, rabbitbrush, gumweed, wild sunflowers and other weeds.

The DC study hives were moved as follows:

- October 2013 to Oakdale and Hughson in California to overwinter.
- Early February 2014 to an almond orchard off of Victory Road near Escalon, CA for pollination.
- Mid-March to northern Utah after almond pollination was completed.
- 18 November 2014 to Oakdale and Hughson in California for overwintering.

Italian race bees were used, with queens purchased from Jackie Park-Burris Queens, Inc. and installed between April and July 2013. Only a single colony was requeened during the study, on 3/24/14. A total of 19 of the 20 colonies were split on 4/20/14, with four frames removed and the original queen left with the colony.

Bees were fed sucrose syrup, corn syrup, and pollen patties as needed. Mite treatments (Taktic, amitraz) were administered on 5/5/2014, 8/29/14, and 9/29/14. Treatment with terramycin as a prophylactic for foulbrood was administered on 9/29/14.

*Appendix A.4. JA Operation*

The JA operation is based in North Central Minnesota in the summer and fall, in an agricultural area where corn, soybeans, alfalfa, oats, barley, sunflowers and additional minor crops are grown. Bees also forage on native plants, tree pollens, dandelions, Dutch white clover, Canadian thistle, yellow and white sweet clover, goldenrod, birdsfoot trefoil and other weeds. The JA study hives were moved as follows:

- November 2013 to Oakdale and Copperopolis in California to overwinter.
- Early February 2014 to an almond orchard off McHenry Road near Escalon, CA for pollination.
- Early March 2014 to a cherry orchard off Copperopolis Road near Stockton for cherry pollination.
- Early April 2014 to Milton, CA.

- Early June 2014 to Eagle Bend in North Central Minnesota.
- 24 November 2014 to California near Copperopolis and Milton for overwintering.

Russian and Carniolan race bees were used, with queens either grafted from larvae taken from other hives in the JA apiaries and Lowman Apiaries, installed in late spring 2013. On 6/10/2014, all 15 colonies remaining in the JA group were found in crisis, with evidence of supersedure cells and dead bees in most colonies. At this point, all 15 colonies were requeened with queens donated by Lohman Apiaries. Only one of the twenty JA colonies was split on 5/26/14, with nine frames removed and the original queen left with the colony.

Bees were fed sucrose syrup and pollen patties as needed. Mite treatments were administered on 4/21/14 (Taktic, amitraz), 9/4/14 (Taktic, amitraz), and 10/20/14 (oxalic acid drench). Treatment with tylosin as a prophylactic for foulbrood was administered on 6/1/14.

*Appendix A.5. SE Operation*

The SE operation is based in West Central Minnesota in the summer and fall, in an agricultural area where corn, soybeans, alfalfa, sugar beets, edible beans, sunflowers and additional minor crops are grown. Bees also forage on native plants, tree pollens, dandelions, Dutch white clover, Canadian thistle, yellow and white sweet clover, goldenrod, birdsfoot trefoil and other weeds. The SE study hives were moved as follows:

- November 2013 to Angel's Camp in California to overwinter.
- Early February 2014 to an almond orchard off McHenry Road near Escalon, CA (SE-1 and SE-4) or to an almond orchard near Madera (SE-2, SE-3, and SE-5) for pollination.
- Late March 2014 to a non-agricultural area in Vallecito, CA in the Sierra foothills, where manzanita and other spring wildflowers were blooming.
- 19 April 2014 to an apiary in an agricultural area between Elbow Lake and Barrett, in West Central Minnesota.
- 23 November 2014 to California and overwintered in Angel's Camp, CA.

Russian and Carniolan race bees and survivor stock were used, with queens from Strachan Apiaries and Lowman Apiaries, installed in late spring 2013. Fifteen of the nineteen remaining colonies were split on 4/10/14, with nine frames removed and the original queen left with the colony. A second split of 9 frames on the same 15 colonies was performed on 5/23/14. On 7/11/2014, all 19 colonies remaining in the SE group were found in crisis, with evidence of supersedure cells. Supersedures were allowed to proceed naturally. No outside queens were introduced.

Bees were fed sucrose syrup and pollen patties, as needed. Mite treatments were administered on 2/1/14 (amitraz), 8/21/14 (oxalic acid drench), 9/9/14 (amitraz), and 10/24/14 (formic acid, half-dose). Treatment with Tylosin as a prophylactic for foulbrood was administered on 9/9/14.

## Appendix B. Measuring the Bee Population Using "Frame Counts"

*Appendix B.1. Frame Counts as a Metric of Colony Health*

A metric of hive health was adopted based on the number of frames within a hive having a minimum number of bees [54,55]. A frame merited one "count" if 2/3 of both sides were covered by bees and/or brood. Photographs of the front and back of each frame were taken during each sample collection to provide a visual data archive, later evaluated by just one team member (RR) to reduce variation in assessments across beekeepers in determining whether the 2/3 criterion was met. This technique incurs the uncertainty of human assessment of "2/3", but it does not inject bias from hive to hive, beekeeper to beekeeper, or over time. In particular, differences in the number of frame counts over time offer a robust metric of changes in colony health. For each colony, frames were counted at each of approximately eight visits (depending on beekeeper) throughout the year, with results given in Tables A1–A3 (below).

*Appendix B.2. Handling of Colony Splits*

A total of 35 of the 60 colonies were split in the spring of 2014. For colonies that were split, the old queen was left with the original study hive, so there was no break in the brood cycle. The split was moved away from the study hive pallets and was no longer considered part of the study, except as described here. Frame counts from splits were accounted for by calculating a split factor (SF), defined as the number of frames remaining in the original hive after the split divided by the total number of frames in the original hive, e.g., for an 18-frame colony from which a 4-frame split was made, SF = 14/18 = 0.78. Corrected frame counts were calculated by dividing the measured frame counts by SF for all frame count measurements made after the split. This correction assumes that the population of the split changes at the same rate as the original colony. Many of the SE hives were split twice, necessitating the use of a second split factor, $SF_2$, defined as the number of frames remaining in the split hive after the second split divided by the total number of frames in the original hive. The corrected frame count was calculated as the measured frame count divided by the product of SF and $SF_2$.

**Table A1.** Frame counts over time for beekeeper DC.

| DC Pallet# | Day 36 | Day 73 | Day 110 | Day 168 | Day 239 | Day 273 | Day 282 | Day 401 |
|---|---|---|---|---|---|---|---|---|
| 1 | 27 | 51 | 68 | 77 | 73 | 53 | 34 | 9 |
| 2 | 41 | 44 | 51 | 63 | 65 | 72 | 56 | 15 |
| 3 | 43 | 65 | 68 | 71 | 60 | 39 | 19 | 5 |
| 4 | 35 | 54 | 68 | 84 | 88 | 81 | 59 | 19 |
| 5 | 41 | 49 | 68 | 78 | 100 | 82 | 73 | 23 |

**Table A2.** Frame counts over time for beekeeper JA.

| JA Pallet | Day 28 | Day 62 | Day 97 | Day 161 | Day 245 | Day 280 | Day 395 |
|---|---|---|---|---|---|---|---|
| 1 | 27 | 58 | 64 | 66 | 52 | 16 | 4 |
| 2 | 30 | 56 | 54 | 51 | 30 | 9 | 7 |
| 3 | 34 | 52 | 68 | 61 | 28 | 18 | 15 |
| 4 | 47 | 60 | 67 | 68 | 34 | 15 | 8 |
| 5 | 34 | 53 | 53 | 43 | 26 | 22 | 13 |

**Table A3.** Frame counts over time for beekeeper SE.

| SE Pallet# | 29 | 68 | 100 | 122 | 143 | 160 | 192 | 237 | 289 | 395 |
|---|---|---|---|---|---|---|---|---|---|---|
| 1 | 43 | 49 | 53 | 101 | 103 | 118 | 75 | 133 | 79 | 42 |
| 2 | 35 | 48 | 61 | 113 | 109 | 124 | 84 | 55 | 49 | 18 |
| 3 | 38 | 46 | 64 | 134 | 128 | 152 | 120 | 100 | 16 | 4 |
| 4 | 41 | 42 | 48 | 92 | 96 | 124 | 120 | 116 | 20 | 12 |
| 5 | 38 | 42 | 60 | 96 | 94 | 132 | 150 | 162 | 12 | 15 |

## Appendix C. Pesticide Data

*Appendix C.1. Pesticide Concentrations, Limits of Detection, and Criteria for Inclusion*

Overall, we found 37 chemicals, listed in Table A4, that exhibited at least one measurement significantly above the LOD. Ten chemicals had either three or fewer significant detections across all samples, namely, captan, cypermethrin, flubendiamide, myclobutanil, pyrimethanil, pyriproxyfen, spirodiclofen, tebufenozide, tetrahydrophthalimde, trifloxystrobin, and trifluralin. We retained these ten chemicals with their trace detections in the analysis that follows, but their integrity and statistical value remain unclear, and they are not included in the major results described below in Figures 2–4. Table A4 contains the sum of chemical exposure across the year for each of the 15 pallets for each of the 37 detected chemicals. Pesticides with measured concentrations at or near the LOD or with three or fewer detections across the 15 pallets were not included in the statistical analysis, including aldrin, bifenthrin, chlorferon, DDE, endosulfan II (beta), fenpropathrin, fenpyroximate, metalaxyl, norflurazon, phorate, quinoxyfen, sethoxydim, and tebuconazole.

**Table A4.** Sum of four measurements (ppb) of 37 chemicals for each pallet.

| Pesticide | DC-1 | DC-2 | DC-3 | DC-4 | DC-5 | JA-1 | JA-2 | JA-3 | JA-4 | JA-5 | SE-1 | SE-2 | SE-3 | SE-4 | SE-5 |
|---|---|---|---|---|---|---|---|---|---|---|---|---|---|---|---|
| 2,4-Dimethylphenyl formamide | 90.2 | 68.7 | 34.3 | 47.3 | 63.7 | 0.0 | 24.4 | 28.1 | 35.2 | 61.3 | 0.0 | 0.0 | 0.0 | 151.0 | 0.0 |
| Acetamiprid | 0 | 0 | 0 | 0 | 0 | 0.77 | 0.76 | 0.92 | 0 | 0 | 0 | 0 | 0 | 0 | 0 |
| Atrazine | 0.0 | 0.0 | 0.0 | 0.0 | 0.0 | 50.1 | 17.1 | 0.0 | 222.0 | 76.2 | 0.0 | 0.0 | 0.0 | 0.0 | 0.0 |
| Azoxystrobin | 0.0 | 0.0 | 0.0 | 0.0 | 0.0 | 10.4 | 0.0 | 107.6 | 16.5 | 8.7 | 0.0 | 0.0 | 0.0 | 0.0 | 0.0 |
| Boscalid | 0.0 | 0.0 | 0.0 | 0.0 | 0.0 | 908.4 | 844.0 | 726.7 | 572.2 | 715.7 | 21.4 | 128.0 | 22.7 | 19.1 | 28.1 |
| Captan | 10.0 | 10.0 | 10.0 | 0.0 | 10.0 | 0.0 | 0.0 | 0.0 | 0.0 | 0.0 | 0.0 | 0.0 | 0.0 | 0.0 | 0.0 |
| Carbendazim | 5.0 | 0.0 | 0.0 | 5.0 | 5.0 | 16.2 | 17.9 | 12.7 | 16.5 | 11.8 | 8.4 | 124.0 | 94.6 | 18.8 | 111.0 |
| Chlorothalonil | 1.0 | 2.0 | 3.0 | 1.0 | 2.0 | 2.0 | 2.0 | 2.5 | 2.5 | 2.5 | 197.0 | 98.5 | 2.0 | 1.0 | 2.0 |
| Chlorpyrifos | 298.0 | 118.0 | 272.9 | 158.4 | 156.9 | 132.5 | 75.2 | 166.7 | 94.3 | 70.1 | 30.8 | 25.7 | 22.0 | 43.2 | 32.4 |
| Clothianidin | 0 | 0 | 0 | 0 | 0 | 0 | 0 | 0 | 0 | 0 | 17.30 | 22.94 | 4.46 | 35.73 | 11.20 |
| Coumaphos | 6.0 | 75.8 | 62.1 | 28.3 | 22.6 | 16.8 | 91.4 | 15.9 | 122.9 | 17.3 | 14.6 | 9.4 | 6.9 | 12.5 | 92.3 |
| Coumaphos oxon | 0.0 | 5.0 | 0.0 | 0.0 | 0.0 | 0.0 | 6.2 | 0.0 | 5.0 | 0.0 | 5.0 | 0.0 | 0.0 | 0.0 | 0.0 |
| Cyhalothrin, lambda | 17.9 | 13.4 | 19.1 | 24.7 | 14.3 | 2.0 | 0.0 | 3.0 | 0.5 | 0.0 | 18.3 | 14.9 | 11.5 | 21.4 | 18.5 |
| Cypermethrin | 0.0 | 0.0 | 0.0 | 0.0 | 0.0 | 0.0 | 4.0 | 0.0 | 0.0 | 0.0 | 0.0 | 0.0 | 0.0 | 0.0 | 0.0 |
| Cyprodinil | 32.5 | 12.8 | 13.2 | 10.2 | 15.5 | 322.4 | 207.2 | 245.1 | 132.7 | 413.0 | 476.0 | 20.1 | 27.4 | 271.0 | 12.5 |
| Diflubenzuron | 0.0 | 0.0 | 0.0 | 0.0 | 68.9 | 0.0 | 0.0 | 0.0 | 0.0 | 0.0 | 163.0 | 213.0 | 294.0 | 209.0 | 717.0 |
| Dinotefuran | 2.5 | 1.8 | 1.1 | 3.0 | 2.1 | 0.0 | 1.6 | 1.1 | 1.1 | 2.3 | 1.6 | 1.8 | 0.0 | 1.3 | 1.1 |
| Esfenvalerate | 0.0 | 0.0 | 0.0 | 0.0 | 0.0 | 9.1 | 0.0 | 1.0 | 3.0 | 1.0 | 2.1 | 2.0 | 2.0 | 4.9 | 2.0 |
| Etoxazole | 0.0 | 0.0 | 1.0 | 1.4 | 0.0 | 15.2 | 1.0 | 0.0 | 0.0 | 0.0 | 0.0 | 0.0 | 0.0 | 0.0 | 0.0 |
| Fenbuconazole | 0.0 | 0.0 | 0.0 | 0.0 | 0.0 | 50.2 | 160.3 | 227.9 | 144.6 | 22.5 | 0.0 | 0.0 | 10.0 | 14.4 | 0.0 |
| Flubendiamide | 0.0 | 863.0 | 629.0 | 182.0 | 0.0 | 0.0 | 0.0 | 0.0 | 0.0 | 0.0 | 0.0 | 0.0 | 0.0 | 0.0 | 0.0 |
| Imidacloprid | 2.7 | 2.0 | 3.2 | 1.7 | 2.9 | 3.0 | 4.9 | 5.8 | 4.9 | 5.5 | 3.9 | 4.5 | 2.7 | 4.9 | 2.7 |
| Methoxyfenozide | 63.4 | 314.0 | 159.2 | 104.0 | 95.5 | 361.0 | 551.0 | 532.0 | 483.5 | 323.1 | 127.0 | 78.3 | 47.8 | 64.5 | 41.3 |
| Myclobutanil | 0.0 | 0.0 | 0.0 | 0.0 | 0.0 | 62.5 | 0.0 | 0.0 | 36.2 | 35.0 | 0.0 | 0.0 | 0.0 | 0.0 | 0.0 |
| Oxyfluorfen | 0.0 | 65.1 | 4.0 | 0.0 | 37.4 | 66.3 | 0.0 | 29.5 | 0.0 | 0.0 | 0.0 | 0.0 | 0.0 | 0.0 | 19.8 |
| Pendimethalin | 107.0 | 113.0 | 172.4 | 104.4 | 196.6 | 837.4 | 399.0 | 397.0 | 428.0 | 176.6 | 117.0 | 247.2 | 176.0 | 160.0 | 156.0 |
| Pyraclostrobin | 0.0 | 0.0 | 0.0 | 0.0 | 0.0 | 591.3 | 597.0 | 559.0 | 387.0 | 464.0 | 19.7 | 181.0 | 22.3 | 15.2 | 22.9 |
| Pyrimethanil | 0.0 | 0.0 | 0.0 | 0.0 | 0.0 | 0.0 | 0.0 | 0.0 | 0.0 | 0.0 | 0.0 | 3.0 | 5.0 | 0.0 | 3.8 |
| Pyriproxyfen | 0.0 | 0.0 | 0.0 | 0.0 | 0.0 | 0.0 | 0.0 | 0.0 | 11.0 | 0.0 | 0.0 | 7.2 | 0.0 | 0.0 | 5.5 |
| Spirodiclofen | 0.0 | 0.0 | 5.0 | 0.0 | 4.8 | 0.0 | 0.0 | 0.0 | 0.0 | 0.0 | 0.0 | 0.0 | 0.0 | 0.0 | 0.0 |
| Tau-fluvalinate | 13.1 | 15.7 | 85.5 | 36.6 | 26.0 | 2.8 | 0.0 | 9.5 | 19.8 | 7.9 | 16.1 | 8.1 | 15.4 | 15.8 | 22.1 |
| Tebufenozide | 0.0 | 0.0 | 0.0 | 0.0 | 0.0 | 0.0 | 29.5 | 0.0 | 0.0 | 0.0 | 0.0 | 0.0 | 0.0 | 0.0 | 0.0 |
| Tetrahydrophthalimide | 50.0 | 0.0 | 0.0 | 0.0 | 0.0 | 0.0 | 0.0 | 0.0 | 0.0 | 0.0 | 0.0 | 0.0 | 0.0 | 0.0 | 0.0 |
| Thiamethoxam | 0.00 | 0.00 | 0.00 | 0.00 | 0.00 | 0.00 | 0.00 | 0.00 | 0.00 | 0.26 | 0.24 | 1.18 | 0.53 | 0.71 | 0.00 |
| Thymol | 0.0 | 0.0 | 0.0 | 100.0 | 50.0 | 150.0 | 100.0 | 616.0 | 839.5 | 236.0 | 1330.0 | 651.0 | 1650.0 | 1057.0 | 2548.0 |
| Trifloxystrobin | 0.0 | 0.0 | 0.0 | 0.0 | 0.0 | 0.0 | 0.0 | 0.0 | 0.0 | 57.0 | 0.0 | 0.0 | 0.0 | 0.0 | 0.0 |
| Trifluralin | 2.0 | 2.0 | 2.0 | 3.0 | 2.0 | 2.0 | 2.0 | 2.0 | 2.0 | 2.0 | 2.0 | 2.0 | 3.0 | 3.0 | 2.0 |

*Appendix C.2. Variability in Pesticide Concentrations Across Pallets*

Although pallets from each beekeeper were placed in different almond orchards in February (with the exception that SE-1 and SE-4 were placed with the JA pallets), the mix of pesticides found in the post-almond pollination samples from each beekeeping operation was similar, dominated by pesticides applied during almond bloom. Analysis of the individual hive samples allowed for an assessment of the variability of pallet pesticide measurements in pollen across hives in the same location and in different locations. Figure A1 shows the variability in concentrations of a commonly detected pesticide, cyprodinil, over the different hives on a pallet and across the different pallets for each operation. The data indicate that while concentrations may vary by a factor of up to 11 across individual hives, pallet concentrations only vary by up to a factor of 3.6.

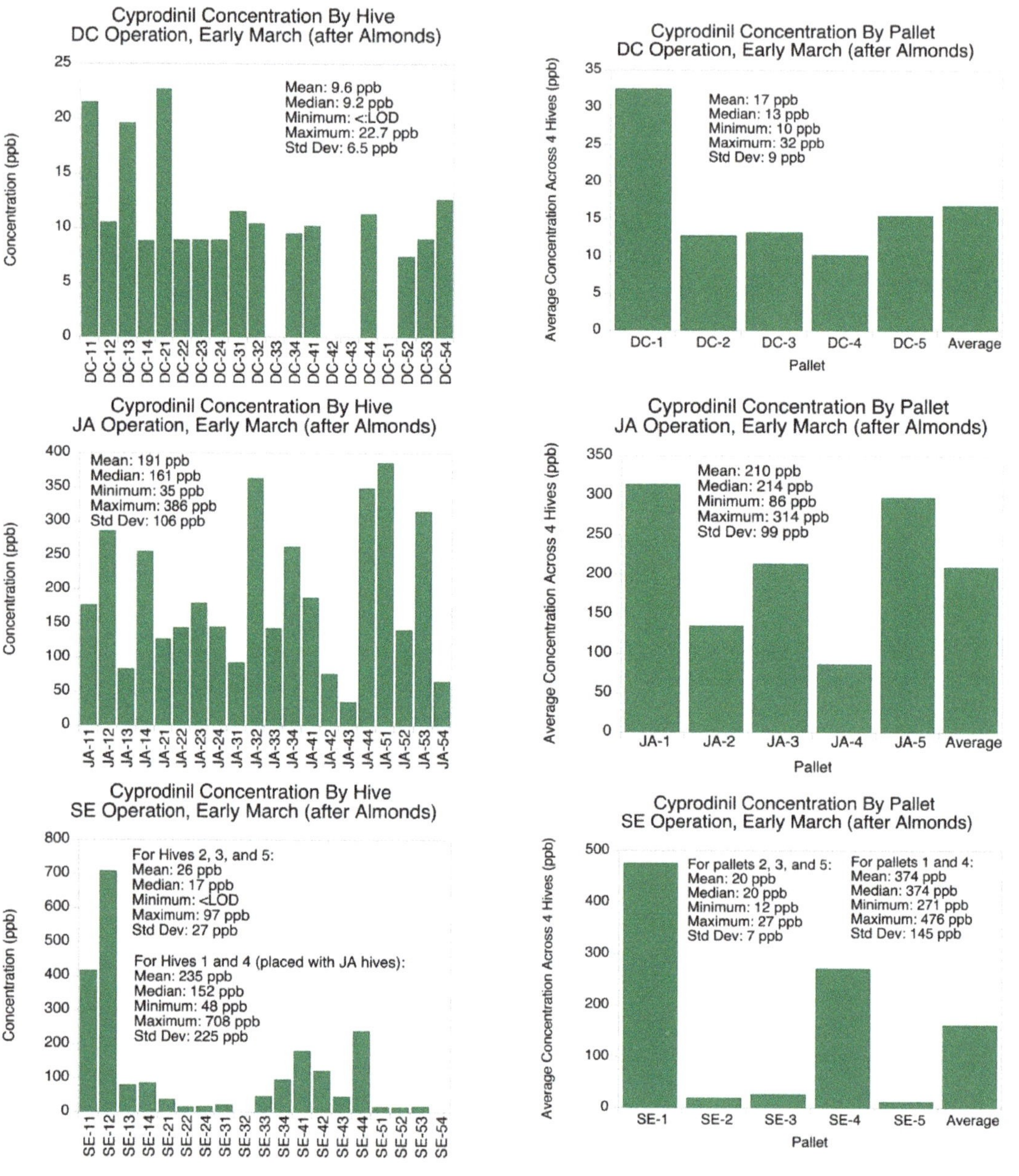


**Figure A1.** Variability in pesticide concentration across hives and pallets for different beekeeping operations. Note that pallets SE-1 and SE-4 were placed with the JA pallets instead of with the rest of the SE hives; thus, the statistics were calculated separately.

## References


1. Steinhauer, N. United States Honey Bee Colony Losses 2022–2023: Preliminary Results from the Bee Informed Partnership. Bee Informed Partnership. 22 June 2023. Available online: https://beeinformed.org/2023/06/22/united-states-honey-bee-colony-losses-2022-23-preliminary-results-from-the-bee-informed-partnership/ (accessed on 22 October 2024).
2. Rucker, R.R.; Thurman, W.N.; Burgett, M. *Economics of the Supply Functions for Pollination and Honey: Marginal Costs and Supply Elasticity*; Economic Research Report Number 246; U.S. Department of Agriculture Economic Research Service: Washington, DC, USA, 2018. Available online: https://www.ers.usda.gov/webdocs/publications/88117/err-246.pdf?v=6612.3 (accessed on 22 October 2024).
3. Van der Zee, R.; Pisa, L.; Andonov, S.; Brodschneider, R.; Charrière, J.-D.; Chlebo, R.; Coffey, M.F.; Crailsheim, K.; Dahle, B.; Gajda, A.; et al. Managed honey bee colony losses in Canada, China, Europe, Israel and Turkey, for the winters of 2008–2009 and 2009–2010. *J. Apic. Res.* **2012**, *51*, 100–114. [CrossRef]
4. Clermont, A.; Eickermann, M.; Kraus, F.; Georges, C.; Hoffmann, L.; Beyer, M. A survey on some factors potentially affecting losses of managed honey bee colonies in Luxembourg over the winters 2010/2011 and 2011/2012. *J. Apic. Res.* **2015**, *53*, 43–56. [CrossRef]
5. Traynor, K.S.; Pettis, J.S.; Tarpy, D.R.; Mullin, C.A.; Frazier, J.L.; Frazier, M.; VanEngelsdorp, D. In-hive Pesticide Exposome: Assessing risks to migratory honey bees from in-hive pesticide contamination in the Eastern United States. *Sci. Rep.* **2016**, *6*, 33207. [CrossRef]
6. Van der Sluijs, J.P.; Simon-Delso, N.; Goulson, D.; Maxim, L.; Bonmatin, J.-M.; Belzunces, L.P. Neonicotinoids, bee disorders and the sustainability of pollinator services. *Curr. Opin. Environ. Sustain.* **2013**, *5*, 293–305. [CrossRef]
7. DiPrisco, G.; Cavaliere, V.; Annoscia, D.; Varricchio, P.; Caprio, E.; Nazzi, F.; Gargiulo, G.; Pennacchio, F. Neonicotinoid clothianidin adversely affects insect immunity and promotes replication of a viral pathogen in honey bees. *Proc. Natl. Acad. Sci. USA* **2013**, *110*, 18466–18471. [CrossRef]
8. VanEngelsdorp, D.; Evans, J.D.; Saegerman, C.; Mullin, C.; Haubruge, E.; Nguyen, B.K.; Frazier, M.; Frazier, J.; Cox-Foster, D.; Chen, Y.; et al. Colony Collapse Disorder: A Descriptive Study. *PLoS ONE* **2009**, *4*, e6481. [CrossRef]
9. Stoner, K.A.; Eitzer, B.D. Using a Hazard Quotient to Evaluate Pesticide Residues Detected in Pollen Trapped from Honey Bees (*Apis mellifera*) in Connecticut. *PLoS ONE* **2013**, *8*, e77550. [CrossRef]
10. Traynor, K.S.; Tosi, S.; Rennich, K.; Steinhauer, N.; Forsgren, E.; Rose, R.; Kunkel, G.; Madella, S.; Lopez, D.; Eversole, H.; et al. Pesticides in honey bee colonies: Establishing a baseline for real world exposure over seven years in the USA. *Environ. Pollut.* **2021**, *279*, 116566. [CrossRef]
11. Runckel, C.; Flenniken, M.L.; Engel, J.C.; Ruby, J.G.; Ganem, D.; Andino, R.; DeRisi, J.L. Temporal Analysis of the Honey Bee Microbiome Reveals Four Novel Viruses and Seasonal Prevalence of Known Viruses, *Nosema*, and *Crithidia*. *PLoS ONE* **2011**, *6*, e20656. [CrossRef] [PubMed]
12. Flenniken, M.L.; Kunitomi, M.; Tassetto, M.; Andino, R. Insect Antiviral Defense; The Antiviral Role of RNA Interference. In *Insect Virology*; Asgari, S., Johnson, K., Eds.; Horizon Scientific Press: Norwich, UK, 2010.
13. Traynor, K.S.; Rennich, K.; Forsgren, E.; Rose, R.; Pettis, J.; Kunkel, G.; Madella, S.; Evans, J.; Lopez, D.; Vanengelsdorp, D. Multiyear survey targeting disease incidence in US honey bees. *Apidologie* **2016**, *47*, 325–347. [CrossRef]
14. Faurot-Daniels, C.; Glenny, W.; Daughenbaugh, K.F.; McMenamin, A.J.; Burkle, L.A.; Flenniken, M.L. Longitudinal monitoring of honey bee colonies reveals dynamic nature of virus abundance and indicates a negative impact of Lake Sinai virus 2 on colony health. *PLoS ONE* **2020**, *15*, e0237544. [CrossRef] [PubMed]
15. Glenny, W.; Cavigli, I.; Daughenbaugh, K.F.; Radford, R.; Kegley, S.E.; Flenniken, M.L. Honey bee (*Apis mellifera*) colony health and pathogen composition in migratory beekeeping operations involved in California almond pollination. *PLoS ONE* **2017**, *12*, e0182814. [CrossRef] [PubMed]
16. Grozinger, C.M.; Flenniken, M.L. Bee Viruses: Ecology, Pathogenicity, and Impacts. *Annu. Rev. Entomol.* **2019**, *64*, 205–226. [CrossRef] [PubMed]
17. Mullin, C.A.; Frazier, M.; Frazier, J.L.; Ashcraft, S.; Simonds, R.; VanEngelsdorp, D.; Pettis, J.S. High Levels of Miticides and Agrochemicals in North American Apiaries: Implications for Honey Bee Health. *PLoS ONE* **2010**, *5*, e9754. [CrossRef]
18. Sanchez-Bayo, F.; Goka, K. Pesticide Residues and Bees—A Risk Assessment. *PLoS ONE* **2014**, *9*, e94482. [CrossRef]
19. Grassl, J.; Holt, S.; Cremen, N.; Peso, M.; Hahne, D.; Baer, B. Synergistic effects of pathogen and pesticide exposure on honey bee (*Apis mellifera*) survival and immunity. *J. Invertebr. Pathol.* **2018**, *159*, 78–86. [CrossRef]
20. Scholer, J.; Krischik, V. Chronic Exposure of Imidacloprid and Clothianidin Reduce Queen Survival, Foraging, and Nectar Storing in Colonies of *Bombus impatiens*. *PLoS ONE* **2014**, *9*, e91573. [CrossRef]
21. Dively, G.P.; Embrey, M.S.; Kamel, A.; Hawthorne, D.J.; Pettis, J.S. Assessment of Chronic Sublethal Effects of Imidacloprid on Honey Bee Colony Health. *PLoS ONE* **2015**, *10*, e0118748. [CrossRef]
22. Straub, L.; Villamar-Bouza, L.; Bruckner, S.; Chantawannakul, P.; Gauthier, L.; Khongphinitbunjong, K.; Retschnig, G.; Troxler, A.; Vidondo, B.; Neumann, P.; et al. Neonicotinoid insecticides can serve as inadvertent insect contraceptives. *Proc. R. Soc. B Biol. Sci.* **2016**, *283*, 20160506. [CrossRef]
23. Fisher, A.I.I.; Rangel, J. Exposure to pesticides during development negatively affects honey bee (*Apis mellifera*) drone sperm viability. *PLoS ONE* **2018**, *13*, e0208630. [CrossRef]

24. Yang, E.C.; Chuang, Y.C.; Chen, Y.L.; Chang, L.H. Abnormal Foraging Behavior Induced by Sublethal Dosage of Imidacloprid in the Honey Bee (Hymenoptera: Apidae). *J. Econ. Entomol.* **2008**, *101*, 1743–1748. [CrossRef] [PubMed]
25. Wright, G.A.; Softley, S.; Earnshaw, H. Low doses of neonicotinoid pesticides in food rewards impair short-term olfactory memory in foraging-age honeybees. *Sci. Rep.* **2015**, *5*, 15322. [CrossRef] [PubMed]
26. Henry, M.; Beguin, M.; Requier, F.; Rollin, O.; Odoux, J.-F.; Aupinel, P.; Aptel, J.; Tchamitchian, S.; Decourtye, A. A Common Pesticide Decreases Foraging Success and Survival in Honey Bees. *Science* **2012**, *336*, 348–350. [CrossRef]
27. Fischer, J.; Müller, T.; Spatz, A.-K.; Greggers, U.; Grünewald, B.; Menzel, R. Neonicotinoids Interfere with Specific Components of Navigation in Honeybees. *PLoS ONE* **2014**, *9*, e91364. [CrossRef] [PubMed]
28. Grillone, G.; Laurino, D.; Manino, A.; Porporato, M. Toxicity of thiametoxam on in vitro reared honey bee brood. *Apidologie* **2017**, *48*, 635–643. [CrossRef]
29. Stuligross, C.; Williams, N.M. Past insecticide exposure reduces bee reproduction and population growth rate. *Proc. Natl. Acad. Sci. USA* **2021**, *118*, e2109909118. [CrossRef]
30. Dolezal, A.G. Carryover insecticide exposure reduces bee reproduction across years. *Proc. Natl. Acad. Sci. USA* **2022**, *119*, e2120128118. [CrossRef]
31. Schuhmann, A.; Schmid, A.P.; Manzer, S.; Schulte, J.; Scheiner, R. Interaction of Insecticides and Fungicides in Bees. *Front. Insect Sci.* **2022**, *1*, 808335. [CrossRef]
32. Rondeau, S.; Raine, N.E. Fungicides and bees: A review of exposure and risk. *Environ. Int.* **2022**, *165*, 107311. [CrossRef]
33. Mussen, E.C.; Lopez, J.E.; Peng, C.Y. Effects of Selected Fungicides on Growth and Development of Larval Honey Bees, *Apis mellifera* L. (Hymenoptera: Apidae). *Envi. Entomol.* **2004**, *33*, 1151–1154. [CrossRef]
34. McArt, S.H.; Urbanowicz, C.; McCoshum, S.; Irwin, R.E.; Adler, L.S. Landscape predictors of pathogen prevalence and range contractions in U.S. bumblebees. *Proc. R. Soc. B Biol. Sci.* **2017**, *284*, 20172181. [CrossRef] [PubMed]
35. Pettis, J.S.; Lichtenberg, E.M.; Andree, M.; Stitzinger, J.; Rose, R.; vanEngelsdorp, D. Crop Pollination Exposes Honey Bees to Pesticides Which Alters Their Susceptibility to the Gut Pathogen *Nosema ceranae*. *PLoS ONE* **2013**, *8*, e70182. [CrossRef] [PubMed]
36. Yoder, J.A.; Jajack, A.J.; Rosselot, A.E.; Smith, T.J.; Yerke, M.C.; Sammataro, D. Fungicide contamination reduces beneficial fungi in bee bread based on an area-wide field study in honey bee, *Apis mellifera*, colonies. *J. Toxicol. Environ. Health Part. A* **2013**, *76*, 587–600. [CrossRef] [PubMed]
37. Johnson, R.M. Honey Bee Toxicology. *Annu. Rev. Entomol.* **2015**, *60*, 141023145309000. [CrossRef] [PubMed]
38. Glavan, G.; Božič, J. The synergy of xenobiotics in honey bee Apis mellifera: Mechanisms and effects. *Acta Biol. Slov.* **2013**, *56*, 1. Available online: http://bijh-s.zrc-sazu.si/ABS/SI/ABS/Cont/56_1/ABS_56-1_2013_11-27.pdf (accessed on 22 October 2024).
39. Thompson, H.M.; Fryday, S.L.; Harkin, S.; Milner, S. Potential impacts of synergism in honeybees (*Apis mellifera*) of exposure to neonicotinoids and sprayed fungicides in crops. *Apidologie* **2014**, *45*, 545–553. [CrossRef]
40. Mao, W.; Schuler, M.A.; Berenbaum, M.R. Disruption of quercetin metabolism by fungicide affects energy production in honey bees (*Apis mellifera*). *Proc. Natl. Acad. Sci. USA* **2017**, *114*, 2538–2543. [CrossRef]
41. Ricke, D.F.; Lin, C.-H.; Johnson, R.M. Pollen Treated with a Combination of Agrochemicals Commonly Applied During Almond Bloom Reduces the Emergence Rate and Longevity of Honey Bee (Hymenoptera: Apidae) Queens. *J. Insect Sci.* **2021**, *21*, 5. [CrossRef]
42. Chandel, R.S.; Gupta, P.R. Toxicity of diflubenzuron and pen- fluron to immature stages of *Apis cerana indica* F and *Apis mellifera* L. *Apidologie* **1992**, *23*, 465–473. [CrossRef]
43. Gupta, P.R.; Chandel, R.S. Effects of diflubenzuron and penfluron on workers of *Apis cerana indica* F. and *Apis mellifera* L. *Apidologie* **1995**, *26*, 3–10. [CrossRef]
44. Pinto, L.Z.; Bitondi, M.M.G.; Simoes, Z.L.P. Inhibition of vitellogenin synthesis in *Apis mellifera* workers by a juvenile hormone analogue, pyriproxyfen. *J. Insect. Physiol.* **2000**, *46*, 153–160. [CrossRef]
45. Fisher, I.I.A.; Colman, C.; Hoffmann, C.; Fritz, B.; Rangel, J. Effects of the Insect Growth Regulators Methoxyfenozide and Pyriproxyfen and the Acaricide Bifenazate on Honey Bee (Hymenoptera: Apidae) Forager Survival. *J. Econ. Entomol.* **2018**, *111*, 510–516. [CrossRef] [PubMed]
46. Mullin, C.A.; Scott, J.G. Biomolecular Basis for Insecticide Resistance: Classification and Comparisons. In *Molecular Mechanisms of Insecticide Resistance*; Mullin, C.A., Scott, J.G., Eds.; American Chemical Society: Washington, DC, USA, 1992; pp. 1–13. Available online: http://pubs.acs.org/doi/abs/10.1021/bk-1992-0505.ch001 (accessed on 22 October 2024).
47. Johnson, R.M.; Dahlgren, L.; Siegfried, B.D.; Ellis, M.D. Acaricide, Fungicide and Drug Interactions in Honey Bees (*Apis mellifera*). *PLoS ONE* **2013**, *8*, e54092. [CrossRef] [PubMed]
48. Han, W.; Yang, Y.; Gao, J.; Zhao, D.; Ren, C.; Wang, S.; Zhao, S.; Zhong, Y. Chronic toxicity and biochemical response of *Apis cerana cerana* (Hymenoptera: Apidae) exposed to acetamiprid and propiconazole alone or combined. *Ecotoxicology* **2019**, *28*, 399–411. [CrossRef] [PubMed]
49. Sgolastra, F.; Medrzycki, P.; Bortolotti, L.; Renzi, M.T.; Tosi, S.; Bogo, G.; Teper, D.; Porrini, C.; Molowny-Horas, R.; Bosch, J. Synergistic mortality between a neonicotinoid insecticide and an ergosterol-biosynthesis-inhibiting fungicide in three bee species. *Pest. Manag. Sci.* **2017**, *73*, 1236–1243. [CrossRef]
50. Tosi, S.; Nieh, J.C. Lethal and sublethal synergistic effects of a new systemic pesticide, flupyradifurone (Sivanto®), on honeybees. *Proc. R. Soc. B* **2019**, *286*, 20190433. [CrossRef]

51. Donley, N. Toxic Concoctions: How the EPA Ignores the Dangers of Pesticide Cocktails. Center for Biological Diversity. July 2016, p. 16. Available online: https://www.biologicaldiversity.org/campaigns/pesticides_reduction/pdfs/Toxic_concoctions.pdf (accessed on 22 October 2024).
52. Philpott, T. Are Your Delicious, Healthy Almonds Killing Bees? *Mother Jones*, 29 April 2014. Available online: https://www.motherjones.com/food/2014/04/california-almond-farms-blamed-honeybee-die/ (accessed on 22 October 2024).
53. Wade, A.; Lin, C.-H.; Kurkul, C.; Regan, E.R.; Johnson, R.M. Combined Toxicity of Insecticides and Fungicides Applied to California Almond Orchards to Honey Bee Larvae and Adults. *Insects* **2019**, *10*, 20. [CrossRef]
54. OSU Extension. Evaluating Honey Bee Colonies for Pollination. Available online: https://catalog.extension.oregonstate.edu/pnw623 (accessed on 4 February 2022).
55. Delaplane, K.S.; van der Steen, J.; Guzman-Novoa, E. Standard methods for estimating strength parameters of *Apis mellifera* colonies. *J. Apic. Res.* **2013**, *52*, 1–12. [CrossRef]
56. Anastassiades, M.; Lehotay, S.J.; Štajnbaher, D.; Schenck, F.J. Quick, Easy, Cheap, Effective, Rugged and Safe (QuEChERS) Method. *J. AOAC Int.* **2003**, *86*, 412. [CrossRef]
57. Chen, M.; Collins, E.M.; Tao, L.; Lu, C. Simultaneous determination of residues in pollen and high-fructose corn syrup from eight neonicotinoid insecticides by liquid chromatography–tandem mass spectrometry. *Anal. Bioanal. Chem.* **2013**, *405*, 9251–9264. [CrossRef]
58. De Jong, D.; De Andrea Roma, D.; Gonçalves, L.S. A comparative analysis of shaking solutions for the detection of *Varroa jacobsoni* on adult honeybees. *Apidologie* **1982**, *13*, 297–306. [CrossRef]
59. Lee, K.V.; Moon, R.D.; Burkness, E.C.; Hutchison, W.D.; Spivak, M. Practical Sampling Plans for Varroa destructor (Acari: Varroidae) in *Apis mellifera* (Hymenoptera: Apidae) Colonies and Apiaries. *J. Econ. Entomol.* **2010**, *103*, 1039–1050. [CrossRef] [PubMed]
60. Tennekes, H.A. The significance of the Druckrey–Küpfmüller equation for risk assessment—The toxicity of neonicotinoid insecticides to arthropods is reinforced by exposure time: Responding to a Letter to the Editor by Drs. C. Maus and R. Nauen of Bayer CropScience AG. *Toxicology* **2011**, *280*, 173–175. [CrossRef]
61. Rondeau, G.; Sánchez-Bayo, F.; Tennekes, H.A.; Decourtye, A.; Ramírez-Romero, R.; Desneux, N. Delayed and time-cumulative toxicity of imidacloprid in bees, ants and termites. *Sci. Rep.* **2014**, *4*, srep05566. [CrossRef] [PubMed]
62. Zhu, Y.C.; Yao, J.; Adamczyk, J.; Luttrell, R. Synergistic toxicity and physiological impact of imidacloprid alone and binary mixtures with seven representative pesticides on honey bee (*Apis mellifera*). *PLoS ONE* **2017**, *12*, e0176837. [CrossRef]
63. Almond Board of California. Honey Bee Best Management Practices for California Almonds: A Guide for Almond Growers. 2018. Available online: https://www.almonds.com/tools-and-resources/grower-tools/pollination-tools (accessed on 22 October 2024).

agronomy

MDPI

Article

# Use of *Beauveria bassiana* and *Bacillus amyloliquefaciens* Strains as *Gossypium hirsutum* Seed Coatings: Evaluation of the Bioinsecticidal and Biostimulant Effects in Semi-Field Conditions

Vasileios Papantzikos [1], Spiridon Mantzoukas [1,*], Alexandra Koutsompina [1], Evangelia M. Karali [1], Panagiotis A. Eliopoulos [2], Dimitrios Servis [3], Stergios Bitivanos [3] and George Patakioutas [1]

[1] Department of Agriculture, Arta Campus, University of Ioannina, 45100 Ioannina, Greece; b.papantzikos@uoi.gr (V.P.); agu16580@uoi.gr (A.K.); agu16574@uoi.gr (E.M.K.); gpatakiu@uoi.gr (G.P.)
[2] Laboratory of Plant Health Management, Department of Agrotechnology, University of Thessaly, Geopolis, 41500 Larissa, Greece; eliopoulos@uth.gr
[3] BASF Hellas S.A., 15125 Marousi, Greece; dimitris.servis@basf.com (D.S.); stergios.bitivanos@basf.com (S.B.)
* Correspondence: sdmantzoukas1979@gmail.com

**Abstract:** There are many challenges in cotton cultivation, which are mainly linked to management practices and market demands. The textile commerce requirements are increasing but the effects of climate change on cotton cultivation are becoming an issue, as its commercial development depends significantly on the availability of favorable climatic parameters and the absence of insect pests. In this research, it was studied whether the use of two commercial strains as cotton seed coatings could effectively contribute to the previous obstacles. The experiment was carried out in semi-field conditions at the University of Ioannina. It used a completely randomized design and lasted for 150 days. The following treatments were tested: (a) coated seeds with a commercial strain of *Beauveria bassiana* (Velifer®); (b) coated seeds with a combination of Velifer® and a commercial strain of *Beauveria bassiana* (Selifer®); and (c) uncoated cotton seeds (control). The biostimulant effect of the two seed coatings was assessed against the growth characteristics of cotton, and the total chlorophyll and proline content. The bioinsecticidal effect was evaluated by measuring the population of *Aphis gossypii* on the cotton leaves. The proline effect increased by 15% in the treated plants, whereas the total chlorophyll was higher in the use of both Velifer® and Velifer®–Selifer® treatments by 32% and 19%, respectively. Aphid populations also decreased in the treated plants compared to the control plants (29.9% in Velifer® and 22.4% in Velifer®–Selifer®). Based on an assessment of the above parameters, it follows that the two seed coatings can significantly enhance the growth performance of cotton and reduce the abundance of *A. gossypii*.

**Keywords:** seed coating; cotton; *Beauveria bassiana*; *Bacillus amyloliquefaciens*; bioinsecticide; biostimulant

## 1. Introduction

Climate change is significantly pressuring cotton cultivation [1], affecting various essential factors for the growth and production of the cotton plant [2,3]. There is a plethora of issues arising due to climate change in cotton cultivation, which are linked to the need for a continuous supply of cotton to the market [4]. One of these factors is the increase in average temperatures, which affects the range of the cotton growing seasons [4], causing problems such as heat stress [5]. Climate change is often associated with changes in the distribution of rain [6], but high water requirements are crucial for cotton as extended periods of drought can affect cotton growth and quality [7]. This can lead to a significant reduction in cotton production [8]. In addition, rising temperatures and changes in the environment may affect the type and life cycle of pathogens and pests [9], thereby affecting the need for pesticides.

The cotton aphid *Aphis gossypii* (Hemiptera: Aphididae) is an insect that affects cotton production in several ways. It is a serious pest and can reduce cotton production [10] by sucking plant juices and consuming important cotton nutrients [11], which can cause severe damage, especially under drought stress [12]. This can lead to nutrient losses and can affect cotton growth and production [13]. It can also reduce the cotton's chlorophyll content [14], affecting photosynthesis and normal growth [15], and causing stress due to juice removal [16]. In addition, they can act as disease vectors [17]: while feeding, insects may sometimes transmit viruses and other pathogens to the plants [18], exacerbating cotton health problems [19]. *Aphis gossypii* secretions can cause the growth of the multi-phytopathogenic fungi, Ascomycetes (Sooty mold), which coats the leaves and blocks sunlight, resulting in low photosynthetic quality [20,21]. The aphid may have developed resistance to some pesticides [20,22,23], making its control an issue for cotton. To deal with the cotton aphid, farmers use a variety of control techniques, including the use of pesticides [24] and the use of natural enemies, such as some Coccinellidae species [25,26]. Also, genetic improvement may play a role in creating cotton varieties that are resistant to the aphid [27].

The above biotic and abiotic issues are directly related to the quality of cotton production and are directly linked to the constant market pressure for high-quality cotton [11,28–30]. Research on microorganisms that have a biostimulant [31] and/or a bioinsecticidal effect [32] enhancing cotton growth could be a sustainable way to address these issues [33].

The term beneficial microorganisms describes those that can live part of their life with certain plant species in a non-parasitic association [34], without adverse effects on plant growth. This type of symbiotic relationship confers many advantages on both sides and greatly benefits the plant's metabolism [35]. Some microorganisms present biostimulant prospects, as they have been reported to enhance the growth of cotton plants [31]. Microorganisms such as *Metarhizium* ssp., transport nutrients and enhance the metabolism of cotton plants [36]. Also, some may reveal bioinsecticidal ability, while simultaneously strengthening the induced systemic resistance (ISR) of plants to pathogenic microorganisms [37].

Seed coatings are an eco-friendly alternative to the adverse effects that cotton cultivation suffers from. In recent years, several organic acid mixtures have been cited as seed coatings, such as proline, glutamate and citric acid [38]. Coating seeds with microorganisms that may have biostimulant and bioinsecticidal properties is an upcoming research field in dealing with biotic and abiotic factors that place stress on cotton cultivation [37,39]. *Beauveria bassiana* (Bals.-Criv.) Vuill. (Hypocreales: Cordycipitaceae) is an endophytic entomopathogenic fungus (EPF) [40–42], which is widely used in integrated pest management (IPM) programs. Its biostimulant effect on cotton cultivation when used as a seed inoculant has been reported [43], as have its improvements in total plant length and biomass [44] in several plant species. At the same time, it exhibits significant entomopathogenic activity against pests such as aphids [45]. The plant-growth-promoting rhizobacterium (PGPR), *Bacillus amyloliquefaciens* (Bacillales: Bacillaceae), is an equally interesting beneficial microorganism in the biological control of many pests for a plethora of crops, including cotton [33,46]. The biostimulant aptitude of *B. amyloliquefaciens* derives from its use as a growth enhancer [47]. Moreover, it has been observed to induce lignin synthesis in cotton seeds [48].

In the present study, we evaluated the biostimulant and bioinsecticidal potential of two seed coatings–one with *B. bassiana* and one with *B. amyloliquefaciens* + *B. bassiana*–on cotton growth characteristics and the biological control of *A. gossypii* nymphs.

## 2. Materials and Methods

### 2.1. Experimental Design

The experiment was performed in semi-field conditions, in a field at the agriculture faculty of the University of Ioannina, in Kostakioi Arta. The biostimulant and the bioinsecticidal effect of two commercial formulations were tested when used as cotton seed coatings in the experiment: *B. bassiana*, strain PPRI 5339 Velifer® OD ($8 \times 10^9$ CFU/mL, 92% Excipients), and *B. amyloliquefaciens*, strain MBI600 Serifel® WP ($5.5 \times 10^{10}$ CFU/g, 95%

Excipients) (BASF SE, Florham Park, NJ, USA). Two seed coating treatments were applied in the experiment: a treatment with Velifer formulation (V) and a treatment with Velifer and Serifel combined (VS). Conventional uncoated cottonseed (C) was used as a control.

### 2.2. Protocol for Coating Seeds with *B. bassiana* and *B. amyloliquefaciens*

Using dispensers and pipettes, the treatment solutions were prepared as the desired combined treatments. The VS treatment was prepared by adding 125 mL of formulated *B. bassiana*/100 Kg cottonseed and 160 mL of formulated *B. amyloliquefaciens*/100 Kg. The strains *B. bassiana* PPRI 5339 Velifer® OD and the *B. amyloliquefaciens* strain MBI600 Serifel® WP (BASF SE, Florham Park, NJ, USA) were added by pipette. The V treatment was prepared using 125 mL of formulated *B. bassiana*/100 Kg cottonseed. The required amount of water was added through the dispenser. Next, the separation of the seed samples was carried out using a precision balance (KERN PES 6200-2M), and, through the use of Wintersteiger Hege 11, the final application of the coating to the seeds was made. The seed sample was then placed in a stainless steel bucket, where stirring began. Finally, using the Eppendorf pipette, an appropriate amount of seed treatment solution was applied. The application time was ~1 min/treatment/sample. In order for the treated seeds to dry completely, they were emptied into a special sample bag, which was left open. The equipment used to apply the treatment solutions and the application bucket were carefully cleaned with dd$H_2O$ when rotating the treatments. The fungal spore concentration in the control treatment was zero.

### 2.3. Experimental Set-Up

A mixture of peat and perlite in 1:1 ratio (*v*/*v*) and in in 9 L pots was used as a growth substrate for the cotton plants (var. Olivia). The experiment was conducted in a completely randomized design. The beginning point (day 0) was 19 May 2023, and the last experimental day was 16 October 2023 (day 150). Each treatment consisted of 21 cotton seeds, each placed in a pot filled with growth substrate. The pots were irrigated daily through a drip system (ARGOS Electronics 2014), automatically controlled by a computer. Irrigation quantity and frequency were based on climatic data taken from temperature and humidity sensors. In addition, to further control the irrigation adequacy, frequent measurements of the relative pot's moisture were made using a soil moisture meter (ΔT-SM150 Kit, Delta T Devices, Cambridge, UK). Initial fertilization was applied to each pot with an N30-P10-K10 fertilizer, which was repeated after 50 days.

### 2.4. Recording of *A. gossypii* Population and Cotton Growth Characteristics

The natural presence of *A. gossypii* on cotton leaves was systematically recorded on a weekly basis from the beginning of the experiment. Cotton growth characteristics such as total plant length (cm); central shoot diameter (mm); and the total number of shoots, internodes, leaves, and cotton bolls were measured. Leaf area ($cm^2$) was calculated from an image analysis of the leaves, using the Image j protocol [49]. At the end of the experiment (day 150), the total plant biomass was measured by dividing each plant into roots, shoots, leaves and seedcotton, taking the values of the fresh weight (g) of each of the previous categories for each plant separately. Following this, dry biomass (g) was recorded after 72 h of 80 °C oven-drying.

### 2.5. Total Chlorophyll Content

Measurement of the total chlorophyll (TCHL) content in the cotton leaves was carried out weekly during the experiment in a non-destructive way, using the SPAD (Minolta Co., Ltd., Tokyo, Japan) instrument method. For the accuracy of the method, the linear correlation of the SPAD method with the chemical method of chlorophyll determination was carried out ($R^2$ = 0.901) in randomly selected cotton leaf samples, according to the protocol of [50], with some modifications: 10 mL of acetone was used as the extraction solvent of 0.04 g cotton leaf tissue, which was crushed in a porcelain mortar with a pestle. Each sample was placed in a 10 mL glass tube, vortexed, and left overnight at 4 °C. The absorbance was measured in a spectrophotometer (Jasco-V630 UV-VIS, JASCO INTERNATIONAL Co., Ltd., Tokyo, Japan), using the equations described by Lichtenthaler and Buschmann [51], and the result was expressed in g of TCHL of fresh leaf per $cm^2$ of cotton leaf area:

$$Ca\ (\mu g/mL) = 11.24 \times A_{661.6} - 2.04 \times A_{644.8}$$

$$Cb\ (\mu g/mL) = 20.13 \times A_{644.8} - 4.19 \times A_{661.6}$$

### 2.6. Proline Content

The total proline content was determined to assess the cotton plants' stress from the environmental conditions according to [52] with some modifications: 4 mL of extractant solution containing 70% ethanol was poured into a mortar containing 0.1 g of fresh cotton leaf plant tissue and was crushed with a pestle until it became a homogeneous mix. The samples were added to glass tubes, which were centrifuged at 4000 g, for 10 min. In a new set of tubes, 1 mL of supernatant extract and 2 mL of fresh acid–ninhydrin solution were placed. Then the samples were vortexed and incubated in a dark water bath at 95 °C for 25 min and the reaction mixture was cooled directly in an ice bath. When at room temperature, the samples were centrifuged for 5 min at 4000 g. The absorbance was determined at 520 nm in a spectrophotometer (Jasco-V630 UV-VIS). The results were reported in μmol of proline, $g^{-1}$ of fresh cotton leaf weight.

### 2.7. Statistical Analysis

One-way ANOVA was performed with Tukey's post hoc test to compare the means of the treatments for the effect on plant growth, proline, total chlorophyll content (TCHL), and cotton fresh and dry mass. Two-way ANOVA was performed for the Insect Population with two variables: population and time. All statistical analyses were conducted using SPSS v. 25 (IBM-SPSS Statistics, Armonk, NY, USA).

## 3. Results

### 3.1. Insect Pest Population

The average population of *A. gossypii* nymphs per treatment in the cotton crop was recorded. More specifically, in all seven samplings, the variation in the average number of *A. gossypii* nymphs was statistically significant among the treatments ($F = 11.881$, $df = 2.511$, $p < 0.001$). Forty days after treatment, the average number of aphids was significantly higher in the control samples than in the treated samples (Figure 1). The average change in the aphid population at the maximum, 56 days, was 12.38 ± 1.35 aphids for the V plants and 13.71 ± 0.38 aphids for the VS plants. In the control plants, the aphid population was 17.68 ± 0.92 aphids (Figure 1). Fewer aphids were almost always counted on the plants treated with V than on the plants inoculated with other strains.

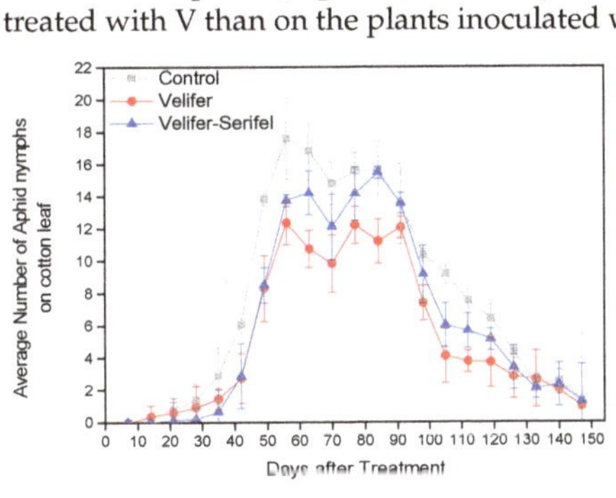


**Figure 1.** The mean number of *A. gossipii* aphids on cotton plant leaves up to 150 days after treatment: **V**—Velifer; **VS**—Velifer–Serifel; and **C**—control.

### 3.2. *Effect on Plant Growth*

The evaluation of the morphological features of the tested plants was based on a recording of the plants' length, shoots, internodes, the number of leaves and cotton bolls, and the stem diameter. In general, V plants and VS plants resulted in better growth than the untreated plants. V plants and VS plants, after 150 days, had statistically more length compared to the control plants (F = 9.553, df = 2, $p < 0.001$) (Figure 2). For the shoots, all the plants, treated and untreated, had the same average growth (F = 1.523, df = 2, $p = 0.951$) (Figure 3).

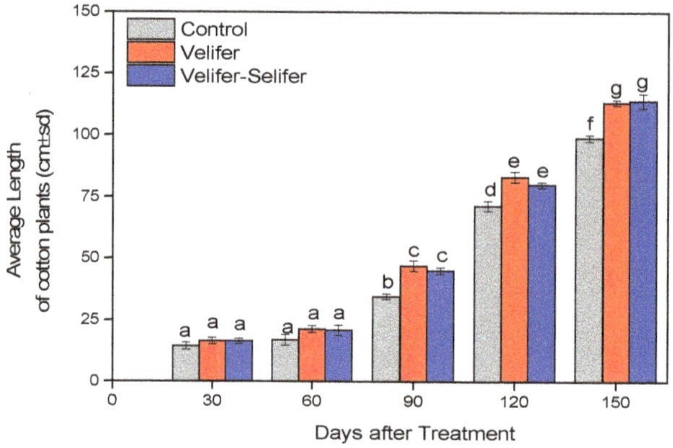


**Figure 2.** The mean length (cm) of cotton plants up to 150 days after treatment: **V**—Velifer; **VS**—Velifer–Serifel; and **C**—control. Different letters among treatments indicate statistically significant differences (Tukey test, $p < 0.05$).

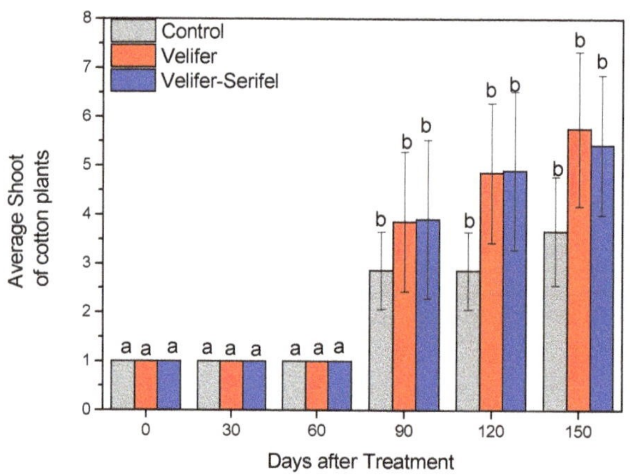


**Figure 3.** The mean shoot number of cotton plants up to 150 days after treatment: **V**—Velifer; **VS**—Velifer–Serifel; and **C**—control. Different letters among treatments indicate statistically significant differences (Tukey test, $p < 0.05$).

The V plants and VS plants caused an increase in the number of internodes (F = 6.759, df = 2, $p = 0.011$) (Figure 4) and cotton bolls (F = 8.759, df = 2, $p = 0.009$) (Figure 5). The leaf areas (cm2) after 150 days were as follows: for the control, 241.15 ± 53.55; for the Velifer treatment, 372.70 ± 44.98; and for the Velifer–Serifel treatment, 397.50 ± 45.81 (F = 11.553, df = 2, $p < 0.001$). The differences proved to be not statistically significant for the number of leaves: F = 2.159, df = 2, $p = 0.870$ (Figure 6). A similar increase (not statistically significant) was recorded for the stem diameter (F = 3.111, df = 2, $p = 0.811$) (Figure 7).

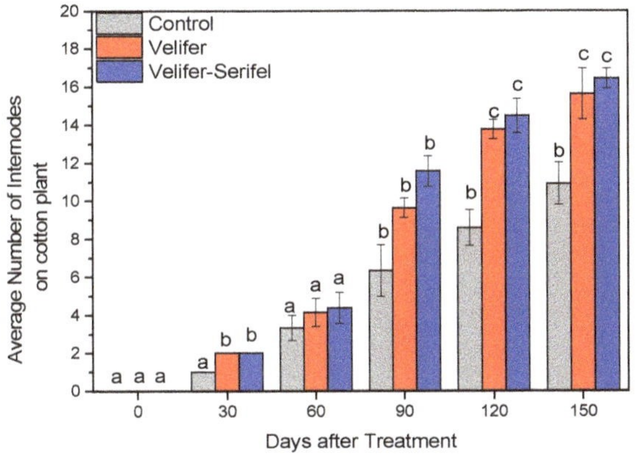


**Figure 4.** The mean number of internodes on cotton plants up to 150 days after treatment: **V**—Velifer; **VS**—Velifer–Serifel; and **C**—control. Different letters among treatments indicate statistically significant differences (Tukey test, $p < 0.05$).

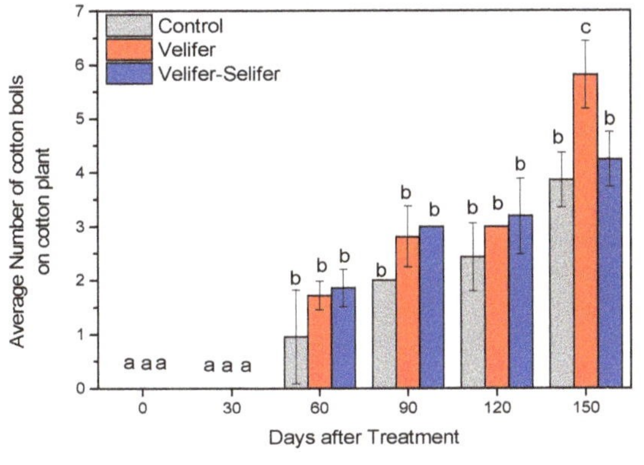


**Figure 5.** The mean number of cotton bolls up to 150 days after treatment: **V**—Velifer; **VS**—Velifer–Serifel; and **C**—control. Different letters among treatments indicate statistically significant differences (Tukey test, $p < 0.05$).

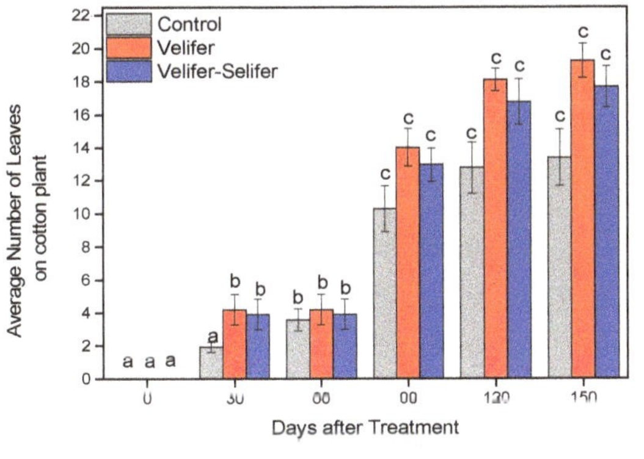


**Figure 6.** The mean number of leaves on cotton plants up to 150 days after treatment: **V**—Velifer; **VS**—Velifer–Serifel; and **C**—control. Different letters among treatments indicate statistically significant differences (Tukey test, $p < 0.05$).

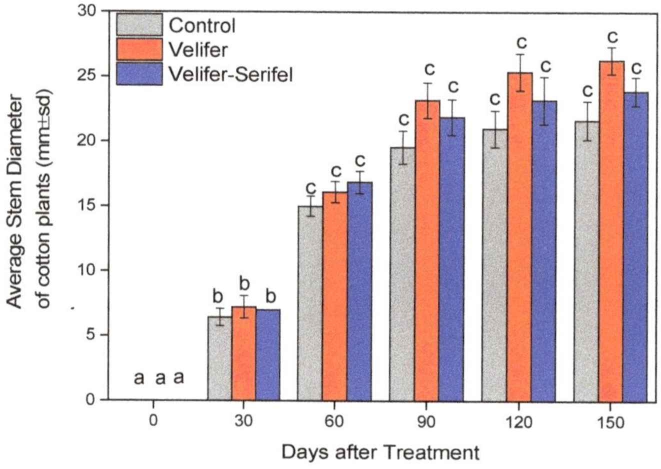


**Figure 7.** The mean stem diameter (mm) of cotton plants up to 150 days after treatment: **V**—Velifer; **VS**—Velifer-Serifel; and **C**—control. Different letters among treatments indicate statistically significant differences (Tukey test, $p < 0.05$).

### 3.3. Effect on Proline and Total Chlorophyll Content (TCHL)

The effect on the proline was found to be not statistically significant in all the plants by the end of the experiment (F = 2.197, df = 2, $p = 0.790$): after 150 days, the proline was the same in all the plants (Figure 8).

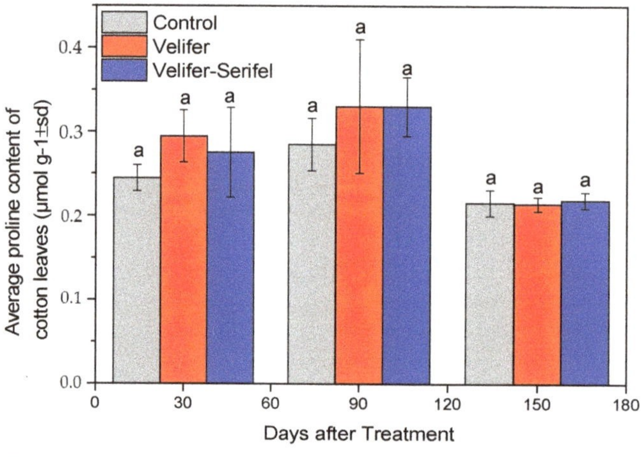


**Figure 8.** The mean proline values (μmol $g^{-1}$) for cotton plants up to 150 days after treatment: **V**—Velifer; **VS**—Velifer–Serifel; and **C**—control. Different letters among treatments indicate statistically significant differences (Tukey test, $p < 0.05$).

The chlorophyll concentration increased in the V and VS plants after 30 days and remained above that of the control plants until the end of the experiment (F = 13.220, df = 2, $p < 0.001$) (Figure 9). The increase in TCHL was attributed to the endophytic effect on the leaves. This effect was especially evident at 56 days due to the low infestation of aphids and high TCHL values for the V plants. The decrease in the TCHL values after 128 days was expected due to the leaf maturation.

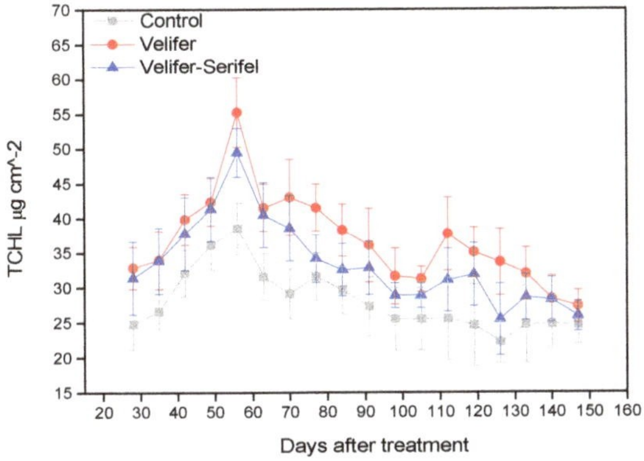


**Figure 9.** The mean values of the leaves' total chlorophyll content (µg $cm^{-2}$) up to 150 days after treatment: **V**—Velifer; **VS**—Velifer-Serifel; and **C**—control.

### 3.4. *Cotton Fresh and Dry Mass*

The fresh and dry mass measured at the final harvest are summarized in Figures 10A and 10B, respectively. The V treatment significantly increased the total fresh mass (F = 34.198, df = 2, $p < 0.001$) and dry mass (F = 25.811, df = 2, $p < 0.001$) of the above-ground part of the cotton plants. The V and VS plants showed significantly increased fresh mass in the roots (F = 24.308, df = 2, $p < 0.001$), and the V treatment significantly increased the roots' dry mass (F = 19.398, df = 2, $p < 0.001$). The V and VS plants significantly increased the shoots' fresh mass (F = 14.118, df = 2, $p < 0.001$) and their dry mass (F = 18.128, df = 2, $p < 0.001$). Also, in the case of leaf weight, the V and VS treatments increased the fresh mass (F = 17.228, df = 2, $p < 0.001$). The leaves' dry mass increased only with the V treatment (F = 8.118, df = 2, $p < 0.001$). Finally, for seedcotton, we did not find a significant difference in the fresh (F = 2.118, df = 2, $p = 0.964$) or the dry mass (F = 1.918, df = 2, $p = 0.916$).

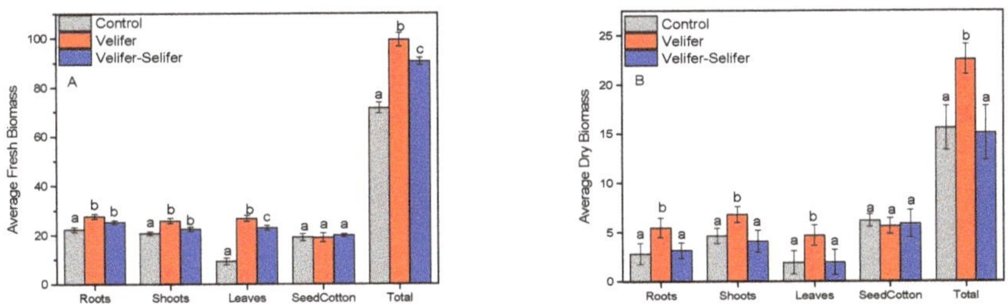


**Figure 10.** The mean fresh (**A**) and dry (**B**) weight (g) of roots, shoots, leaves, and seedcotton of *G. hirsutum* plants up to 150 days after treatment: **V**—Velifer; **VS**—Velifer–Serifel; and **C**—control. Different letters among treatments indicate statistically significant differences (Tukey test, $p < 0.05$).

## 4. Discussion

Some EPFs and PGPRs can be used as seed dressings for a wide range of plant species [37,53,54]. Cottonseed arouses an interest in experimentation on such microorganisms as seed coatings in order to improve the plant's metabolism, as it is a high-demand crop. Due to its increased use for commercial purposes [55], it is important to find eco-friendly solutions to combat its insect pests, such as *A. gossypii*. In this study, the bioinsecticidal activity of fungi and bacteria coatings was demonstrated by the significant reduction in the *A. gossypii* population on the cotton plants. This outcome conforms with our last work [56], where the inoculated cottonseed coatings (with the same *B. bassiana* strain) was effective in minimizing the *A. gossypii* population. The coating with *B. bassiana* showed a higher bioinsecticidal effect, with a longer duration than its co-inoculated application with *B. amyloliquefaciens*. In field conditions, a *B. bassiana* coating on maize *Zea mays* (Poales: Poaceae) seed has shown bioinsecticidal action on *Spodoptera frugiperda* (Lepidoptera: Noctuidae) [57], with no negative effects on beneficial insects, such as honeybees. A similar result was presented in the study by Mishra et al., 2013 [58], where a *B. bassiana* formulated strain was effective on seeds via encapsulation, against *Musa domestica* (Diptera: Muscidae).

In this study, the application of seed coatings using *B. Bassiana* (V) and the combined application of *B. Bassiana* and *B. amyloliquefaciens* (VS), demonstrated a beneficial effect on cotton's growth and metabolism, and the same effect has been found in other crops [59–62]. The growth characteristics of cotton, such as fresh and dry biomass, were increased in the coated treatments, especially in the case of *B. bassiana*. This result has also been observed in the work of [41], where the fresh weight of the roots and shoots of coated *Vicia faba* (Fabales: Fabaceae) seeds was increased. Moreover, in the work of Sánchez-Rodríguez et al., 2018 [63], endophytic *B. bassiana,* used as a seed dressing, increased the dry weight and total grain weight of bread wheat, *Triticum aestivum* (Poales: Poaceae). It also effectively controlled cotton leafworm larvae, *Spodoptera littoralis* (Lepidoptera: Noctuidae). The number of internodes was higher than on the untreated cotton seed. An equivalent result was exhibited in the study by Canassa et al., 2019 [39], where coated bean seeds (with *B. Bassiana)* strengthened plant growth. A similar effect on the growth of *Z. mays* seeds coated with *B. bassiana* was also observed in the study of Rivas-Franco et al., 2019 [64], in which a significant increase in the vegetation length was noted, something that was not observed to a significant extent in this experiment.

The use of *B. bassiana* as a seed coating for *Phaseolus vulgaris* L. (Fabales: Fabaceae) showed an increased number of leaves [46], while in other studies, an enlarged leaf area was observed [57]. A similar occurrence was noted in this study regarding treatment with V. The number of cotton bolls were also greater when treated with V. This has been observed in other *B. bassiana* application experiments [65]. In a lot of studies, the beneficial impact of seed coatings has been mentioned, regarding several metabolic traits [31,66]. The higher the amount of chlorophyll in the leaves, the more beneficial it is for the plant [67] because it is involved in various metabolic processes that relate to enhancing its robustness. The single-inoculated treatment containing *B. bassiana* had a higher performance on chlorophyll, compared to the co-inoculated treatment with *B. bassiana* + *B. amyloliquefaciens*. Of course, both the coating treatments showed a more significant effect compared to the uncoated seeds. A similar effect of *B. bassiana* has been observed on rice, *Oryza sativa* (Poales: Poaceae) [68]; on barley, *Hordeum vulgare* (Poales: Poaceae) [69]; on chili, *Capsicum annuum* L. (Solanales: Solanaceae) [70]; and on cucumber, *Cucumis sativus* L. (Cucurbitales: Cucurbitaceae) [71]. However, the low chlorophyll in the control treatment may also be related to the fact that the sucking damage caused by *A. gossypii* was higher.

Proline is an indicator of abiotic stress [72] and has been detected as enriched in some *B. Bassiana* application experiments [59], which contrasts with our experiment, where proline levels were not affected, as the same levels were presented in all treatments.

The above led us to ascertain that the coating composition using *B. bassiana* had a biostimulant effect due to the enhancement of growth characteristics and chlorophyll in the cotton crop. On the contrary, we could not identify the same strong biostimulant and bioinsecticidal properties in the co-inoculation treatment (VS), as this did not result in the same vigorous impact in all growth parameters. This may be attributed to the antagonistic interactions that evolved between the two microorganisms [73,74]. However, in the work of Prabhukarthikeyan et al., 2014, the combined application of *Bacillus* ssp. and *Bassiana ssp.* was effective in combating *Fusarium oxysporum* f., sp., lycopersici's (Hypocreales: Nectriaceae) wilt and the fruit borer *Helicoverpa armigera* (Lepidoptera: Noctuidae) in tomato plants, *Solanum lycopersicum* (Solanales: Solanaceae) [75]. This was achieved without competing to the detriment of the plant's metabolism. The application of *Beauveria* and *Bacillus* strains in a simultaneous mixture has been referred to as highly effective in the control of greenhouse whitefly, *Trialeurodes vaporariorum* (Hemiptera: Aleyrodidae) [76]; red palm weevil, *Rhynchophorus ferrugineus* Olivier (Coleoptera: Curculionidae) [77]; and tobacco cutworm, *Spodoptera litura* (Lepidoptera: Noctuidae) [78]. However, this performance was not repeated in our results. In addition, in a study by Wang et al., 2015 [79], co-inoculated plants were weaker than plants with individual inoculations, but no harmful effects on their growth were observed. This fact has also been reflected in other studies [80]. Quantitative PCR studies have shown that the presence of beneficial bacteria within plant tissues suppressed fungal colonization. This may explain the reduction in biomass of the co-inoculated plants, compared to the single-inoculated plants. Two potential ways could lead to the observed decrease in fungal colonization mentioned in this context. Firstly, the decline in fungal population density might be caused directly by certain substances, which are produced by bacteria that inhibit fungal growth. Research has demonstrated that alkaloids from beneficial microorganisms and the defensive chemicals produced by the host in response to endophytic invasion can reduce the colonization of fungi [81,82]. Secondly, the coexistence of symbiotic microorganism reforms how the host distributes its resources, consequently, impacting fungal colonization [83,84]. The aforementioned concept of antagonism between *B. bassiana* and *B. amyloliquefaciens* in the co-inoculated coated treatment (VS) also derives from the study of metabolic parameters, such as chlorophyll, which was higher in the treatment with *B. bassiana* (V) than in the co-inoculated treatment. Moreover, a similar effect was presented by the reduced population of *A. gossypii* in the same treatment. However, this hypothesis needs further study to define the interactions between the two microorganisms: initially, by studying the degree of their interaction during the endophytic growth of the co-inoculated coated cotton seed and, next, establishing by which factors the synergism and viability of each strain can be affected.

The typical functioning and development of a plant is generally unaffected by symbiotic microorganisms [85]. Nevertheless, EPF and PGPR may occasionally boost the host's ability to withstand challenging environmental factors, such as drought [86] and lack of nutrients [87], or fortify the host's defenses against pests [88–90]. For instance, due to EPF, plants amplify their resistance to insect feeding damage and encounter fewer biomass losses [91].

## 5. Conclusions

In our experiment, the seed coatings created by EPF and PGPR are of interest when focusing on their application in IPM cotton cultivation programs. The cotton seed coating treatment with *B. bassiana* and the co-inoculation treatment with *B. amyloliquefaciens* and *B. bassiana* showed bioinsecticidal properties, reducing the population of the cotton aphid, *A. gossypii*, for a long time. The application of *B. bassiana* enhanced the growth of cotton in several parameters, such as the total fresh and dry biomass. In addition, the larger leaf area and the higher amount of TCHL in the leaves were shown to be an effect of *B. bassiana* and its co-inoculation with *B. amyloliquefaciens*, which improved the robustness of the cotton crop. In addition, the results of our research suggest that the combined application of the two microorganisms was not as successful as the single inoculation treatment with *B. bassiana*. The data present, once again, the biostimulant and bioinsecticidal effect of beneficial EPF and PGPR on the growth and metabolic traits of cotton. In order to shed light on the potential antagonistic interactions between the two microorganisms, it would be relevant to explore the precise mode of action of the endophytic coatings that co-inoculate EPF *B. bassiana* and PGPR *B. amyloliquefaciens*. Our research on utilizing EPF and PGPR bacteria as seed coatings for commercially important plant species has provided viable pathways towards reducing the excessive use of agrochemicals, while boosting the metabolic resilience of plants. Considering these promising opportunities, more research must be conducted on the use of endophytic seed coatings containing *B. amyloliquefaciens* and *B. bassiana* in cotton production under realistic circumstances. The positive results of our research highlight the significance of these initiatives.

**Author Contributions:** Conceptualization, V.P. and S.M.; methodology, V.P.; software, V.P. and S.M.; validation, V.P., S.M., A.K., E.M.K., P.A.E., D.S., S.B. and G.P.; formal analysis, V.P., A.K. and E.M.K.; investigation, V.P., A.K. and E.M.K.; resources, V.P. and S.M.; data curation, V.P., S.M. and P.A.E.; writing—original draft preparation, V.P.; writing—review and editing, V.P., S.M., P.A.E., D.S. and S.B.;

visualization, V.P. and S.M.; supervision, V.P., SM. and G.P.; project administration, V.P., S.M. and G.P. All authors have read and agreed to the published version of the manuscript.

**Funding:** The authors declare that this study received funding from the University of Ioannina (grant number 61392). The funder had the following involvement in the study: funding and material supplies for the experiment.

**Data Availability Statement:** The data presented in this study are available on request from the corresponding authors, V.P. and S.M.

**Acknowledgments:** The authors gratefully acknowledge the Department of Agriculture of the University of Ioannina for providing the necessary facilities to carry out the experiments.

**Conflicts of Interest:** Authors Dimitrios Servis and Stergios Bitivanos were employed by the company BASF Hellas S.A. The remaining authors declare that the research was conducted in the absence of any commercial or financial relationships that could be construed as a potential conflict of interest.

## References

1. Aslan, R.; Aygun, Y.Z.; Mert, M. Yield and Fiber Quality Characteristics of Some Cotton (*Gossypium hirsutum* L.) Cultivars Grown in the Southeastern Anatolian Conditions. *Turk. J. Field Crops* **2022**, *27*, 285–292. [CrossRef]
2. Namrata, P.B.; Hardik, L.S.; Ramani, H.; Rajkumar, B.K. Screening of Cotton (*Gossypium hirsutum* L.) Genotypes for Drought Tolerance. *Pharma Innov. J.* **2022**, *11*, 1634–1639.
3. Ertekin, I.; Atis, I.; Aygun, Y.Z.; Yilmaz, S.; Kizilsimsek, M. Effects of Different Nitrogen Doses and Cultivars on Fermentation Quality and Nutritive Value of Italian Ryegrass (*Lolium multiflorum* Lam.) Silages. *Anim. Biosci.* **2022**, *35*, 39–46. [CrossRef] [PubMed]
4. Li, N.; Lin, H.; Wang, T.; Li, Y.; Liu, Y.; Chen, X.; Hu, X. Impact of Climate Change on Cotton Growth and Yields in Xinjiang, China. *Field Crops Res.* **2020**, *247*, 107590. [CrossRef]
5. Saleem, M.A.; Malik, W.; Qayyum, A.; Ul-Allah, S.; Ahmad, M.Q.; Afzal, H.; Amjid, M.W.; Ateeq, M.F.; Zia, Z.U. Impact of Heat Stress Responsive Factors on Growth and Physiology of Cotton (*Gossypium hirsutum* L.). *Mol. Biol. Rep.* **2021**, *48*, 1069–1079. [CrossRef] [PubMed]
6. Tabari, H. Climate Change Impact on Flood and Extreme Precipitation Increases with Water Availability. *Sci. Rep.* **2020**, *10*, 13768. [CrossRef] [PubMed]
7. Ul-Allah, S.; Rehman, A.; Hussain, M.; Farooq, M. Fiber Yield and Quality in Cotton under Drought: Effects and Management. *Agric. Water Manag.* **2021**, *255*, 106994. [CrossRef]
8. Zafar, S.; Afzal, H.; Ijaz, A.; Mahmood, A.; Ayub, A.; Nayab, A.; Hussain, S.; UL-Hussan, M.; Sabir, M.A.; Zulfiqar, U.; et al. Cotton and Drought Stress: An Updated Overview for Improving Stress Tolerance. *S. Afr. J. Bot.* **2023**, *161*, 258–268. [CrossRef]
9. Raza, M.M.; Bebber, D.P. Climate Change and Plant Pathogens. *Curr. Opin. Microbiol.* **2022**, *70*, 102233. [CrossRef]
10. Nikam, T.A.; Latpate, C.B.; Bhosale, B.B. Estimation of Yield Losses Due to Sucking Pests of Bt Cotton under High Density Planting System. *J. Cotton Res. Dev.* **2019**, *33*, 273–280. [CrossRef]
11. Liu, J.; Wang, C.; Desneux, N.; Lu, Y. Impact of Temperature on Survival Rate, Fecundity, and Feeding Behavior of Two Aphids, Aphis Gossypii and Acyrthosiphon Gossypii, When Reared on Cotton. *Insects* **2021**, *12*, 565. [CrossRef] [PubMed]
12. Liu, J.; Wang, C.; Li, H.; Gao, Y.; Yang, Y.; Lu, Y. Bottom-Up Effects of Drought-Stressed Cotton Plants on Performance and Feeding Behavior of Aphis Gossypii. *Plants* **2023**, *12*, 2886. [CrossRef] [PubMed]
13. Ramalho, F.S.; Fernandes, F.S.; Nascimento, A.R.B.; Nascimento, J.L.; Malaquias, J.B.; Silva, C.A.D. Feeding Damage from Cotton Aphids, Aphis Gossypii Glover (Hemiptera: Heteroptera: Aphididae), in Cotton with Colored Fiber Intercropped with Fennel. *Ann. Entomol. Soc. Am.* **2012**, *105*, 20–27. [CrossRef]
14. XiaoXia, D.; HaiLan, J.; Jun, P.; ZeMin, H.; TianWen, M.; JunGang, W. Physiological Responses of Cotton to Feeding by Aphis Gossypii during the Flower-Boiling Stage. *Chin. J. Appl. Entomol.* **2013**, *50*, 161–166.
15. Shannag, H.K.; Thorvilson, H.; El-Shatnawi, M.K. Changes in Photosynthetic and Transpiration Rates of Cotton Leaves Infested with the Cotton Aphid, Aphis Gossypii: Unrestricted Infestation. *Ann. Appl. Biol.* **1998**, *132*, 13–18. [CrossRef]
16. Zhang, P.; Zhang, X.; Zhao, Y.; Ren, Y.; Mu, W.; Liu, F. Efficacy of Granular Applications of Clothianidin and Nitenpyram against Aphis Gossypii (Glover) and Apolygus Lucorum (Meyer-Dür) in Cotton Fields in China. *Crop Prot.* **2015**, *78*, 27–34. [CrossRef]
17. Mahas, J.W.; Hamilton, F.B.; Roberts, P.M.; Ray, C.H.; Miller, G.L.; Sharman, M.; Conner, K.; Bag, S.; Blythe, E.K.; Toews, M.D.; et al. Investigating the Effects of Planting Date and Aphis Gossypii Management on Reducing the Final Incidence of Cotton Leafroll Dwarf Virus. *Crop Prot.* **2022**, *158*, 106005. [CrossRef]
18. Escriu, F.; Perry, K.L.; Garcia-Arenal, F. Transmissibility of Cucumber Mosaic Virus by Aphis Gossypii Correlates with Viral Accumulation and Is Affected by the Presence of Its Satellite RNA. *Phytopathology* **2000**, *90*, 1068–1072. [CrossRef]
19. Tarazi, R.; Vaslin, M.F.S. The Viral Threat in Cotton: How New and Emerging Technologies Accelerate Virus Identification and Virus Resistance Breeding. *Front. Plant Sci.* **2022**, *13*, 851939. [CrossRef]
20. Li, R.; Cheng, S.; Liang, P.; Chen, Z.; Zhang, Y.; Liang, P.; Zhang, L.; Gao, X. Status of the Resistance of Aphis Gossypii Glover, 1877 (Hemiptera: Aphididae) to Afidopyropen Originating from Microbial Secondary Metabolites in China. *Toxins* **2022**, *14*, 750. [CrossRef]

21. Srivastava, R.; Shukla, A.C. Fusarium Pallidoroseum: A Potential Entomopathogenic Agent for the Biological Management of Aphis Gossypii. *J. Appl. Nat. Sci.* **2021**, *13*, 775–785. [CrossRef]
22. Ahmad, M.; Arif, M.I.; Denholm, I. High Resistance of Field Populations of the Cotton Aphid Aphis Gossypii Glover (Homoptera: Aphididae) to Pyrethroid Insecticides in Pakistan. *J. Econ. Entomol.* **2003**, *96*, 875–878. [CrossRef]
23. Wang, K.Y.; Guo, Q.L.; Xia, X.M.; Wang, H.Y.; Liu, T.X. Resistance of Aphis Gossypii (Homoptera: Aphididae) to Selected Insecticides on Cotton from Five Cotton Production Regions in Shandong, China. *J. Pestic. Sci.* **2007**, *32*, 372–378. [CrossRef]
24. Gaber, A.S.; Abd-Ella, A.A.; Abou-Elhagag, G.H.; Abdel-Rahman, Y.A. Field Efficiency and Selectivity Effects of Selected Insecticides on Cotton Aphid, Aphis Gossypii Glover (Homoptera: Aphididea) and Its Predators. *J. Phytopathol. Dis. Manag.* **2015**, *2*, 22–35.
25. Chamuene, A.; Araújo, T.A.D.; Lopes, M.C.; Ramos Pereira, R.; Berger, P.G.; Picanço, M.C. Investigating the Natural Mortality of Aphis Gossypii (Hemiptera: Aphididae) on Cotton Crops in Tropical Regions Using Ecological Life Tables. *Environ. Entomol.* **2020**, *49*, 66–72. [CrossRef]
26. Eid, A.E.; El-Heneidy, A.H.; Hafez, A.A.; Shalaby, F.F.; Adly, D. On the Control of the Cotton Aphid, Aphis Gossypii Glov. (Hemiptera: Aphididae), on Cucumber in Greenhouses. *Egypt. J. Biol. Pest. Control* **2018**, *28*, 64. [CrossRef]
27. Mbiza, N.I.T.; Hu, Z.; Zhang, H.; Zhang, Y.; Luo, X.; Wang, Y.; Wang, Y.; Liu, T.; Li, J.; Wang, X.; et al. GhCalS5 Is Involved in Cotton Response to Aphid Attack through Mediating Callose Formation. *Front. Plant Sci.* **2022**, *13*, 892630. [CrossRef]
28. Kerns, D.L.; Yates, J.A.; Baugh, B.A. Economic Threshold for Cotton Aphid (Hemiptera: Aphididae) on Cotton in the Southwestern United States. *J. Econ. Entomol.* **2015**, *108*, 1795–1803. [CrossRef]
29. Abang, A.F.; Srinivasan, R.; Hanna, R.; Fotso, A.K.; Kekeunou, S.; Tenkouano, A.; Bilong, C.F.B. Productivity and Resistance of Okra (*Abelmoschus* spp.) to the Cotton Aphid Aphis Gossypii Glover (Hemiptera: Aphididae) under Tropical Conditions. *Int. J. Trop. Insect Sci.* **2021**, *41*, 197–208. [CrossRef]
30. Karamchedu, A. Dried up Bt Cotton Narratives: Climate, Debt and Distressed Livelihoods in Semi-Arid Smallholder India. *Clim. Dev.* **2023**, *16*, 289–300. [CrossRef]
31. Sánchez-Rodríguez, A.R.; Del Campillo, M.C.; Quesada-Moraga, E. Beauveria Bassiana: An Entomopathogenic Fungus Alleviates Fe Chlorosis Symptoms in Plants Grown on Calcareous Substrates. *Sci. Hortic.* **2015**, *197*, 193–202. [CrossRef]
32. Dannon, H.F.; Dannon, A.E.; Douro-Kpindou, O.K.; Zinsou, A.V.; Houndete, A.T.; Toffa-Mehinto, J.; Elegbede, I.A.T.M.; Olou, B.D.; Tamò, M. Toward the Efficient Use of Beauveria Bassiana in Integrated Cotton Insect Pest Management. *J. Cotton Res.* **2020**, *3*, 24. [CrossRef]
33. Lopez, D.C.; Sword, G.A. The Endophytic Fungal Entomopathogens Beauveria Bassiana and Purpureocillium Lilacinum Enhance the Growth of Cultivated Cotton (*Gossypium hirsutum*) and Negatively Affect Survival of the Cotton Bollworm (*Helicoverpa zea*). *Biol. Control* **2015**, *89*, 53–60. [CrossRef]
34. Kanasagra, J.R.; Valu, M.; Raval, L.J.; Rupapara, S. Heterosis, Combining Ability and Gene Action for Seed Cotton Yield and Its Contributing Characters in Cotton (*Gossypium hirsutum* L.). *Pharma Innov. J.* **2022**, *11*, 2050–2056.
35. Khashaba, E.H.K. Inoculation and Colonization of Isolated Entomopathogenic Fungi Beauveria Bassiana in Rice Plants, *Oryza sativa* L. through Seed Immersion Method. *Egypt. J. Biol. Pest. Control* **2021**, *31*, 92. [CrossRef]
36. Nawaz, A.; Razzaq, F.; Razzaq, A.; Gogi, M.D.; Fernández-Grandon, G.M.; Tayib, M.; Ayub, M.A.; Sufyan, M.; Shahid, M.R.; Qayyum, M.A.; et al. Compatibility and Synergistic Interactions of Fungi, Metarhizium Anisopliae, and Insecticide Combinations against the Cotton Aphid, Aphis Gossypii Glover (Hemiptera: Aphididae). *Sci. Rep.* **2022**, *12*, 4843. [CrossRef]
37. Dara, S.K. Non-Entomopathogenic Roles of Entomopathogenic Fungi in Promoting Plant Health and Growth. *Insects* **2019**, *10*, 277. [CrossRef]
38. Kasim Ongun, A.; Karademir, C. The Effect of Application of Different Doses of Acid Mixture Seed Coating Method on Cotton (*Gossypium hirsutum* L.) Yield and Fiber Technological Characteristics. Technology and Engineering Management. *J. Agron. Technol. Eng. Manag.* **2022**, *2022*, 841. [CrossRef]
39. Canassa, F.; Tall, S.; Moral, R.A.; de Lara, I.A.R.; Delalibera, I.; Meyling, N.V. Effects of Bean Seed Treatment by the Entomopathogenic Fungi Metarhizium Robertsii and Beauveria Bassiana on Plant Growth, Spider Mite Populations and Behavior of Predatory Mites. *Biol. Control* **2019**, *132*, 199–208. [CrossRef]
40. Quesada-Moraga, E.; López-Díaz, C.; Landa, B.B. The Hidden Habit of the Entomopathogenic Fungus Beauveria Bassiana: First Demonstration of Vertical Plant Transmission. *PLoS ONE* **2014**, *9*, e89278. [CrossRef]
41. Jaber, L.R.; Enkerli, J. Effect of Seed Treatment Duration on Growth and Colonization of Vicia Faba by Endophytic Beauveria Bassiana and Metarhizium Brunneum. *Biol. Control* **2016**, *103*, 187–195. [CrossRef]
42. Brownbridge, M.; Reay, S.D.; Nelson, T.L.; Glare, T.R. Persistence of Beauveria Bassiana (Ascomycota: Hypocreales) as an Endophyte Following Inoculation of Radiata Pine Seed and Seedlings. *Biol. Control* **2012**, *61*, 194–200. [CrossRef]
43. Rasool, S.; Vidkjær, N.H.; Hooshmand, K.; Jensen, B.; Fomsgaard, I.S.; Meyling, N.V. Seed Inoculations with Entomopathogenic Fungi Affect Aphid Populations Coinciding with Modulation of Plant Secondary Metabolite Profiles across Plant Families. *New Phytol.* **2021**, *229*, 1715–1727. [CrossRef]
44. Liu, Y.; Yang, Y.; Wang, B. Entomopathogenic Fungi Beauveria Bassiana and Metarhizium Anisopliae Play Roles of Maize (*Zea mays*) Growth Promoter. *Sci. Rep.* **2022**, *12*, 15706. [CrossRef]
45. Mantzoukas, S.; Daskalaki, E.; Kitsiou, F.; Papantzikos, V.; Servis, D.; Bitivanos, S.; Patakioutas, G.; Eliopoulos, P.A. Dual Action of Beauveria Bassiana (Hypocreales; Cordycipitaceae) Endophytic Stains as Biocontrol Agents against Sucking Pests and Plant Growth Biostimulants on Melon and Strawberry Field Plants. *Microorganisms* **2022**, *10*, 2306. [CrossRef]

46. Afandhi, A.; Widjayanti, T.; Emi, A.A.L.; Tarno, H.; Afiyanti, M.; Handoko, R.N.S. Endophytic Fungi Beauveria Bassiana Balsamo Accelerates Growth of Common Bean (*Phaeseolus vulgaris* L.). *Chem. Biol. Technol. Agric.* **2019**, *6*, 11. [CrossRef]
47. El Husseini, M.M.; Bochow, H.; Junge, H. The Biofertilising Effect of Seed Dressing with PGPR Bacillus Amyloliquefaciens FZB 42 Combined with Two Levels of Mineral Fertilising in African Cotton Production. *Arch. Phytopathol. Plant Prot.* **2012**, *45*, 2261–2271. [CrossRef]
48. Irizarry, I.; White, J.F. Bacillus Amyloliquefaciens Alters Gene Expression, ROS Production and Lignin Synthesis in Cotton Seedling Roots. *J. Appl. Microbiol.* **2018**, *124*, 1589–1603. [CrossRef]
49. Bakr, E.M. A New Software for Measuring Leaf Area, and Area Damaged by Tetranychus Urticae Koch. *J. Appl. Entomol.* **2005**, *129*, 173–175. [CrossRef]
50. Priya, S.; Ghosh, R. Monitoring Effects of Heavy Metal Stress on Biochemical and Spectral Parameters of Cotton Using Hyperspectral Reflectance. *Environ. Monit. Assess.* **2023**, *195*, 112. [CrossRef]
51. Lichtenthaler, H.K.; Buschmann, C. Chlorophylls and Carotenoids: Measurement and Characterization by UV-VIS Spectroscopy. *Curr. Protoc. Food Anal. Chem.* **2001**, *1*, F4.3.1–F4.3.8. [CrossRef]
52. Carillo, P.; Gibon, Y. PROTOCOL: Extraction and Determination of Proline. *Prometh. Wiki* **2011**, *2011*, 1–5.
53. Mnasri, N.; Chennaoui, C.; Gargouri, S.; Mhamdi, R.; Hessini, K.; Elkahoui, S.; Djébali, N. Efficacy of Some Rhizospheric and Endophytic Bacteria in Vitro and as Seed Coating for the Control of Fusarium Culmorum Infecting Durum Wheat in Tunisia. *Eur. J. Plant Pathol.* **2017**, *147*, 501–515. [CrossRef]
54. García-Espinoza, F.; García, M.J.; Quesada-Moraga, E.; Yousef-Yousefa, M. Entomopathogenic Fungus-Related Priming Defense Mechanisms in Cucurbits Impact Spodoptera Littoralis (Boisduval) Fitness. *Appl. Environ. Microbiol.* **2023**, *89*, e00940-23. [CrossRef]
55. Faiz, A.; Faiz, A.; Wang, W.; Bennett, C. Sustainable Rural Roads for Livelihoods and Livability. *Procedia Soc. Behav. Sci.* **2012**, *53*, 1–8. [CrossRef]
56. Mantzoukas, S.; Papantzikos, V.; Katsogiannou, S.; Papanikou, A.; Koukidis, C.; Servis, D.; Eliopoulos, P.; Patakioutas, G. Biostimulant and Bioinsecticidal Effect of Coating Cotton Seeds with Endophytic Beauveria Bassiana in Semi-Field Conditions. *Microorganisms* **2023**, *11*, 2050. [CrossRef]
57. Kuzhuppillymyal-Prabhakarankutty, L.; Ferrara-Rivero, F.H.; Tamez-Guerra, P.; Gomez-Flores, R.; Rodríguez-Padilla, M.C.; Ek-Ramos, M.J. Effect of Beauveria Bassiana-Seed Treatment on *Zea mays* L. Response against Spodoptera Frugiperda. *Appl. Sci.* **2021**, *11*, 2887. [CrossRef]
58. Mishra, S.; Kumar, P.; Malik, A. Preparation, Characterization, and Insecticidal Activity Evaluation of Three Different Formulations of Beauveria Bassiana against Musca Domestica. *Parasitol. Res.* **2013**, *112*, 3485–3495. [CrossRef]
59. Batool, R.; Umer, M.J.; Wang, Y.; He, K.; Zhang, T.; Bai, S.; Zhi, Y.; Chen, J.; Wang, Z. Synergistic Effect of Beauveria Bassiana and Trichoderma Asperellum to Induce Maize (*Zea mays* L.) Defense against the Asian Corn Borer, Ostrinia Furnacalis (Lepidoptera, Crambidae) and Larval Immune Response. *Int. J. Mol. Sci.* **2020**, *21*, 8215. [CrossRef]
60. Luo, L.; Zhao, C.; Wang, E.; Raza, A.; Yin, C. Bacillus Amyloliquefaciens as an Excellent Agent for Biofertilizer and Biocontrol in Agriculture: An Overview for Its Mechanisms. *Microbiol. Res.* **2022**, *259*, 127016. [CrossRef]
61. Bisht, N.; Mishra, S.K.; Chauhan, P.S. Bacillus Amyloliquefaciens Inoculation Alters Physiology of Rice (*Oryza sativa* L. Var. IR-36) through Modulating Carbohydrate Metabolism to Mitigate Stress Induced by Nutrient Starvation. *Int. J. Biol. Macromol.* **2020**, *143*, 937–951. [CrossRef] [PubMed]
62. Kramski, D.J.; Nowinski, D.; Kowalczuk, K.; Kruszyński, P.; Radzimska, J.; Greb-Markiewicz, B. Beauveria Bassiana Water Extracts' Effect on the Growth of Wheat. *Plants* **2023**, *12*, 326. [CrossRef] [PubMed]
63. Sánchez-Rodríguez, A.R.; Raya-Díaz, S.; Zamarreño, Á.M.; García-Mina, J.M.; del Campillo, M.C.; Quesada-Moraga, E. An Endophytic Beauveria Bassiana Strain Increases Spike Production in Bread and Durum Wheat Plants and Effectively Controls Cotton Leafworm (*Spodoptera littoralis*) Larvae. *Biol. Control* **2018**, *116*, 90–102. [CrossRef]
64. Rivas-Franco, F.; Hampton, J.G.; Morán-Diez, M.E.; Narciso, J.; Rostás, M.; Wessman, P.; Jackson, T.A.; Glare, T.R. Effect of Coating Maize Seed with Entomopathogenic Fungi on Plant Growth and Resistance against Fusarium Graminearum and Costelytra Giveni. *Biocontrol Sci. Technol.* **2019**, *29*, 877–900. [CrossRef]
65. Shaalan, R.S.; Gerges, E.; Habib, W.; Ibrahim, L. Endophytic Colonization by Beauveria Bassiana and Metarhizium Anisopliae Induces Growth Promotion Effect and Increases the Resistance of Cucumber Plants against Aphis Gossypii. *J. Plant Prot. Res.* **2021**, *61*, 358. [CrossRef]
66. Qin, X.; Zhao, X.; Huang, S.; Deng, J.; Li, X.; Luo, Z.; Zhang, Y. Pest Management via Endophytic Colonization of Tobacco Seedlings by the Insect Fungal Pathogen Beauveria Bassiana. *Pest. Manag. Sci.* **2021**, *77*, 2007–2018. [CrossRef]
67. Martins, T.; Barros, A.N.; Rosa, E.; Antunes, L. Enhancing Health Benefits through Chlorophylls and Chlorophyll-Rich Agro-Food: A Comprehensive Review. *Molecules* **2023**, *28*, 5344. [CrossRef]
68. Akter, T.; Mimma, A.A.; Haque, M.A.; Hossain, M.M.; Ghosh, T.K.; Zinan, N.; Chowdhury, M.Z.H.; Islam, S.M.N. Seed Priming with Beauveria Bassiana Improves Growth and Salt Stress Response in Rice. *Environ. Exp. Bot.* **2023**, *213*, 105427. [CrossRef]
69. Veloz-Badillo, G.M.; Riveros-Ramírez, J.; Angel-Cuapio, A.; Arce-Cervantes, O.; Flores-Chávez, B.; Espitia-López, J.; Loera, O.; Garza-López, P.M. The Endophytic Capacity of the Entomopathogenic Fungus Beauveria Bassiana Caused Inherent Physiological Response in Two Barley (*Hordeum vulgare*) Varieties. *3 Biotech.* **2019**, *9*, 12. [CrossRef]

70. Saragih, M.; Trizelia; Nurbailis; Yusniwati. Endophytic Colonization and Plant Growth Promoting Effect by Entomopathogenic Fungus, Beauveria Bassiana to Red Chili (*Capsicum annuum* L.) with Different Inoculation Methods. *IOP Conf. Ser. Earth Environ. Sci.* **2019**, *305*, 12070. [CrossRef]
71. Homayoonzadeh, M.; Esmaeily, M.; Talebi, K.; Allahyari, H.; Reitz, S.; Michaud, J.P. Inoculation of Cucumber Plants with Beauveria Bassiana Enhances Resistance to Aphis Gossypii (Hemiptera: Aphididae) and Increases Aphid Susceptibility to Pirimicarb. *Eur. J. Entomol.* **2022**, *119*, 1–11. [CrossRef]
72. Zulfiqar, F.; Ashraf, M. Proline Alleviates Abiotic Stress Induced Oxidative Stress in Plants. *J. Plant Growth Regul.* **2023**, *42*, 4629–4651. [CrossRef]
73. Card, S.D.; Bastías, D.A.; Caradus, J.R. Antagonism to Plant Pathogens by Epichloë Fungal Endophytes—A Review. *Plants* **2021**, *10*, 1997. [CrossRef] [PubMed]
74. Schulz, B.; Römmert, A.K.; Dammann, U.; Aust, H.J.; Strack, D. The Endophyte-Host Interaction: A Balanced Antagonism? *Mycol. Res.* **1999**, *103*, 1275–1283. [CrossRef]
75. Prabhukarthikeyan, R.; Saravanakumar, D.; Raguchander, T. Combination of Endophytic Bacillus and Beauveria for the Management of Fusarium Wilt and Fruit Borer in Tomato. *Pest. Manag. Sci.* **2014**, *70*, 1742–1750. [CrossRef]
76. Komagata, Y.; Sekine, T.; Oe, T.; Kakui, S.; Yamanaka, S. Simultaneous Use of Beauveria Bassiana and Bacillus Subtilis-Based Biopesticides Contributed to Dual Control of Trialeurodes Vaporariorum (Hemiptera: Aleyrodidae) and Tomato Powdery Mildew without Antagonistic Interactions. *Egypt. J. Biol. Pest. Control* **2024**, *34*, 18. [CrossRef]
77. Malik, M.A.; Ahmad, S.J.N.; Ahmad, J.N.; Abbasi, A.; Sufyan, M.; Arif, M.J. Efficacy of Bacillus Thuringiensis and Beauveria Bassiana against Red Palm Weevil Rhynchophorus Ferrugineus Olivier (Coleoptera: Curculionidae). *Afr. Entomol.* **2019**, *27*, 386–394. [CrossRef]
78. Sharma, A.; Thakur, N.; Hashem, A.; Dawoud, T.M.; Abd_Allah, E.F. Insecticidal Potential of Bacillus Thuringiensis, Beauveria Bassiana and Metarhizium Anisopliae Individually and Their Synergistic Effect with Barazide against Spodoptera Litura. *Heliyon* **2024**, *10*, e37175. [CrossRef]
79. Wang, X.M.; Yang, B.; Wang, H.W.; Yang, T.; Ren, C.G.; Zheng, H.L.; Dai, C.C. Consequences of Antagonistic Interactions between Endophytic Fungus and Bacterium on Plant Growth and Defense Responses in Atractylodes Lancea. *J. Basic. Microbiol.* **2015**, *55*, 659–670. [CrossRef]
80. Alquéres, S.; Meneses, C.; Rouws, L.; Rothballer, M.; Baldani, I.; Schmid, M.; Hartmann, A. The Bacterial Superoxide Dismutase and Glutathione Reductase Are Crucial for Endophytic Colonization of Rice Roots by Gluconacetobacter Diazotrophicus PAL5. *Mol. Plant-Microbe Interact.* **2013**, *26*, 937–945. [CrossRef]
81. Mack, K.M.L.; Rudgers, J.A.; Mack, K.M.L.; Rudgers, J.A. Balancing Multiple Mutualists: Asymmetric Interactions among Plants, Arbuscular Mycorrhizal Fungi, and Fungal Endophytes. *Oikos* **2008**, *117*, 310–320. [CrossRef]
82. Rasmussen, S.; Parsons, A.J.; Bassett, S.; Christensen, M.J.; Hume, D.E.; Johnson, L.J.; Johnson, R.D.; Simpson, W.R.; Stacke, C.; Voisey, C.R.; et al. High Nitrogen Supply and Carbohydrate Content Reduce Fungal Endophyte and Alkaloid Concentration in Lolium Perenne. *New Phytol.* **2007**, *173*, 787–797. [CrossRef] [PubMed]
83. Liu, Q.; Parsons, A.J.; Xue, H.; Fraser, K.; Ryan, G.D.; Newman, J.A.; Rasmussen, S. Competition between Foliar *Neotyphodium Lolii* Endophytes and Mycorrhizal *Glomus* spp. Fungi in *Lolium perenne* Depends on Resource Supply and Host Carbohydrate Content. *Funct. Ecol.* **2011**, *25*, 910–920. [CrossRef]
84. Omacini, M.; Eggers, T.; Bonkowski, M.; Gange, A.C.; Jones, T.H. Leaf Endophytes Affect Mycorrhizal Status and Growth of Co-Infected and Neighbouring Plants. *Funct. Ecol.* **2006**, *20*, 226–232. [CrossRef]
85. Russo, M.L.; Scorsetti, A.C.; Vianna, M.F.; Allegrucci, N.; Ferreri, N.A.; Cabello, M.N.; Pelizza, S.A. Effects of Endophytic Beauveria Bassiana (Ascomycota: Hypocreales) on Biological, Reproductive Parameters and Food Preference of the Soybean Pest Helicoverpa Gelotopoeon. *J. King Saud. Univ. Sci.* **2019**, *31*, 1077–1082. [CrossRef]
86. Zimmermann, G. The 'Galleria Bait Method' for Detection of Entomopathogenic Fungi in Soil. *J. Appl. Entomol.* **1986**, *102*, 213–215. [CrossRef]
87. Kannadan, S.; Rudgers, J.A. Endophyte Symbiosis Benefits a Rare Grass under Low Water Availability. *Funct. Ecol.* **2008**, *22*, 706–713. [CrossRef]
88. Akello, J.; Dubois, T.; Gold, C.S.; Coyne, D.; Nakavuma, J.; Paparu, P. Beauveria Bassiana (Balsamo) Vuillemin as an Endophyte in Tissue Culture Banana (*Musa* spp.). *J. Invertebr. Pathol.* **2007**, *96*, 34–42. [CrossRef]
89. Jaber, L.R.; Vidal, S. Fungal Endophyte Negative Effects on Herbivory Are Enhanced on Intact Plants and Maintained in a Subsequent Generation. *Ecol. Entomol.* **2010**, *35*, 25–36. [CrossRef]
90. Wagner, B.L.; Lewis, L.C. Colonization of Corn, *Zea mays*, by the Entomopathogenic Fungus Beauveria Bassiana. *Appl. Environ. Microbiol.* **2000**, *66*, 3468. [CrossRef]
91. McKinnon, A.C.; Saari, S.; Moran-Diez, M.E.; Meyling, N.V.; Raad, M.; Glare, T.R. Beauveria Bassiana as an Endophyte: A Critical Review on Associated Methodology and Biocontrol Potential. *BioControl* **2017**, *62*, 1–17. [CrossRef]

agronomy

MDPI

Article

# Using Multispectral Data from UAS in Machine Learning to Detect Infestation by *Xylotrechus chinensis* (Chevrolat) (Coleoptera: Cerambycidae) in Mulberries

**Christina Panopoulou [1], Athanasios Antonopoulos [1], Evaggelia Arapostathi [1], Myrto Stamouli [1], Anastasios Katsileros [2] and Antonios Tsagkarakis [1,*]**

[1] Laboratory of Sericulture and Apiculture, Agricultural University of Athens, 11855 Athens, Greece; stud118114@aua.gr (C.P.); antonopoulos@aua.gr (A.A.); e.arapostathi@aua.gr (E.A.); myrtostamouli@aua.gr (M.S.)

[2] Laboratory of Plant Breeding and Biometry, Agricultural University of Athens, 11855 Athens, Greece; katsileros@aua.gr

* Correspondence: atsagarakis@aua.gr; Tel.: +30-210-529-4578

**Abstract:** The tiger longicorn beetle, *Xylotrechus chinensis* Chevrolat (Coleoptera: Cerambycidae), has posed a significant threat to mulberry trees in Greece since its invasion in 2017, which may be associated with global warming. Detection typically relies on observing adult emergence holes on the bark or dried branches, indicating severe damage. Addressing pest threats linked to global warming requires efficient, targeted solutions. Remote sensing provides valuable, swift information on vegetation health, and combining these data with machine learning techniques enables early detection of pest infestations. This study utilized airborne multispectral data to detect infestations by *X. chinensis* in mulberry trees. Variables such as mean NDVI, mean NDRE, mean EVI, and tree crown area were calculated and used in machine learning models, alongside data on adult emergence holes and temperature. Trees were classified into two categories, infested and healthy, based on *X. chinensis* infestation. Evaluated models included Random Forest, Decision Tree, Gradient Boosting, Multi-Layer Perceptron, K-Nearest Neighbors, and Naïve Bayes. Random Forest proved to be the most effective predictive model, achieving the highest scores in accuracy (0.86), precision (0.84), recall (0.81), and F-score (0.82), with Gradient Boosting performing slightly lower. This study highlights the potential of combining remote sensing and machine learning for early pest detection, promoting timely interventions, and reducing environmental impacts.

**Keywords:** climate change; early detection; machine learning; multispectral imagery; mulberry tree random forest; remote sensing; tiger longicorn beetle; UAS

## 1. Introduction

Global warming exerts strong effects on various organisms, including insects. Abiotic factors such as solar radiation, relative humidity, rainfall, and temperature define the suitable ecosystem for each species to inhabit. Insect populations, physiology, phenology, and abundance tend to be susceptible to changes as climate conditions become warmer [1]. Thus, temperature increases, notably in mid to high latitudes, are responsible for insects' rapid development, high reproductive potential, survival, and population rearrangement [2,3]. Given that plants cannot spread out in a short period of time, climatic warming may enlarge the range of herbivores poleward, into previously uninhabited, regions of their host plants' distributions [1].

The tiger longicorn beetle, *Xylotrechus chinensis* Chevrolat (Coleoptera: Cerambycidae), is a newly invasive woodborer species in Greece that originates from Eastern Asia. It causes severe damage to the mulberry trees (*Morus* spp.), although there are references for damage in species of the Rosaceae family (*Malus* spp., *Pyrus* spp.) and *Vitis vinifera* in its area of origin [4–6]. However, in Europe, it has been shown that this pest so far only attacks *Morus* spp., with a previous study proving that *X. chinensis* has not used *Vitis vinifera* as an alternative host plant under laboratory conditions [6]. Concerning its presence in Europe, initially it has been imported due to international commerce and mainly through wooden packaging materials and objects [4,7,8]. It is possible that its establishment in Spain, France, Italy, and Greece in the EPPO region (Figure 1) was facilitated by the warmer temperature conditions in the Mediterranean area. The first recorded infestation of *X. chinensis* in Greece was recorded in Heraklion, Crete, in 2017, where 200 infected mulberry trees were observed, and approximately 15% of the mulberry tree population eventually succumbed to the pest's infestation. Considering the insect's life cycle, its introduction to Greece probably occurred around 2014–2015 [5,8]. Until 2019, the distribution of the insect in Greece had reached considerable proportions, notably affecting the mulberry tree population in Athens, where 1300 out of 20,000 trees displayed severe damage [4].

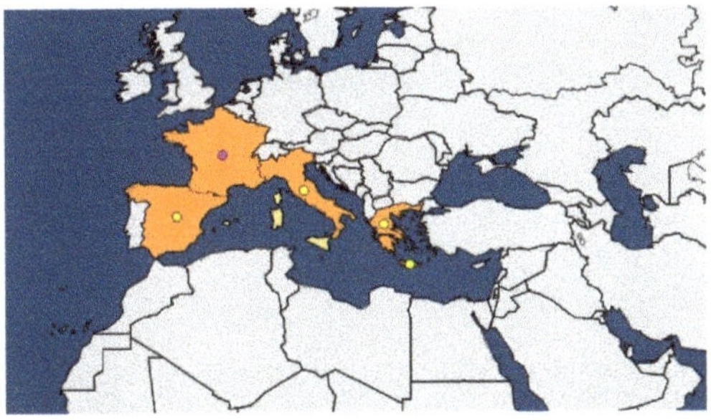

**Figure 1.** Distribution of *X. chinensis* in Europe. The species is present in Spain, Italy, and Greece (yellow dot), while in France it is transient (purple dot) (EPPO Global Database, 2023, www.gd.eppo.int, accessed on 30 June 2024).

Generally, the impacts of the climatic variation on insect phenology are associated with life cycle alterations. In particular, voltinism, emergence, and the duration of the life cycle are likely to be affected [1]. Regarding the influence of temperature rise, especially on bark beetles, the decrease in the duration of the life cycle from 2 years to 1 in the case of *Dendroctonus rufipennis* is demonstrated (Coleoptera: Curculionidae) [9]. Current data in Greece do not testify any differences concerning the duration of *X. chinensis'* life cycle in the areas in which it occurs. Additionally, warm climate conditions correlate with distribution expansion. In the EFSA Panel on Plant Health (2021), after comparing the eight Köppen–Geiger climate types occurring in countries where *X. chinensis* has been reported and the climate types occurring in the European Union, it is concluded that the pest may be established in the largest part of EU territory [7]. Nevertheless, there is a risk that temperature warming may provoke a population surge in northern regions or may decrease the required period that a generation needs to be completed. Taking into consideration the insects' outbreaks, [10] note that a warming trend of 3.3 °C in minimum winter temperatures for more than 40 years generated a population burst of *Dendroctonus frontalis* (Coleoptera: Curculionidae).

Currently, the insect's biological cycle consists of four stages: egg, larva, pupa, and adult. Adults (Figure 2) emerge between May and June, feed on leaves and stems of mulberries and after their sexual reproduction, the females deposit their eggs on the bark of the trunk of the trees [4–6]. During midsummer, the larvae penetrate the trunk and bore elongated tunnels within the phloem while they are feeding on them. They overwinter as developed larvae and pupate into the xylem [4–6,8].

**Figure 2.** Adult of *X. chinensis* on the trunk of a mulberry tree in Agricultural University of Athens.

The result of the feeding activity of the larvae is the gradual destruction of the xylem tissue of the trees, so water and inorganic compounds cannot be transported and distributed through it. During spring, adult emergence holes are visible on the bark of the trees in the middle and late stages of the infestation. Eventually, the decline and death of the tree occur (Figure 3) [6,8].

**Figure 3.** Symptoms of the pest damage in mulberry trees in the orchard of the Agricultural University of Athens, Greece. (**A**). Adult emergence holes of X. *chinensis* on the bark of a mulberry tree (**B**). Bark discoloration by the activity of the larvae of the pest (**C**). Dried sprouts on a mulberry tree in the orchard of Agricultural University of Athens.

Apart from insects, global warming poses a significant threat for plants. Environmental change, particularly temperature increase, can affect plants' species range, abundance, and phenology [11,12]. Climatic alteration, primarily temperature augmentation, often leads to drought events and consequently to water deprivation. Water stress restricts hydraulic functions inside the plants and destroys the production, transport, and availability of nonstructural carbohydrates (NSC) [13]. These NSCs are related to plant defense compounds, and as their production decreases, trees are likely to be more vulnerable to insect attacks [13–19]. Spring drought conditions have been shown to increase infestations of certain insects, such as *Ips typographus* (Coleoptera: Curculionidae) [14]. Climate warming is also linked to reduced precipitation. Mulberries in Greece are used, mainly, as ornamental plants on the streets or for their shadow-providing role. In these cases, where no economic interest is involved, trees receive little to no irrigation, making rainwater their exclusive water source. Due to drought events and as the water stress rises, mulberries are more likely to be attacked by *X. chinensis.*

The timing of *X. chinensis* attacks on mulberry trees does not synchronize with the appearance of the symptoms. Due to the severe damage and the stealth action of the larvae, early detection of the infestation is crucial to take timely control measures and limit the spread [15,16]. This is feasible by using remote sensing for the recording of the spectral signature of a plant. A plant's spectral signature is the reflectance of electromagnetic radiation by its tissues [17]. Normally, a healthy plant absorbs the radiation of the blue and red bands of the electromagnetic spectrum, using it in photosynthesis, while it reflects the radiation of the green and NIR (near infra-red) radiation of the spectrum [17,18]. Combining the reflectance in different spectral bands, phytosanitary differences can be highlighted. Mathematical relations using the reflectance in different spectral bands, called vegetation indices, have been created, and they provide information on the state of the health of the plants [17–20].

Remote sensing has already been used for phytosanitary monitoring to minimize crop loss. Besides that, it can forecast favorable conditions that are responsible for pest outbreaks caused by climate warming, thus contributing to pest control optimization [21].

This study presents the methodology for early detection of infestations of tiger longicorn beetle, *X. chinensis*, in mulberries at the Agricultural University of Athens using remote sensing data and machine-learning models. Multispectral data were captured by attaching a multispectral sensor to a UAV. Mean NDVI, mean NDRE, mean EVI, and tree crown area were calculated and used as predictors in machine learning models, along with mean temperature data and data of the adult emergence holes, to determine whether detection of the infestation by this pest can be successfully achieved.

## 2. Materials and Methods

### *2.1. UAV Flight Schedule and On-Site Observations*

The study was conducted during 2022 and 2023 in the mulberry orchards of the Agricultural University of Athens. The flight schedule was devised following the biological cycle of *X. chinensis*, targeting the stages of adult emergence and active larvae. Late spring and early summer corresponded with pupation and adult emergence, marked by the appearance of emergence holes. Summer involved reproduction and oviposition, while late summer and autumn corresponded to the active larval phase, characterized by severe infestation symptoms like weakening and drying of leaves and branches due to sap flow disruption.

Considering the biological cycle and the appearance of the symptoms, four aerial missions with UAV were executed, with two flights per year (July and September of 2022, June and September of 2023). The first flight of each year coincided with the pivotal stage of adult emergence, while the second aligned with the active larvae. Flights were conducted at 13:00 to ensure optimal lighting and minimize shadows. In addition, during the midday hours, the maximum daily temperature was recorded, which increases the evapotranspiration of the plant, amplifying stress symptoms in mulberry trees.

The UAV used was the 2021 model "Mera" (UcanDrone S.A., Koropi Attica, Greece), an EASA Class C2 quadcopter designed for aerial photography and mapping. The RedEdge-MX multispectral camera manufactured by MicaSense (AgEagle Aerial Systems Inc., Wichita, KS, USA) was attached to the "Mera" UAV to acquire the multispectral data from the mulberry trees (Figure 4). UAV data were preferred over satellite imagery due to the high cost and low accuracy of the latter. UAVs offer real-time capabilities, cost efficiency, precise data collection, and extensive data acquisition [22]. All the flight missions were designed and executed in the Mission Planner program (ArduPilot Development Team, New York, NY, USA).

**Figure 4.** Quadcopter "Mera" (UcanDrone S.A., Koropi Attica, Greece) with the attached multispectral camera, MicaSense RedEdge MX (AgEagle Aerial Systems Inc., Wichita, KS, USA).

Flight parameters for the four missions included an altitude of 70 m, a speed of 5 m/s, and an 80% image overlap in both forward and backward directions. The RedEdge MX camera (AgEagle Aerial Systems Inc., Wichita, KS, USA) captured multispectral images at one frame every 2 s. All flights were conducted by a certified A2 category pilot, adhering to existing legislation.

Field scouting in the mulberry orchard occurred concurrently with each flight, aiming to observe and document the health status of individual trees, identify potential disease and pest symptoms, and document prevalent weed species. The field scouting extended beyond typical pest and disease considerations to encompass various stress factors affecting the mulberry trees, such as symptoms of nutrient deficiencies. During scouting sessions, adult specimens of *X. chinensis*, adult emergence holes, galleries, and cracks on the trunk were observed. The emergence holes were tallied for each tree during on-site sessions.

The health status of mulberry trees was categorized into two classes during on-site observations:

1. Healthy tree: maintaining full turgor pressure in leaves during midday, showing no discernible indications of debilitation or abnormal growth.
2. Tree with *X. chinensis* infestation: displaying generalized weakening, dried shoots, yellow leaves, and adult emergence holes.

### 2.2. Multispectral Data Processing

Multispectral imagery was processed using Metashape (Version 1.7.3) (Agisoft LLC., St. Petersburg, Russia) for photogrammetric processing and 3D model generation. Two-dimensional pre-georeferenced images were imported, and image alignment discerned shared points for precise relative positioning, resulting in a Dense Point Cloud (X, Y, Z coordinates).

The Dense Point Cloud facilitated the generation of a Digital Surface Model (DSM), representing the Earth's surface and reflective features. The DSM is a comprehensive representation of the Earth's surface, encompassing all reflective elements, whether natural or man-made. These elements include vegetation, water bodies, buildings, roads, bridges, and more. The DSM serves as a depiction of the Earth's topography, providing elevation data for both terrestrial and aquatic features, as well as vegetative and man-made components. Additionally, each pixel of the DSM is embedded with georeferencing, enhancing its geographical and spatial accuracy.

Utilizing the Dense Point Cloud once more, the Digital Terrain Model (DTM) is generated, which delineates the Earth's surface, providing elevation information pertaining to the topographic features, excluding both natural vegetation and man-made structures. This model represents the elevation characteristics of the bare terrain while imparting additional topographical information, including slope, orientation, and both horizontal and vertical curvature of the terrain.

Orthomosaic maps (Figure 5), created from the DSM and DTM, comprised distortion-free orthophotos for detailed area representation. These maps are composed of smaller orthophotos and are constructed utilizing the spectral bands of the visible spectrum of electromagnetic radiation, specifically the Red, Green, and Blue (RGB). Each small orthophoto is precisely georeferenced, resulting in each pixel in the final map corresponding to a specific geographic point on the Earth's surface with well-defined coordinates (X, Y, Z). The orthophotos essentially represent aerial images that have undergone correction processes to eliminate distortions induced by various factors, including lens characteristics, capture angles, and the Earth's topographical variations. Therefore, the orthomosaic maps provide a highly detailed and accurate representation of the Earth's surface, encompassing both natural and man-made features, all presented with high spatial resolution.

**Figure 5.** Orthomosaic map of the mulberry orchard (flight of 28 June 2023).

The DSM, DTM, and orthomosaic maps were integrated into QGIS (Version 3.32.0) (QGIS Development Team, London, UK) for further processing and map creation. QGIS enhanced the exhibition and representation of data, making geospatial information more accessible and interpretable.

For mulberry tree health assessment, the orthomosaic map was analyzed using Object-Based Classification (OBIA) in the QGIS Orfeo Toolbox (Figure 6). Three vegetation indices (VIs) were applied: Normalized Difference Vegetation Index (NDVI), Normalized Difference Red Edge Index (NDRE), and Enhanced Vegetation Index (EVI). These indices provided insights into tree health based on spectral characteristics.

**Figure 6.** Classified output of the Object-Based Classification for the airborne data of the 28 June 2023 flight.

The NDVI is derived through the mathematical expression [23–25]:

$$NDVI = (R_{NIR} - R_{RED})/(R_{NIR} + R_{RED})$$

In this equation, $R_{NIR}$ represents the reflection of solar radiation within the near infrared band, at 842 nm, while $R_{RED}$ expresses the reflection of solar radiation within the red band, specifically at 668 nm. The NDVI values resulting from this computation range from −1 to +1. Negative values indicate non-vegetated surfaces, encompassing areas such as bare ground, rocky or sandy terrains, water bodies, and urban spaces. A zero NDVI denotes very sparse or stressed vegetation, including dried or aged plants. Positive NDVI values approaching +1 correlate with increasingly healthier and more robust vegetation. Conversely, low positive values suggest vegetation that is sparse or compromised. This calculated NDVI serves as a quantitative metric, offering a precise and standardized measure for the evaluation of vegetation health and distribution based on the unique spectral characteristics of the near infrared and red bands.

The NDRE is calculated using the mathematical formula [24,26,27]:

$$NDRE = (R_{NIR} - R_{RED\ EDGE})/(R_{NIR} + R_{RED\ EDGE})$$

In this formula, $R_{NIR}$ denotes the reflection of solar radiation within the NIR band, specifically at 842 nm, while $R_{RED\ EDGE}$ signifies the reflection of solar radiation within the red edge band, at 717 nm. The NDRE also ranges from −1 to +1. Higher values indicate good health of the vegetation, while lower NDRE values indicate stressed vegetation.

The EVI is computed through the mathematical expression [28]:

$$EVI = 2.5 \times (R_{NIR} - R_{RED}/(R_{NIR} + 6 \times R_{RED} - 7.5 \times R_{BLUE}) + 1)$$

In this mathematical expression, $R_{NIR}$ represents the reflection of solar radiation within the near Infrared band, at 842 nm and $R_{RED}$ represents the reflection of solar radiation within the red band, at 668 nm. The blue represents the reflection of solar radiation within the blue band, at 475 nm. This index is suitable for high biomass areas characterized by a dense canopy. The range of values for the EVI spans from −1 to 1, with values indicative of robust vegetation within the range of 0.20 to 0.80.

### 2.3. Statistical Analysis

Machine learning algorithms were used to construct a predictive model for the identification of the tiger longicorn beetle infestation, *X. chinensis*, by considering variables such as mean NDVI, mean NDRE, mean EVI, mean temperature, number of adult emergence holes, and the tree crown area (canopy). For the performance evaluation of the model on the aforementioned independent data, preprocessing was performed on the variable data, such as cleaning and scaling. The evaluated models were the following: Random Forest, Decision Tree, Gradient Boosting, Multi-Layer Perceptron, K-Nearest Neighbors, and Naïve Bayes. Training was performed for each model in 70% of the observations, utilizing the remaining 30% of the observations for validation purposes. The performance of the models was evaluated using the following metrics: accuracy, precision, recall, and F-measure.

Accuracy was calculated to determine which model was the most appropriate for utilization in the present study and is defined as follows [29–31]:

$$\text{Accuracy} = \text{TP} + \text{TN}/\text{TP} + \text{TN} + \text{FP} + \text{FN}$$

where TP = true positives, TN = true negatives, FP = false positives, and FN = false negatives.

Precision measures the accuracy of positive predictions made by a model and is defined as [29–31]:

$$\text{Precision} = \text{TP}/\text{TP} + \text{FP}$$

Recall is a metric that measures the ability of a model to capture all the relevant instances of a particular class and is defined as [29–31]:

$$\text{Recall} = \text{TP}/\text{TP} + \text{FN}$$

The F-measure is a metric that combines both precision and recall and is defined as [29–31]:

$$\text{F-measure} = 2 \times ((\text{Precision} \times \text{Recall})/(\text{Precision} + \text{Recall}))$$

Models were constructed and evaluated using Python (Version 3.10.10) and the libraries Pandas (Version 1.5.1), Scikit-learn (Version 1.1.3), Matplotlib (Version 3.6.2), Numpy (Version 1.23.4), Seaborn (Version 0.13.0), and XGBoost (Version 2.0.1).

The Decision Tree (DT) algorithm in machine learning constructs a tree-like model by recursively selecting the most informative features to partition the input data. It aims to maximize information gain for classification. The process continues until a predefined stopping criterion is met, producing a hierarchical structure of decision rules. Decision trees are interpretable and handle various data types well but may be prone to overfitting, which can be addressed through techniques like pruning [32].

Random Forest (RF) is a supervised machine learning classifier that uses multiple decision trees to make predictions. It is an ensemble learning method that corrects for decision trees' tendency to overfit their training set. Random Forest is widely used in pest and disease prediction due to its excellent classification results and fast implementation [33,34].

Gradient Boosting (GB) is an ensemble method that iteratively improves predictive performance by combining weak learners, usually decision trees, and correcting errors made by the ensemble. It uses the negative gradient of the loss function to guide the learning process [35].

Naïve Bayes (NB) is a probabilistic classification algorithm that uses Bayes' theorem with a simplified assumption of feature independence. It calculates class probabilities based on prior information and likelihood of observing features and assigns new instances to the class with the highest probability [36].

Multi-Layer Perceptron (MLP) is a neural network used in machine learning, consisting of input, hidden, and output layers, with nodes applying activation functions to weighted inputs. Trained via backpropagation and optimized using algorithms like stochastic gradient descent, MLPs excel in tasks like image recognition by capturing complex patterns in data [37].

K-Nearest Neighbors (KNN) is a basic machine-learning algorithm for classification and regression. It predicts the outcome for a new data point by considering the majority or average of its K closest neighbors in the training dataset, using a chosen distance metric like Euclidean distance. KNN is a simple and effective algorithm, but its performance can be sensitive to the choice of K and distance metric [38].

## 3. Results

### 3.1. Descriptive Statistics

Table 1 provides a summary of the descriptive statistics related to the predictor variables in this study, including the number of holes, mean NDVI, mean NDRE, mean EVI, mean T, and canopy size ($m^2$). Upon an initial inspection of variable values, conducted through both graphical representations and statistical tests, it was observed that there were no missing, duplicate, or extreme outlier values (Table 1).

**Table 1.** Descriptive statistics of the six predictors that were used in the models.

| | Holes | Canopy | Mean NDVI | Mean NDRE | Mean EVI | MeanT (°C) |
|---|---|---|---|---|---|---|
| Mean | 5.13 | 2.22 | 0.86 | 0.43 | 2.28 | 25.70 |
| Std | 9.97 | 1.63 | 0.08 | 0.10 | 2.96 | 2.00 |
| Min | 0 | −1.30 | 0.03 | −0.03 | −87.59 | 23.06 |
| 25% | 0 | 1.28 | 0.84 | 0.38 | 2.14 | 24.34 |
| Median | 2 | 1.89 | 0.89 | 0.45 | 2.34 | 27.23 |
| 75% | 6 | 2.67 | 0.91 | 0.50 | 2.44 | 27.94 |
| Max | 95 | 17.40 | 0.94 | 0.66 | 100.08 | 27.94 |
| N | 2081 | 2081 | 2081 | 2081 | 2081 | 4 |

### 3.2. Correlation Analysis

Figure 7 illustrates the correlation matrix, showing the relationships among the variables. The most notable statistically significant correlation was identified between the variables mean NDVI and mean NDRE (r = 0.74, $p$-value < 0.001). Following this, the Pearson correlation coefficients between the canopy and mean NDVI and mean NDRE were 0.27 and 0.25, respectively, both demonstrating statistical significance ($p$-value < 0.001). The absence of significantly strong correlations (greater than 0.9) among the predictors in a machine learning context is crucial because such strong correlations can lead to multicollinearity, which can negatively impact the performance of algorithms like Random Forest (Figure 7).

### 3.3. Model Performance

Figure 8 displays the metrics of the various algorithms employed. In terms of accuracy, the Random Forest model demonstrated the highest performance at 0.86, closely followed by the Gradient Boosting model with an accuracy of 0.85. The Decision Tree and Multi-Layer Perceptron models achieved accuracy values of 0.82 and 0.79, respectively. For the F-1 score, both the Random Forest and Gradient Boosting models achieved the top score at 0.82, while the Decision Tree model scored 0.77. Regarding precision, the Random Forest and Gradient Boosting models led with the highest values of 0.84, followed by the Decision Tree model at 0.80. According to the recall metric, the Random Forest model attained the highest value at 0.81, followed by the Gradient Boosting and Decision Tree models at 0.80 and 0.79, respectively. The K-nearest Neighbors and Naïve Bayes models exhibited the lowest metric values compared to the other models (Figure 8).

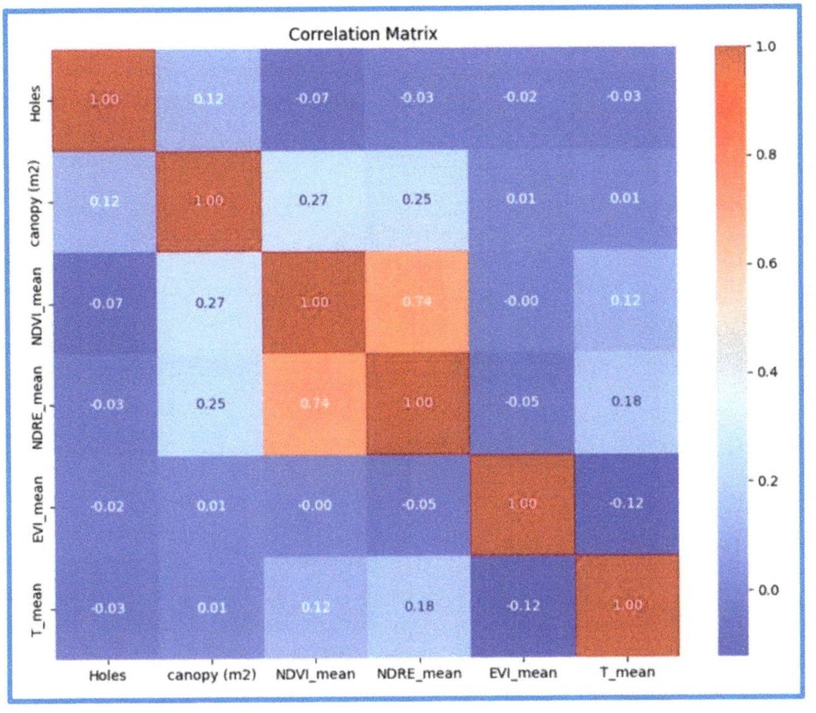


**Figure 7.** Correlation matrix of the variables depicting the relationship between them. The most statistically significant linear correlation is found between the two vegetation indices, NDVI and NDRE (r = 0.74).

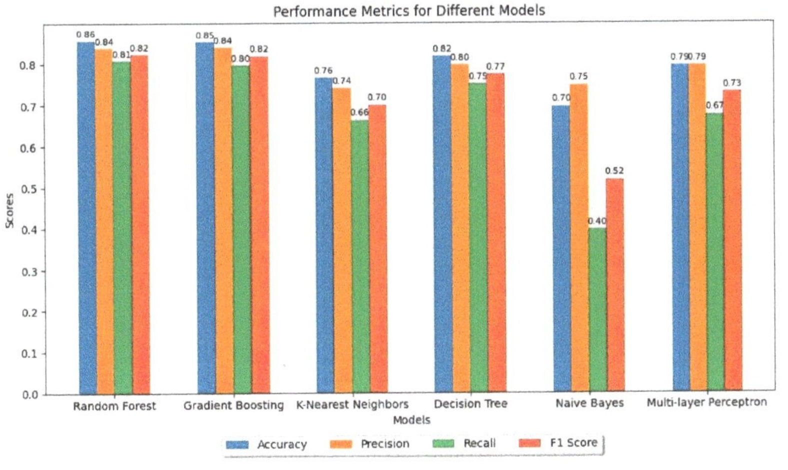


**Figure 8.** Evaluation of the six algorithms based on accuracy, precision, recall, and F1 Score.

### 3.4. Variable Importance

The plots in Figure 9 suggest that the number of holes and mean temperature are the most crucial explanatory variables across all six models. When examining the Random Forest model, a significant degree of importance is attributed to all the variables used in the prediction, in contrast to the Gradient Boosting model, where only the number of holes and mean temperature were considered particularly important (Figure 9).

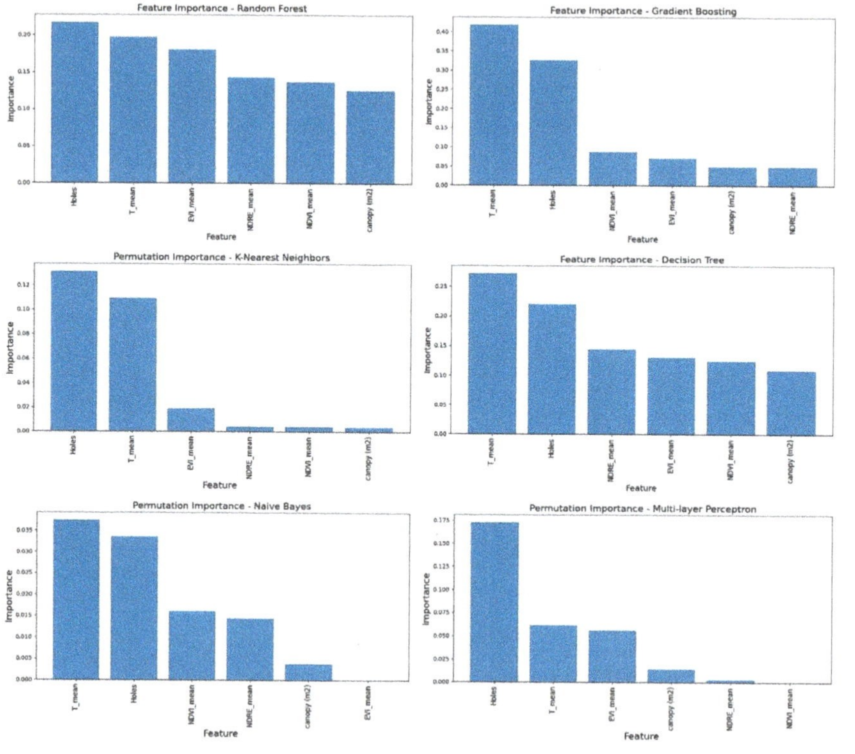


**Figure 9.** Importance of variables per learning algorithm based on the training data.

### 3.5. Confusion Matrices

To assess the performance of the classification model, confusion matrices were employed (Figure 10). In these matrices, we observe that the percentages of the evaluated models ranged for true negatives from 49.16 to 53.24, while true positives ranged from 16.31 to 33.57. In both cases, the Random Forest and Gradient Boosting models exhibited the highest percentages.

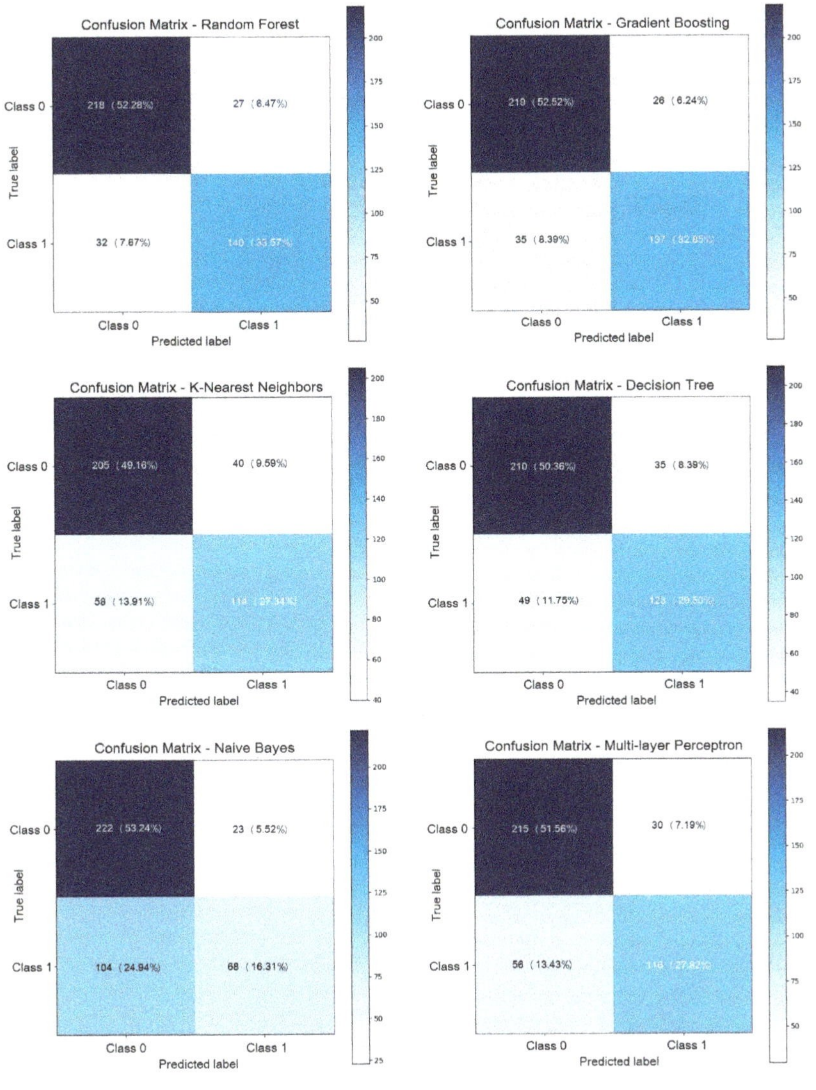


**Figure 10.** Confusion matrices for the machine-learning algorithms.

### 3.6. ROC Curves

To assess and compare the performance of classifiers, Receiving Operating Characteristic (ROC) curves are employed (Figure 11). The closer the ROC curve is to the upper left corner of the graph, the higher the accuracy of the test because the greater the area underneath it, with the ideal ROC curve having Area Under the Curve (AUC) = 0.1. The dotted line shows the ROC curve of a random classification so below this line, the performance of the test is worse than random. In the graphical plots, it is apparent that the Random Forest and Gradient Boosting models demonstrated the highest area with a value of 0.93, followed by the Multi-Layer Perceptron and K-Nearest Neighbors models with areas of 0.86 and 0.84, respectively. The Decision Tree and Naïve Bayes models exhibit the smallest areas with 0.79 and 0.77, respectively.

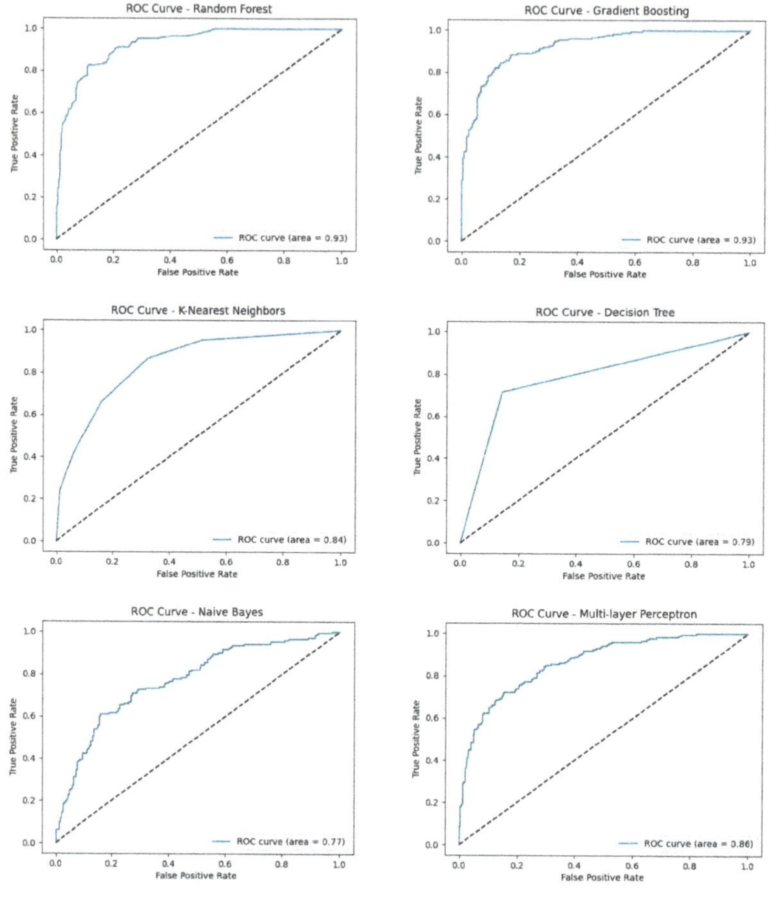


**Figure 11.** ROC curves of the six models.

## 4. Discussion

According to the *Annual Global Climate Report* of the National Oceanic and Atmospheric Administration (NOAA), the average global surface temperature in 2023 was the highest since 1850 and 1.18 °C above the average of the 20th century. In Europe, 2023 was the second warmest year on record. These increased temperatures were often accompanied by extreme weather phenomena, such as hurricanes, wildfires, and cyclones [39]. This ever-deteriorating situation has unpredictable effects on ecosystems and their organisms. The effects on crops and their pests are complex and far-reaching. The geographical distribution and biological cycle of both crops, crop pests, and their natural enemies, as well as their multiple interactions, will become increasingly difficult to predict to prevent future infestations and outbreaks [1,11,12].

Moreover, pests that complete most of their life cycle within their hosts, such as woodborers, are even more challenging to detect in time before they do irreversible damage to crops and spread significantly over an area. The tiger longicorn beetle, *X. chinensis*, is one such insect that attacks mulberry trees, destroying their vascular bundles while they are feeding on them [4–6,8]. Considering that this insect has invaded the European continent, attacking mulberries in four European Mediterranean countries and causing the death of many trees, early detection is vital to limit its spread across Europe and to control it in time.

Phytosanitary surveillance has been facilitated, expedited, and become more effective owing to the utilization of remote sensing technology. Through remote sensing, changes in absorption and reflectance of the electromagnetic radiance by the plants can be detected even before visible symptoms of the infestation appear, minimizing crop losses [17,21]. Remote sensing has proven effective in early pest detection by identifying subtle changes in plant health indicators such as chlorophyll content and water stress [16,24].

The prediction of the existence of infestation by *X. chinensis* in mulberries utilizing multispectral data, deriving from UAV remote sensing, was the aim of the present study. Six machine learning models were used for the detection of the infestation, with Random Forest and Gradient Boosting models having the best performance with slight differences. Random Forest has proven to be a very effective model for the prediction of pest infestations, achieving high overall performance [16,40–43].

In this study, mean temperature and the number of adult emergence holes were crucial factors for all six prediction models. This emphasizes that ground truth data are crucial for the construction of a prediction model to achieve high classification accuracy. Adult emergence holes are a symptom of either an ongoing or a past infestation by *X. chinensis*. Like other woodboring pest infestations, such as by *Capnodis tenebrionis* (Coleoptera: Buprestidae), where weakened or already infested trees are more susceptible to new infestations by woodborers, infestations by *X. chinensis* are also more likely to be repeated in trees that have been previously infected by the insect [44]. Therefore, the results show that even past adult emergence hole data can be used as an important predictor for future infestations.

Remote sensing data from which vegetation indices were computed were also important factors for the prediction models. Among the models, EVI was the most important vegetation index predictor in three of them (Random Forest, K-Nearest Neighbors, Multi-Layer Perceptron), NDVI was the second most important in two of the models (Gradient Boosting, Naïve Bayes), and NDRE was the most important vegetation index predictor only in one model (Decision Tree). The main disadvantage of NDVI is its saturation in areas with high biomass and dense vegetation, which does not occur with EVI, making it more useful in such areas like the one in this study [45,46]. NDRE is more useful for detecting subtle changes in chlorophyll content, but NDVI is more efficient in large areas with dense vegetation [47]. In the Random Forest model, all the vegetation indices (EVI, NDVI, and NDRE) played an important role, without significant differences from the mean temperature and adult emergence hole factors. Moreover, the factor of tree crown area (canopy) had few differences from the NDVI factor. NDVI is correlated with crop biomass, so it makes sense that if NDVI plays an important role as a predictor in a model, the same will apply to the tree crown area ($m^2$ of canopy) [40].

UAV remote sensing data have been used successfully for the detection of infestation by other woodborers or wood-eating pests in previous studies, mostly for pest infestations in forests [15,42,48–50]. Unfortunately, there are no studies about *X. chinensis*, which seems to be a rising threat to mulberry trees in Mediterranean countries of Europe. This study fills a crucial gap in the literature and provides a foundation for further research into the early detection and control of this invasive pest.

Climate change unpredictably affects the geographical distribution, life cycle, and hosts of many pests. Hence, frequent, easy, and quick monitoring of plants is more essential than ever. Remote sensing is a promising technology that can be used for the early detection of pest infestations and timely control of the spread of the pests. This approach aligns with the principles of Integrated Pest Management (IPM), which aims to reduce pesticide use and mitigate environmental impact by facilitating early intervention and precise pest control measures [17,21].

## 5. Conclusions

In summary, this study aimed to detect infestations by *X. chinensis* in mulberries using UAV remote sensing multispectral data and six classification models. The key factors identified were mean temperature, number of adult emergence holes, and mean EVI.

Among the models, Random Forest demonstrated the best performance, closely followed by Gradient Boosting. Our findings confirm that classifying healthy and infested mulberry trees is feasible through remote sensing data. However, the inclusion of ground truth data is crucial for achieving high classification accuracy. Models such as Random Forest and Gradient Boosting achieved high overall performance (0.93), making them ideal for this application. Early detection of pest infestations facilitates timely interventions, leading to reduced pesticide use and mitigating negative environmental impacts, aligning with the principles of Integrated Pest Management (IPM). This study lays important groundwork for the use of UAV remote sensing in pest detection, but further research is needed to refine these methods and differentiate between various pest infestations more accurately.

**Author Contributions:** Conceptualization, A.T.; methodology, C.P., E.A. and A.A.; software, E.A., C.P. and A.A.; validation, C.P. and E.A.; formal analysis, A.K., A.A. and E.A.; investigation, E.A., A.A. and C.P.; data curation, C.P., E.A. and A.A.; writing—original draft preparation, C.P., E.A., A.A., M.S. and A.T.; writing—review and editing, C.P., E.A., A.A. and A.T.; visualization, C.P. and E.A.; supervision, A.T.; project administration, A.T. and E.A. All authors have read and agreed to the published version of the manuscript.

**Funding:** This research received no external funding.

**Data Availability Statement:** The data presented in this study are available on request from the corresponding author. The data are not publicly available due to also forming part of an ongoing study.

**Conflicts of Interest:** The authors declare no conflicts of interest.

## References


1. Stange, E.E.; Ayres, M.P. *Climate Change Impacts: Insects*; Wiley: Hoboken, NJ, USA, 2010. [CrossRef]
2. Gillooly, J.; West, G.; Savage, V.; Brown, J. Effects of Size and Temperature on Developmental Time. *Nature* **2002**, *417*, 70–73. [CrossRef] [PubMed]
3. Sharpe, P.J.H.; DeMichele, D.W. Reaction Kinetics of Poikilotherm Development. *J. Theor. Biol.* **1977**, *64*, 649–670. [CrossRef]
4. Kavallieratos, N.G.; Boukouvala, M.C.; Skourti, A.; Nika, E.P.; Papadoulis, G.T. Trunk Injection with Insecticides Manages *Xylotrechus chinensis* (Chevrolat) (Coleoptera: Carambycidae). *Insects* **2022**, *13*, 1106. [CrossRef]
5. Leivadara, E.; Leivadaras, I.; Vontas, I.; Trichas, A.; Simoglou, K.; Roditakis, E.; Avtzis, D.N. First Record of *Xylotrechus chinensis* (Coleoptera: Cerambycidae) in Greece and in the EPPO Region. *EPPO Bull.* **2018**, *48*, 277–280. [CrossRef]
6. Sarto i Monteys, V.; Torras i Tutusaus, G. A New Alien Invasive Longhorn Beetle, *Xylotrechus chinensis* (Cerambycidae), Is Infesting Mullberries in Catalonia (Spain). *Insects* **2018**, *9*, 52. [CrossRef]
7. Bragard, C.; Baptista, P.; Chatzivassiliou, E.; Di Serio, F.; Gonthier, P.; Jaques Miret, J.A.; Justesen, A.F.; Magnusson, C.S.; Milonas, P.; Navas-Cortes, J.A.; et al. Pest Categorisation of *Xylotrechus chinensis*. *EFSA J.* **2021**, *19*, e07022. [CrossRef]
8. Sarto i Monteys, V.; Costa Ribes, A.; Savin, I. The Invasive Longhorn Beetle *Xylotrechus chinensis*, Pest of Mullberries, in Europe: Study on Its Local Spread and Efficacy of Abamectin Control. *PLoS ONE* **2021**, *16*, e0245527. [CrossRef] [PubMed]
9. Logan, J.A.; Régnière, J.; Powell, J.A. Assessing the Impacts of Global Warming on Forest Pest Dynamics. *Front. Ecol. Environ.* **2003**, *1*, 130–137. [CrossRef]
10. Trần, J.K.; Ylioja, T.; Billings, R.F.; Régnière, J.; Ayres, M.P. Impact Of Minimum Winter Temperatures On The Population Dynamics of *Dendroctonus frontalis*. *Ecol. Appl.* **2007**, *17*, 882–899. [CrossRef]
11. Peñuelas, J.; Filella, I.; Comas, P. Changed Plant and Animal Life Cycles from 1952 to 2000 in the Mediterranean Region. *Glob. Chang. Biol.* **2002**, *8*, 531–544. [CrossRef]
12. Sykes, M.T. Climate Change Impacts: Vegetation. In *Encyclopedia of Life Sciences*; Wiley: Hoboken, NJ, USA, 2009; ISBN 978-0-470-01617-6.
13. McDowell, N.G. Mechanisms Linking Drought, Hydraulics, Carbon Metabolism, and Vegetation Mortality. *Plant Physiol.* **2011**, *155*, 1051–1059. [CrossRef] [PubMed]
14. Faccoli, M. Effect of Weather on *Ips typographus* (Coleoptera Curculionidae) Phenology, Voltinism, and Associated Spruce Mortality in the Southeastern Alps. *Environ. Entomol.* **2009**, *38*, 307–316. [CrossRef]
15. Lehmann, J.R.; Nieberding, F.; Prinz, T.; Knoth, C. Analysis of Unmanned Aerial System-Based CIR Images in Forestry—A New Perspective to Monitor Pest Infestation Levels. *Forests* **2015**, *6*, 594–612. [CrossRef]
16. Luo, Y.; Huang, H.; Roques, A. Early Monitoring of Forest Wood-Boring Pests with Remote Sensing. *Annu. Rev. Entomol.* **2023**, *68*, 277–298. [CrossRef]
17. Iost Filho, F.H.; Heldens, W.B.; Kong, Z.; De Lange, E.S. Drones: Innovative Technology for Use in Precision Pest Management. *J. Econ. Entomol.* **2020**, *113*, 1–25. [CrossRef]

18. Roman, A.; Ursu, T. Multispectral Satellite Imagery and Airborne Laser Scanning Techniques for the Detection of Archaeological Vegetation Marks. In *Landscape Archaeology on the Northern Frontier of the Roman Empire at Porolissum: An Interdisciplinary Research Project*; Mega Publishing House: Cluj-Napoca, Romania, 2016; pp. 141–152.
19. Acharya, M.C.; Thapa, R.B. Remote Sensing and Its Applications in Agricultural Pest Management. *J. Agric. Environ.* **2015**, *16*, 43–61. [CrossRef]
20. Wójtowicz, M.; Wójtowicz, A.; Piekarczyk, J. Application in Remote Sensing Methods in Agriculture. *Commun. Biometry Crop Sci.* **2016**, *11*, 31–50.
21. Prasad, Y.G.; Prabhakar, M. Pest Monitoring and Forecasting. In *Integrated Pest Management: Principles and Practice*; Abrol, D.P., Shankar, U., Eds.; CABI: Wallingford, UK, 2012; pp. 41–57. ISBN 978-1-84593-808-6.
22. Yue, J.; Lei, T.; Li, C.; Zhu, J. The Application of Unmanned Aerial Vehicle Remote Sensing in Quickly Monitoring Crop Pests. *Intell. Autom. Soft Comput.* **2012**, *18*, 1043–1052. [CrossRef]
23. Bannari, A.; Morin, D.; Bonn, F.; Huete, A. A Review of Vegetation Indices. *Remote Sens. Revies* **1995**, *13*, 95–120. [CrossRef]
24. Dash, J.P.; Watt, M.S.; Pearse, G.D.; Heaphy, M.; Dungey, H.S. Assessing Very High Resolution UAV Imagery for Monitoring Forest Health during a Simulated Disease Outbreak. *ISPRS J. Photogramm. Remote Sens.* **2017**, *131*, 1–14. [CrossRef]
25. Rouse, J.W.; Haas, R.H.; Schell, J.A.; Deering, D.W. Monitoring Vegetation Systems in the Great Plains with ERTS. *NASA Spec. Publ.* **1974**, *351*, 309.
26. Gitelson, A.; Merzlyak, M.N. Spectral Reflectance Changes Associated with Autumn Senescence of *Aesculus hippocastanum* L. and *Acer platanoides* L. Leaves. *J. Plant Physiol.* **1994**, *143*, 286–292. [CrossRef]
27. Sims, D.A.; Gamon, J.A. Relationships between Leaf Pigment Content and Spectral Reflectance across a Wide Range of Species, Leaf Structures and Developmental Stages. *Remote Sens. Environ.* **2002**, *81*, 337–354. [CrossRef]
28. Justice, C.O.; Vermote, E.; Townshend, J.R.; Defries, R.; Roy, D.P.; Hall, D.K.; Salomonson, V.V.; Privette, J.L.; Riggs, G.; Strahler, A.; et al. The Moderate Resolution Imaging Spectroradiometer (MODIS): Land Remote Sensing for Global Change Research. *IEEE Trans. Geosci. Remote Sens.* **1998**, *36*, 1228–1249. [CrossRef]
29. Altman, D.G.; Bland, J.M. Diagnostic Tests 1: Sensitivity and Specificity. *BMJ* **1994**, *308*, 1552. [CrossRef] [PubMed]
30. Altman, D.G.; Bland, J.M. Diagnostic Tests 2: Predictive Values. *BMJ* **1994**, *309*, 102. [CrossRef]
31. Velez, D.R.; White, B.C.; Motsinger, A.A.; Bush, W.S.; Ritchie, M.D.; Williams, S.M.; Moore, J.H. A Balanced Accuracy Function for Epistasis Modeling in Imbalanced Datasets Using Multifactor Dimensionality Reduction. *Genet. Epidemiol. Off. Publ. Int. Genet. Epidemiol. Soc.* **2007**, *31*, 306–315. [CrossRef]
32. Quinlan, J.R. Induction of Decision Trees. *Mach. Learn.* **1986**, *1*, 81–106. [CrossRef]
33. Belgiu, M.; Drăguţ, L. Random Forest in Remote Sensing: A Review of Applications and Future Directions. *ISPRS J. Photogramm. Remote Sens.* **2016**, *114*, 24–31. [CrossRef]
34. Skawsang, S.; Nagai, M.K.; Tripathi, N.; Soni, P. Predicting Rice Pest Population Occurence with Satellite-Derived Crop Phenology, Ground Meterological Observation, and Machine Learning: A Case Study for the Central Plain of Thailand. *Appl. Sci.* **2019**, *9*, 4846. [CrossRef]
35. Friedman, J.H. Greedy Function Approximation: A Gradient Boosting Machine. *Ann. Stat.* **2001**, *29*, 1189–1232. [CrossRef]
36. Webb, G.I. Naïve Bayes. In *Encyclopedia of Machine Learning*; Sammut, C., Webb, G.I., Eds.; Springer: Boston, MA, USA, 2010; pp. 713–714. ISBN 978-0-387-30164-8.
37. Haykin, S.S. *Neural Networks and Learning Machines*, 3rd ed.; Prentice-Hall: New York, NY, USA; Munich, Germany, 2009; ISBN 978-0-13-147139-9.
38. Mucherino, A.; Papajorgji, P.J.; Pardalos, P.M. K-Nearest Neighbor Classification. In *Data Mining in Agriculture*; Mucherino, A., Papajorgji, P.J., Pardalos, P.M., Eds.; Springer Optimization and Its Applications; Springer: New York, NY, USA, 2009; pp. 83–106. ISBN 978-0-387-88615-2.
39. NOAA. *Annual 2023 Global Climate Report*; NOAA National Centers for Environmental Information: Washington, DC, USA, 2024.
40. Arapostathi, E.; Panopoulou, C.; Antonopoulos, A.; Katsileros, A.; Karellas, K.; Dimopoulos, C.; Tsagkarakis, A. Early Detection of Potential Infestation by *Capnodis tenebrionis* (L.) (Coleoptera: Buprestidae), in Stone and Pome Fruit Orchards, Using Multispectral Data from a UAV. *Agronomy* **2024**, *14*, 20. [CrossRef]
41. Bozzini, A.; Francini, S.; Chirici, G.; Battisti, A.; Faccoli, M. Spruce Bark Beetle Outbreak Prediction through Automatic Classification of Sentinel-2 Imagery. *Forests* **2023**, *14*, 1116. [CrossRef]
42. Junttila, S.; Näsi, R.; Koivumäki, N.; Imangholiloo, M.; Saarinen, N.; Raisio, J.; Holopainen, M.; Hyyppä, H.; Hyyppä, J.; Lyytikäinen-Saarenmaa, P.; et al. Multispectral Imagery Provides Benefits for Mapping Spruce Tree Decline Due to Bark Beetle Infestation When Acquired Late in the Season. *Remote Sens.* **2022**, *14*, 909. [CrossRef]
43. Sumido, E.C.; Feliscuzo, L.S.; Aliac, C.J.G. Pest classification and prediction: Analyzing the impact of weather to pest occurrence through machine learning. *J. Eng. Sci. Technol.* **2023**, *18*, 124–138.
44. Rodríguez-González, Á.; Sánchez-Maíllo, E.; Peláez, H.J.; González-Núñez, M.; Hall, D.R.; Casquero, P.A. Field Evaluation of 3-Hydroxy-2-Hexanone and Ethanol as Attractants for the Cerambycid Beetle Pest of Vineyards, Xylotrechus Arvicola. *Pest Manag. Sci.* **2017**, *73*, 1598–1603. [CrossRef]
45. Huete, A. Vegetation Indices, Remote Sensing and Forest Monitoring. *Geogr. Compass* **2012**, *6*, 513–532. [CrossRef]
46. Huete, A.; Didan, K.; Miura, T.; Rodriguez, E.P.; Gao, X.; Ferreira, L.G. Overview of the Radiometric and Biophysical Performance of the MODIS Vegetation Indices. *Remote Sens. Environ.* **2002**, *83*, 195–213. [CrossRef]

47. Boiarskii, B.; Hasegawa, H. Comparison of NDVI and NDRE Indices to Detect Differences in Vegetation and Chlorophyll Content. *J. Mech. Contin. Math. Sci.* **2019**, *4*, 20–29. [CrossRef]
48. Cessna, J.; Alonzo, M.G.; Foster, A.C.; Cook, B.D. Mapping Boreal Forest Spruce Beetle Health Status at the Individual Crown Scale Using Fused Spectral and Structural Data. *Forests* **2021**, *12*, 1145. [CrossRef]
49. Safonova, A.; Tabik, S.; Alcaraz-Segura, D.; Rubtsov, A.; Maglinets, Y.; Herrera, F. Detection of Fir Trees (*Abies sibirica*) Damaged by the Bark Beetle in Unmanned Aerial Vehicle Images with Deep Learning. *Remote Sens.* **2019**, *11*, 643. [CrossRef]
50. Stoyanova, M.; Kandilarov, A.; Koutev, V.; Nitcheva, O.; Dobreva, P. Potential of Multispectral Imaging Technology for Assessment Coniferous Forests Bitten by a Bark Beetle in Central Bulgaria. *MATEC Web Conf.* **2018**, *145*, 01005. [CrossRef]

agronomy

MDPI

*Article*

# Stress Response of Citrus Leaves under Mechanical Damage and Huanglongbing Disease Infection Using Plasmonic $TiO_2$ Nanotube Substrate-Based Imprinting Mass Spectrometry Imaging

**Yaming Sun †, Dong Chen †, Xiran Chen and Xinzhou Wu ***

National Key Laboratory of Green Pesticide and Key Laboratory of Natural Pesticide and Chemical Biology of the Ministry of Education, South China Agricultural University, Guangzhou 510642, China

* Correspondence: wuxz@scau.edu.cn

† The authors contributed equally to this work.

**Abstract:** Mapping the molecular signatures and metabolic regulation of plant tissues under biotic/abiotic stresses and defensive responses has become a subject of increasing interest in plant biology and systems biology, but determining when and where specialized metabolites are produced and accumulated currently remains a somewhat elusive goal. Herein, we demonstrated the use of a $TiO_2$ nanotube-based composite substrate modified with plasmonic gold nanoparticles and hydrophobic polydopamine (AuNP-hPDA-TDNT) for surface-assisted laser desorption/ionization mass spectrometry (SALDI-MS) analysis of a wide range of pesticides and for visualizing the stress-responsive metabolites of citrus leaves during various plant defense processes. This method enabled the visualization of non-uniform and tissue-specific distribution patterns of functional metabolites of citrus leaves that were mechanically damaged, fed to larvae, and infected by Huanglongbing disease. Interestingly, some specialized metabolites exhibited different accumulation and regulation patterns for mechanical damage and larval feeding, suggesting that plant-derived secondary metabolites exercise specific defensive functions with respect to various damage processes. Moreover, the early diagnosis and detection of HLB disease-associated biomarkers can facilitate the prevention of citrus HLB diseases. Overall, this imprinting MS imaging strategy will expand the scope of MS techniques in plant biology, providing more biologically relevant insights into the biosynthesis, accumulation, and defensive role of bioactive metabolites in economically important plants.

**Keywords:** Huanglongbing disease; leaf imprinting; mass spectrometry imaging; plant stress response; surface-assisted laser desorption/ionization

## 1. Introduction

Citrus (*Citrus japonica* Thunb.) Huanglongbing (HLB) is a disease of worldwide incidence that causes considerable economic losses during the development, growth, and production stages of citrus plants [1,2]. Citrus HLB is caused by infection with *Liberbacter asianticum* jagoueix, which is propagated mainly by the citrus psyllid and grafting [3]. HLB has an incubation period during which newly infected trees do not show symptoms, posing a challenge for the early detection of HLB [4,5]. As the main pest of young citrus shoots, the citrus psyllid is a propagation medium for citrus HLB. The adults mostly lay eggs on the susceptible young treetops and begin to suck the sap of young shoots after the emergence of larvae. The adult citrus psyllid can then fly to new plants to transmit the citrus HLB. Currently, there is no completely effective treatment for citrus HLB, but early detection and treatment are the preferred strategies [6,7]. Thus, the early detection of HLB and the control of citrus psyllids are the keys to the prevention and control of HLB. Additionally, insights into the stress-induced secondary metabolites of plant tissues that are mechanically damaged and fed larvae will help researchers evaluate their defensive roles and decipher how plants deploy defenses for their maximum benefit [8]. However, the interactions between defensive responses and secondary metabolites produced by plants remain largely unclear.

Given that primary metabolites are closely associated with the growth and development of plant systems, plant secondary metabolites have received a significant amount of attention because of their specific biological functions under biotic/abiotic stresses [9–11]. Thus, gas chromatography or liquid chromatography coupled with mass spectrometry (GC−/LC−MS) has been well established for direct component analysis [12,13]. Unfortunately, spatial distribution information on minute amounts of metabolites where stress responses occur is normally missing during tissue homogenization and metabolite extraction processes. To circumvent these problems, mass spectrometry imaging (MSI) has evolved as a promising and molecule-specific tool for spatially and temporally characterizing a wide range of metabolites in complex plant systems [14–19]. Among the available techniques, three mainstream types of MSI techniques, namely, matrix-assisted laser desorption/ionization (MALDI), desorption electrospray ionization (DESI), and laser ablation electrospray ionization (LA-ESI), have become the most routinely used ionization sources of choice for plant metabolomics studies [20–22]. Despite being extensively used, several key challenges are currently facing conventional MALDI-MSI, such as background interference at $m/z < 500$ from organic matrices, poor reproducibility resulting from "sweet spots", and potential imaging artifacts due to inhomogeneous co-crystallization [23–25]. Even worse, unlike plant stem or root tissues, which can be frozen-sectioned for MALDI-MSI, leaf and flower tissues cannot be sliced, which makes it exceedingly difficult to perform MSI analysis [26,27]. Although direct leaf and petal MSI can be obtained via DESI techniques, the MSI results might be compromised to a large extent because of wax layer protection and fluctuating surfaces. To this end, several porous materials have been proposed for use in imprinting strategies in DESI-MSI applications [28–30]. However, the limited spatial resolution of ambient MSI methods limits their ability to acquire high-quality images. Thus, there is a high demand for developing matrix-free LDI approaches that are compatible with plant leaves and flowers for capturing molecular signatures under biotic/abiotic stresses.

Inspired by the pioneering efforts of Gary Siuzdak's group [31], a plethora of ultraviolet (UV)-absorbing nanostructured substrates, such as semiconductor-based [32–34], metal-based [35,36], carbon-based [37,38], and silicon-based [33,39] substrates, have enabled matrix-free LDI for MS analysis, termed surface-assisted laser desorption/ionization (SALDI). More valuably, these SALDI substrates are very suitable for MSI analysis of plant leaves and flowers and can be simply imprinted onto nanostructured material surfaces, regardless of matrix deposition and tissue sectioning. Benefiting from the advantages of high light absorption, a high surface area-to-volume ratio, and low thermal conductivity [40,41], several recent examples of these methods include visualizing endogenous glycoalkaloids, flavonoids, and carbohydrates within *Catharanthus roseus* flowers and spearmint leaves by imprinting SALDI-MSI [42,43]. Recently, we constructed a $TiO_2$ nanotube (TDNT)-based composite material with plasmonic gold nanoparticles (AuNPs) and hydrophobic polydopamine (PDA) modification (AuNP-hPDA-TDNT) for primary and secondary metabolite mapping in a wide range of plant tissues and fruits [34,44]. The synergistic effects between n-type semiconductor TDNTs and plasmonic AuNPs make this SALDI-MSI strategy effective and sensitive. Despite considerable efforts, the great potential of imprinting MSI in real-case applications with respect to the stress response of plant tissues to disease infection and mechanical damage remains to be explored.

In this study, we demonstrated the potential of AuNP-hPDA-TDNT-based nanostructured substrates for SALDI-MS analysis of a wide range of pesticides and exploited the imprinting MSI capability in mapping secondary metabolites during various plant defense processes. Moreover, various nonuniform and tissue-specific distribution patterns of functional metabolites can be clearly visualized in plant leaves that are mechanically damaged, fed to larvae, and infected by Huanglongbing disease via AuNP-hPDA-TDNT-based SALDI-MSI. Taken together, these results demonstrate that this imprinting MSI method has great potential for visualizing plant defense-derived metabolites, providing insights into the biosynthesis, accumulation, and defensive role of a diverse variety of bioactive metabolites in economically important plants.

## 2. Materials and Methods

### 2.1. Chemicals and Materials

Titanium sheets (99.95% purity) were purchased from Qingyuan Metal Materials Ltd. (Shijiazhuang, China). Chloroauric acid ($HAuCl_4$, 98% purity and 47.8% Au content), sodium citrate (98% purity), sodium fluoride (99% purity), and anhydrous sodium sulfate (99% purity) were purchased from Energy Chemistry Ltd. (Shanghai, China). Dopamine hydrochloride (98% purity) and 1H,1H,2H,2H-perfluorooctyltrichlorosilane (FOTS, 97% purity) were purchased from Shanghai Macklin Biochemical Co., Ltd. (Shanghai, China). α-Cyano-4-hydroxycinnamic acid (CHCA, 99.0% purity) and Tris buffer were purchased from Sigma-Aldrich (St. Louis, MO, USA). Acephate, dinotefuran, thiamethoxam, spirotetramat, rotenone, azoxystrobin, cyantraniliprole, chlorantraniliprole, and abamectin were purchased from J&K Scientific Ltd. (Beijing, China). HPLC-grade acetonitrile was purchased from Tedia Company, Inc. (Fairfield, OH, USA). All the chemicals were used without further purification.

### 2.2. Preparation of the Pesticide Standards and AuNPs

Thiamethoxam, rotenone, chlorobenzamide, and abamectin were dissolved in acetone. Spirotetramat, dinotefuran, acephate, cyantraniliprole, and azoxystrobin were dissolved in an acetonitrile/water (1:1, $v/v$) mixture. All the analytes were dissolved in the stock solution at a concentration of 10 mM. Then, the standard individual solution and mixed solution were prepared by progressively diluting the stock solution to the desired concentration. Then, 0.1 μL of standard solution was manually deposited onto the AuNP-hPDA-TDNT substrate, which was naturally air-dried prior to MS analysis. AuNPs were synthesized via thermal treatment according to previous reports [17,27].

### 2.3. Preparation of the AuNP-hPDA-TDNT Substrate

The composite nanostructured substrate was prepared via an electrochemical anodization method according to previous reports [34,44]. Briefly, anhydrous $Na_2SO_4$ and NaF were dissolved in deionized water as the electrolyte solution, a Ti sheet was adopted as the anode, and a Pt electrode was adopted as the cathode. The oxidation reaction proceeded at a voltage of 20 V at room temperature for 1 h. To avoid adverse side effects, the voltage was increased slowly. Then, $TiO_2$ nanotube (TDNT) materials were produced and washed ultrasonically with deionized water for 1 min. After air drying at room temperature, the TDNT substrates were stored in clean containers until future use.

The TDNT substrates were placed on a mixed solution containing dopamine hydrochloride and Tris-HCl buffer (pH 8.5) that was heated with stirring at 600 rpm in a 90 °C water bath for 1 h to prepare the PDA-TDNT substrate. The sample was subjected to ultrasonic cleaning twice with ethanol and deionized water to remove surface residues. Then, the hPDA-TDNT substrate was obtained by spraying 10 times with 0.25% ($v/v$) FOTS at a flow rate of 1 mL/min onto the PDA-TDNT surface in a high-performance ultrasonic sprayer (UAM4000, Hangzhou, China). Finally, a 0.1 mg/mL AuNP suspension was evenly sprayed on the hPDA-TDNT substrate surface 200 times at a flow rate of 0.03 mL/min to prepare the AuNP-hPDA-TDNT substrate. The detailed workflow of the AuNP-hPDA-TDNT substrate fabrication is shown in Figure S1.

### 2.4. Preparation of Imprinted Citrus Leaves That Were Infected by Huanglongbing Disease, Mechanically Damaged, and Fed Larvae for MSI Analysis

The experimental citrus (*Citrus japonica* Thunb.) variety was sour orange (a native wild species in Hainan), and susceptible citrus plants were provided by South China

Agricultural University. Four- to five-month-old leaves were chosen for infection with HLB (*Liberbacter asianticum* jagoueix), which is an early pathogen. Susceptible leaves with obvious diseased spots were selected. Additionally, one- to two-month-old leaves that were mechanically damaged and fed larvae were chosen for imprinting MSI analysis. Specifically, the mechanically damaged leaves were subsequently crushed with needle-nose pliers and then harvested after they had grown for 2 h, 12 h, and 24 h. The leaves that were fed with larvae were placed on young leaves to eat and then collected after they had grown for 12 h. After the leaves were collected, the leaf surface was simply wiped with a water-soaked paper towel and then placed on the AuNP-hPDA-TDNT substrate for imprinting, and pressure was slowly applied at 4 MPa, where the duration of the imprinting process was 20 min. Then, the leaves were removed, and the leaf-imprinted substrates were transferred to room temperature for drying and subjected to MSI analysis.

### 2.5. Metabolite Extraction of Citrus Leaves for Ultra-Performance Liquid Chromatography Cou-Pled with Mass Spectrometry (UPLC–MS) Experiments

For metabolite extraction from citrus plant leaves, 50 mg of each sample was weighed and placed into centrifuge tubes, where 800 μL of precooled extraction solvent containing methanol and water (7:3, $v/v$) at −20 °C was added for metabolite extraction. After the tissue was ground with two small steel balls at 50 Hz for 5 min, it was placed in an ultrasonic water bath at 4 °C for 30 min and then transferred to −20 °C for 1 h. Following centrifugation at 25,000 rpm for 15 min, 600 μL of the supernatant was collected through a 0.22 μm filter and analyzed via UPLC-MS.

### 2.6. UPLC-MS Conditions for Metabolite Identification in Citrus Leaves

For metabolite identification in citrus leaves, a two-dimensional UPLC (Waters, Milford, MA, USA) coupled to a high-resolution Orbitrap mass spectrometer (Q Exactive HF, Thermo Fisher Scientific, Waltham, MA, USA) was used. In brief, 0.1% formic acid in water (Solvent A) and 0.1% formic acid in methanol (Solvent B) composed the mobile phases. The gradient elution conditions were as follows: 0–1 min, 2% Solvent B; 1–9 min, 2–98% Solvent B; 9–12 min, 98% Solvent B; 12–12.1 min, 98–2% Solvent B; and 12.1–15 min, 2% Solvent B. The flow rate and injection volume were set at 0.35 mL/min and 5 μL, respectively. The column oven was maintained at 45 °C. LC-eluted fractions were subjected to MS through electrospray ionization. In this study, the positive ion mode for electrospray ionization (ESI) was used with the spray voltage set at 3.80 kV, where 40 L/min of sheath gas and 10 L/min of auxiliary gas at 350 °C were used. The ion transfer capillary temperature was set at 320 °C.

Mass spectra were acquired in the $m/z$ range from 70 to 1050 Da with a mass resolution of 70,000 at $m/z$ 200. The automatic gain control (AGC) target value was set to $3 \times 10^6$ for full MS mode. For MS/MS analysis, data-dependent acquisition (DDA) mode was employed with a mass resolution of 17,500 and an AGC setting of $1 \times 10^5$. Step-increase collision voltages of 20−40−60 eV were adopted to activate and fragment the precursor ions of interest. Metabolite identification was performed by matching the measured $m/z$ values and fragmental patterns with the Human Metabolome Database (HMDB) (http://www.hmdb.ca, accessed on 5 October 2022).

### 2.7. SALDI-MS and Imprinting MSI Experiments

For AuNP-hPDA-TDNT-based SALDI-MS analysis, 0.1 μL of sample solution was spotted onto the AuNP-hPDA-TDNT substrate. After it was naturally air-dried, the AuNP-hPDA-TDNT substrate was mounted on a stainless steel target (MTP Slide Adapter II, Bruker Daltonics, Billerica, MA, USA). All SALDI-MS analyses were performed on an UltrafleXtreme mass spectrometer (Bruker Daltonics, Billerica, MA, USA) equipped with a Nd:YAG laser at a wavelength of 355 nm, a laser frequency of 1000 Hz, and a laser energy of 45% in position ion mode. The mass spectral data were acquired with a mass range from 100 to 1000 Da. The spot size was selected as "Ultra", and mass axis calibration was performed using the signals of the $Au_n^+$ ions ($n = 1$ to 5) generated by the AuNPs. Each SALDI mass spectrum was obtained by accumulating the MS signals from 400 laser shots. For MSI analysis, the laser power was set at 80–90%, and the mass spectrum at each pixel was obtained by accumulating 200 laser shots unless otherwise indicated. The optimized voltage parameters, such as a reflector voltage of 20.84 kV, a lens voltage of 11.00 kV, an ion source voltage of 20.00 kV, and an extraction delay time of 100 ns, were adopted throughout the MSI experiments.

### 2.8. Data Analysis

FlexControl 5.0 and flexAnalysis 5.0 software were used to acquire and analyze the data. For MSI data analysis, the raw data were visualized via flexImaging 5.0 software and then imported into SCiLS Lab 2016b software for further processing. The acquired raw mass spectra were normalized to the total ion count for each image. The statistical data were processed using SPSS software (version 19.0; SPSS, Inc., Chicago, IL, USA). One-way analysis of variance (ANOVA) and Tukey's test were performed to exhibit statistical significance.

## 3. Results and Discussion

### 3.1. SALDI-MS Analysis of Various Pesticide Molecules

Previous results have demonstrated that the AuNP-hPDA-TDNT material shows great promise as a SALDI substrate for primary and secondary metabolites in plant tissues, whereas the great potential of this SALDI-MS method in pesticide mixture detection remains to be explored. To this end, a total of nine pesticides were selected for SALDI-MS analysis via a conventional MALDI matrix and four SALDI substrates at different construction stages. Information on the detected pesticides and their ion forms is listed in Tables S1 and S2. As shown in Figure 1, only three pesticides could be detected when CHCA was used as the matrix (Figure 1a), whereas all of the pesticides could be clearly detected using the AuNP-hPDA-TDNT substrate (Figure 1e). With the same laser energy, the other substrates, including TDNT, PDA-TDNT, and hPDA-TDNT, afforded three, four, and eight pesticides, respectively. Notably, the excellent photothermal conversion efficiency and enhanced ion production contributed to the improved desorption/ionization performance in the analysis of agrichemicals from the AuNP-hPDA-TDNT substrate compared with those from other substrates or conventional MALDI matrices, which was in accordance with previous studies [34,44]. More specifically, the deposited PDA layer and AuNPs facilitate enhanced photothermal conversion efficiency and surface plasmon excitation for high-density charges [45,46]. These results suggest that the enhanced desorption/ionization performance of these molecules of interest can be greatly enhanced by step-by-step modifications of the TDNT substrate.

Additionally, we further assessed the repeatability, linearity, and sensitivity of this method for detecting pesticides. Thus, azoxystrobin, thiamethoxam, and rotenone were chosen for the assessment of the AuNP-hPDA-TDNT material, as the relative standard deviation (RSD) values were lower than 6% for both the spot-to-spot and batch-to-batch homogeneity investigations (Figures S2–S4). This good reproducibility can be attributed to the fact that this SALDI substrate afforded uniform nanoscale microregions on the surface, facilitating subsequent MSI analysis regardless of the presence of imaging artifacts. Given that the stability of the SALDI material is crucial for practical MSI applications, the day-to-day stability of these composite materials was also investigated, and the results indicated that all of the RSD values were lower than 6.2% for the three selected species (Figures S5–S7). Furthermore, good linearities of the calibration curves with correlation coefficient ($R^2$) values better than 0.98 for all the model pesticides can be acquired (Figures S8–S10). We further performed SALDI-MS analysis of azoxystrobin, rotenone, and thiamethoxam at absolute concentrations of ~10 fmol, ~100 fmol, and ~100 fmol, respectively, at each spot (Figure S11). These results indicate that methodological limits of detection below 100 fmol can be reached for some pesticides.

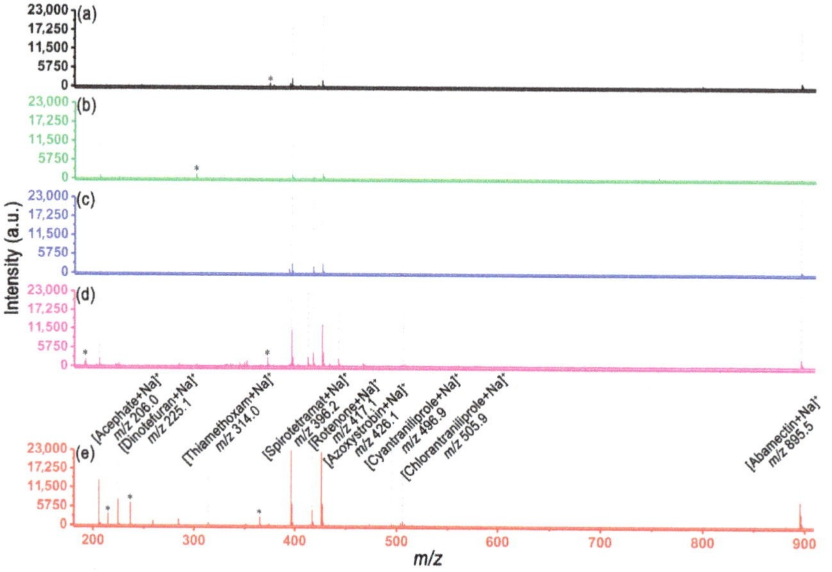


**Figure 1.** SALDI mass spectra of nine pesticide mixtures in positive ion mode. (**a**) Comparison of mass spectra of a mixture of nine pesticides, including acephate, thiamethoxam, spirotetramat, rotenone, azoxystrobin, cyantraniliprole, chlorantraniliprole, and abamectin, using (**a**) CHCA and four SALDI substrates, including (**b**) TDNT, (**c**) PDA-TDNT, (**d**) hPDA-TDNT, and (**e**) AuNP-hPDA-TDNT. The concentration of each analyte was 1 mM, and * represents the background interference peaks.

### 3.2. Imprinting SALDI-MSI of Citrus Leaves

Histologic sectioning has been widely applied to plant roots and stems, but this strategy is incompatible with plant leaves and petals because of their inability to be sliced. Thus, prior to performing imprinting MSI of citrus leaves during various plant defense processes, we visualized the spatial distribution of primary and secondary metabolites within a citrus leaf. Previous studies have revealed that citrus leaves are the main source of botanical drugs, so visualizing their intrinsic distribution patterns is highly important [47]. As shown in Figure 2, some tentatively assigned metabolites, such as leucine ($[M+Na]^+$, $m/z$ 154.1), 7-hydroxycoumarin ($[M+K]^+$, $m/z$ 201.0), 8-methoxypsoralen ($[M+H]^+$, $m/z$ 217.1), scopoletin ($[M+K]^+$, $m/z$ 231.0), scoparone ($[M+K]^+$, $m/z$ 633.1), apigenin ($[M+Na]^+$, $m/z$ 293.0), oxypeucedanin ($[M+Na]^+$, $m/z$ 309.1), astragalin ($[M+H]^+$, $m/z$ 449.1), and kaempferitrin ($[M+K]^+$, $m/z$ 617.1), were found to be homogeneously distributed across the whole leaf. In contrast, cathinone ($[M+H]^+$, $m/z$ 150.1), caffeic acid ($[M+K]^+$, $m/z$ 219.0), and hesperidin ($[M+K]^+$, $m/z$ 649.2) were co-localized with a tissue-specific distribution pattern and accumulated near the leaf stalk. A representative mass spectrum acquired via SALDI-MSI is shown in Figure S12. Some typical tandem mass spectra of secondary metabolites are displayed in Figure S13. Taken together, these results support the idea that leaf-imprinted MSI can be used for the investigation of bioactive compounds in some functional plants.

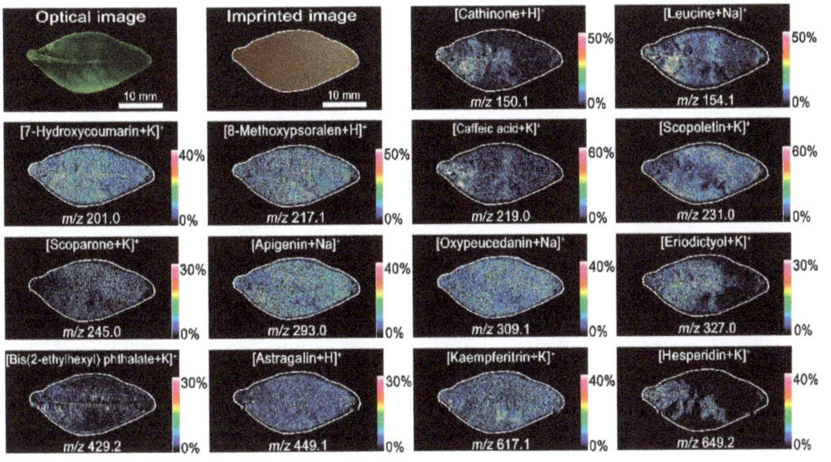


**Figure 2.** Spatial distribution of various small-molecule metabolites resulting from an imprinted citrus leaf. Optical image of a citrus leaf and the corresponding image imprinted onto the surface of the AuNP-hPDA-TDNT substrate and representative ion images of a citrus leaf obtained via SALDI-MSI. The colored bars indicate the relative signal intensity. Scale bar: 10 mm.

### 3.3. Stress Response of Citrus Leaves to Insect Feeding and Mechanical Damage

Having optimized the AuNP-hPDA-TDNT-based SALDI-MSI, we investigated its potential for mapping stress-responsive metabolites of citrus leaves in real cases. As a proof-of-concept study, we visualized the spatial distribution of some secondary metabolites in citrus leaves subjected to insect feeding and mechanical damage by imprinting MSI based on the AuNP-hPDA-TDNT substrate. Figure 3 shows that when citrus leaves were damaged by insect feeding and mechanical damage for 12 h, some metabolic marker substances exhibited specific distributions. Specifically, the levels of protocatechuic acid ($[M+Na]^+$, $m/z$ 177.0), trigonelline ($[M+Na]^+$, $m/z$ 160.0), and caffeic acid ($[M+Na]^+$, $m/z$ 203.0) decreased after citrus leaves were subjected to mechanical damage, but increased to a great extent after insect feeding. Figure 3 shows that these compounds significantly accumulated at insect feeding sites, but did not significantly change at mechanically damaged sites or in the CK group. Previous studies have reported that an increase in the levels of protocatechuic acid and caffeic acid may be the main stress defense substance produced in response to insect mouth injury [4]. Interestingly, trigonelline is a plant alkaloid that is closely related to coffee and fenugreek, the main components of which have antioxidant, anti-inflammatory, and neuroprotective effects. This study suggested that trigonelline has important antioxidant effects in response to damage caused by insect feeding. On the other hand, oxypeucedanin ($[M+K]^+$, $m/z$ 325.0) obviously accumulated at sites of mechanical damage, which may be closely related to tissue damage and repair. Additionally, given that the levels of most amino acids also tend to increase in leaves damaged by larval bites [48], proline ($[M+Na]^+$, $m/z$ 138.1) significantly accumulated at sites of larval damage, but there was no significant change at sites of mechanical damage.

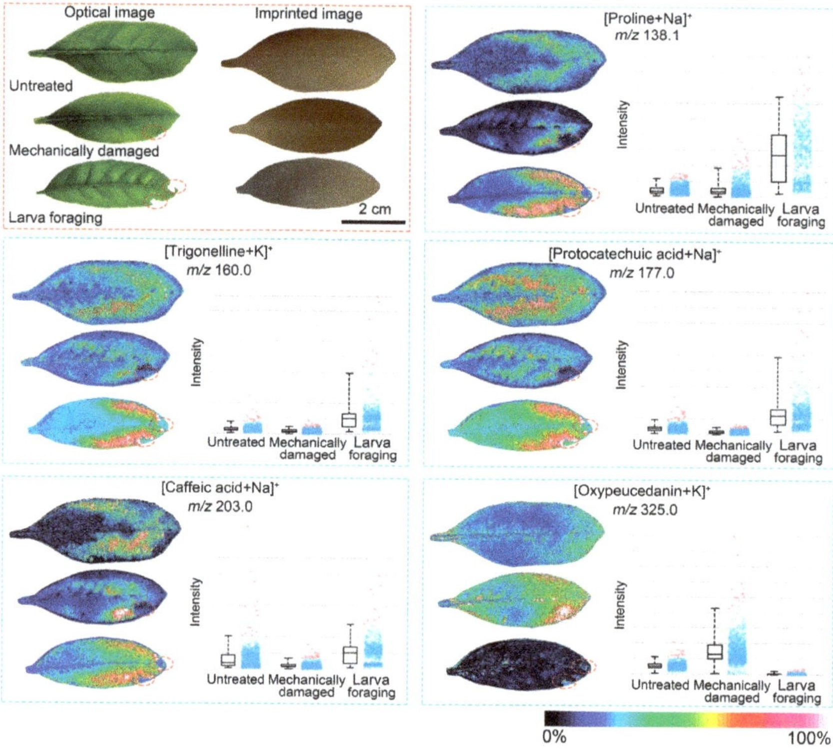


**Figure 3.** Mass spectrometry images of mechanically damaged and larva-fed citrus leaves. Optical images of citrus leaves, corresponding imprinted images of the surface of the AuNP-hPDA-TDNT substrate, and representative ion images of the citrus leaves obtained via SALDI-MSI with a pixel size of 160 µm. The red dashed ellipses indicate the specific locations that are mechanically damaged and larva foraged. The blue and red data points indicate the reasonable values and outliers, respectively. Scale bar: 20 mm.

### 3.4. *Stress Response of Citrus after Mechanical Damage for Different Repair Durations*

In addition to the static spatial distribution variation in mechanically damaged leaves, we further investigated the dynamic stress response of citrus leaves in the face of mechanical damage after different repair durations (i.e., 2 h, 12 h, and 24 h). Using the AuNP-hPDA-TDNT-based imprinting MSI method, we visualized several metabolic markers with specific spatial distributions in citrus leaves at different repair times (Figure 4). The results revealed that the levels of various amino acids, such as leucine ($[M+Na]^+$, $m/z$ 132.1) and methionine ($[M+Na]^+$, $m/z$ 172.0), increased, and that the methionine content in the CK group was greater than that at different repair times, indicating that the level decreased sharply after mechanical damage and then gradually increased. Although the defensive role of the amino acid methionine in mechanical damage is not yet fully understood, the dynamic variations in methionine levels for different durations of repair are interesting. Several studies have revealed that Met, a fundamental metabolite in plant cells, controls the levels of several essential metabolites, such as polyamines, biotin, ethylene, and phytosiderophores, after its conversion to *S*-adenosylmethionine (SAM) [49–51]. SAM, which serves as a primary methyl group donor, can regulate key processes, including the formation of chlorophyll and the cell wall, as well as the synthesis of many secondary metabolites [52]. Additionally, Met can also be catabolized to produce a group of plant secondary metabolites (e.g., glucosinolates), which exhibit repellent activity against herbivorous insects and pathogens [8,53]. When a leaf is mechanically damaged, the cell wall can be severely destroyed so that Met levels can decrease within damaged regions. As the repair mechanism is triggered, the level of the amino acid methionine gradually recovers to a normal level. More specifically, the leucine content increased significantly at the injury site after mechanical damage, while its contents at 2 h, 12 h, and 24 h were not significantly different, which was in agreement with previous studies showing that tryptophan and serine are common biomarkers of wound stress in citrus plants [54]. In sharp contrast, *N*-formylglycine ($[M+H]^+$, $m/z$ 104.1) increased slowly after mechanical damage and peaked at 12 h but began to decrease slowly from 12 h to 24 h. Notably, both the oxobutanedioic acid ($[M+K]^+$, $m/z$ 171.0) and citric acid ($[M+H]^+$, $m/z$ 193.0) levels increased after mechanical damage, which is part of the TCA cycle. Taken together, these results suggest that the TCA cycle produces increased amounts of energy in response to the threat posed by mechanical damage.

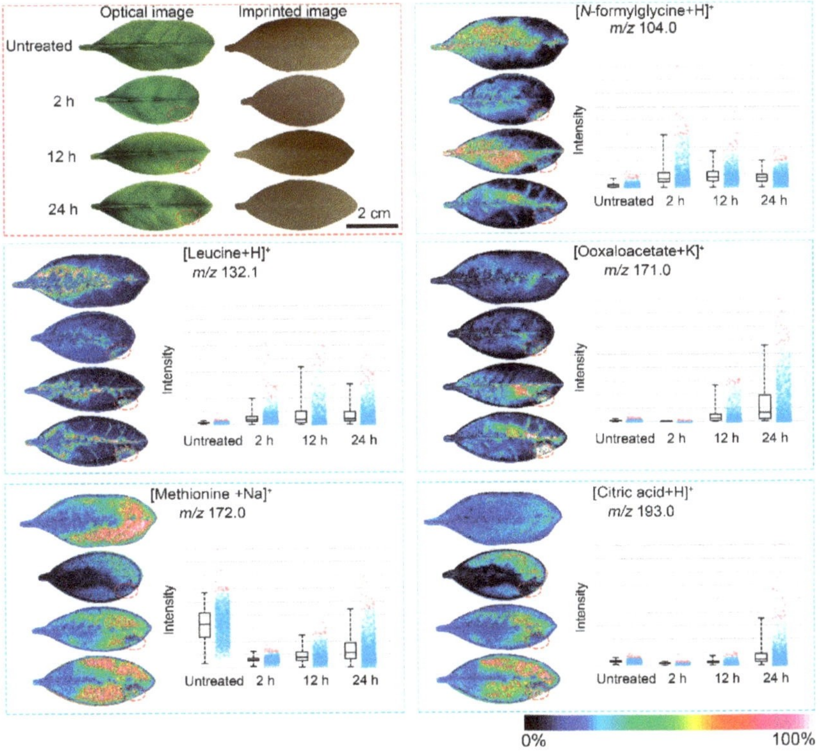


**Figure 4.** Imprinting MS images of mechanically damaged citrus leaves for different durations after repair. Optical images of citrus leaves, corresponding imprinted images of the surface of the AuNP-hPDA-TDNT substrate, and representative ion images of the citrus leaves obtained via SALDI-MSI with a pixel size of 160 µm. The red dashed ellipses indicate the specific locations that are mechanically damaged. The blue and red data points indicate the reasonable values and outliers, respectively. Scale bar: 20 mm.

### *3.5. Stress Response of Citrus Leaves during Infection with HLB*

In this study, in addition to insect feeding and mechanical damage, the spatial distribution of some secondary metabolites in citrus leaves before and after HLB infection was also visualized by imprinting MSI based on the AuNP-hPDA-TDNT substrate. We found that some metabolic markers related to defense response processes were generated when citrus leaves were infected by HLB and exhibited a specific distribution on the leaves

(Figure 5). The MSI results revealed that the contents of several metabolites, such as guaiacol ($[M+Na]^+$, $m/z$ 147.0), quinic acid ($[M+K]^+$, $m/z$ 231.0), dehydroabietic acid ($[M+Na]^+$, $m/z$ 323.2), mignonette glycosides ($[M+K]^+$, $m/z$ 487.1), and hesperidin ($[M+K]^+$, $m/z$ 649.2), were greater in the HLB-infected citrus leaves than in the healthy leaves. Notably, the concentration of quinic acid markedly increased throughout the infected leaves, whereas guaiacol was present throughout the whole leaf except in the vein. Previous studies have demonstrated that guaiacol and quinic acid can be candidate biomarkers of HLB disease infection [55], but their specific spatial distributions are poorly understood. Specifically, Hijaz et al. and Jones et al. demonstrated that the organic acid quinic acid can be detected at a relatively high level in symptomatic or *C*Las-infected leaves, corroborating our imprinting MSI results [56,57]. As a key metabolite associated with plant stress and defense, quinic acid has been reported to exhibit a significantly different profile in asymptomatic leaves of *Citrus sinensis* [54,55], suggesting that quinic acid can be a vital biomarker candidate with potential for HLB infection recognition, even in the asymptomatic stage. Additionally, dehydroabietic acid, a natural product isolated from the nonvolatile residue of citric acid essential oil, was markedly increased in symptomatic citrus leaves, suggesting that it might be another candidate biomarker that is directly associated with HLB disease [55]. Given that HLB is a disease of worldwide incidence that could result in great losses to the citrus industry, the early diagnosis and detection of HLB disease-associated biomarkers can facilitate the prevention of citrus HLB diseases.

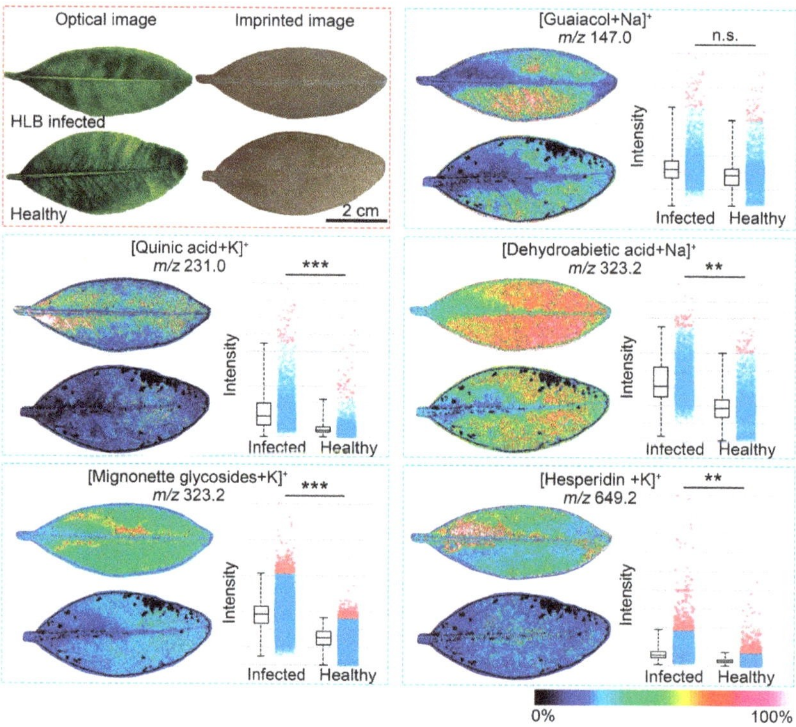


**Figure 5.** Imprinting MS images of leaves infected with citrus HLB and healthy leaves. Optical images of citrus leaves, corresponding imprinted images of the surface of the AuNP-hPDA-TDNT substrate, and representative ion images of the citrus leaves obtained via SALDI-MSI with a pixel size of 160 μm. The blue and red data points indicate the reasonable values and outliers, respectively. The colored bars indicate the relative signal intensity. The asterisks indicate the level of significance: ** $p < 0.01$, *** $p < 0.001$, and $p > 0.05$ (not significant, n.s.). Scale bar, 20 mm.

## 4. Conclusions

In summary, this work demonstrates a AuNP-hPDA-TDNT-based SALDI-MS method for the detection of a wide range of pesticides and the spatial visualization of primary and secondary metabolites during various plant defense processes by imprinting MSI. This nanostructured substrate, with the synergistic advantages of imprinting capability toward plant tissues and enhanced ionization efficiency, enables simplified sample preparation in real-case MSI applications, particularly for plant leaves that are incompatible with conventional histologic sectioning procedures. Moreover, the results presented here revealed that various nonuniform and tissue-specific distribution patterns of stress-induced metabolites can be clearly visualized in citrus leaves facing mechanical damage, insect feeding, and HLB disease infection via AuNP-hPDA-TDNT-based SALDI-MSI. Intriguingly, several potential biomarkers associated with the stress response of citrus leaves, such as quinic acid, could be visualized during mechanical damage and HLB disease infection, opening new avenues for a deeper understanding of crucial metabolites that participate in plant defense and infection processes. Efforts will be made in the future to decipher the biosynthesis, accumulation, and defensive role of various bioactive metabolites in economically important plants.

**Supplementary Materials:** The following supporting information can be downloaded at: https://www.mdpi.com/article/10.3390/agronomy14081797/s1. Figure S1: Workflow of the AuNP-hPDA-TDNT substrate fabrication. Figure S2: Repeatability test of the AuNP-hPDA-TDNT substrates from 3 batches for SALDI-MS analysis of azoxystrobin. Figure S3: Repeatability test of the AuNP-hPDA-TDNT substrates from 3 batches for SALDI-MS analysis of thiamethoxam. Figure S4: Repeatability test of the AuNP-hPDA-TDNT substrates from 3 batches for SALDI-MS analysis of rotenone. Figure S5: Stability test of the AuNP-hPDA-TDNT substrates for SALDI-MS analysis of azoxystrobin on different days. Figure S6: Stability test of the AuNP-hPDA-TDNT substrates for SALDI-MS analysis of thiamethoxam on different days. Figure S7: Stability test of the AuNP-hPDA-TDNT substrates for SALDI-MS analysis of rotenone on different days. Figure S8: Calibration curve of azoxystrobin using the AuNP-hPDA-TDNT substrate. Figure S9: Calibration curve of thiamethoxam using the AuNP-hPDA-TDNT substrate. Figure S10: Calibration curve of rotenone using the AuNP-hPDA-TDNT substrate. Figure S11: AuNP-hPDA-TDNT-based SALDI mass spectra of three pesticides with 10 amol amount. Figure S12: Representative mass spectrum acquired from an imprinted *Citrus madurensis* leaf using the AuNP-hPDA-TDNT substrate. Figure S13: MS/MS spectra of some typical metabolites from *Citrus madurensis* leaves in the positive ion mode. Table S1: Information on the pesticides used in the experiments. Table S2: Ion forms of standards of pesticides observed in SALDI-MS.

**Author Contributions:** Conceptualization, D.C. and X.W.; methodology, D.C.; validation, Y.S. and X.C.; formal analysis, Y.S.; investigation, Y.S. and X.C.; visualization, Y.S. and X.C.; writing—original draft preparation, Y.S., D.C. and X.C.; writing—review and editing, X.W.; supervision, X.W.; project administration, X.W.; funding acquisition, X.W. All authors have read and agreed to the published version of the manuscript.

**Funding:** This research was funded by the Spring Sunshine Programme by the Ministry of Education of China (No. HZKY20220113) and the Natural Science Foundation of Guangdong Province (No. 2023A1515030241). This work was also supported by the Original Innovation Cultivation Project of South China Agricultural University (No. 8400-223377).

**Data Availability Statement:** Data are contained within the article and Supplementary Materials.

**Acknowledgments:** We gratefully acknowledge the contributions of Yuhui Zhao and Wenlin Wen to data curation for this work.

**Conflicts of Interest:** The authors declare no conflict of interest.

## References


1. Gottwald, T.R. Current epidemiological understanding of citrus Huanglongbing. *Annu. Rev. Phytopathol.* **2010**, *48*, 119–139. [CrossRef] [PubMed]
2. Ma, W.; Pang, Z.; Huang, X.; Xu, J.; Pandey, S.S.; Li, J.; Achor, D.S.; Vasconcelos, F.N.C.; Hendrich, C.; Huang, Y.; et al. Citrus Huanglongbing is a pathogen-triggered immune disease that can be mitigated with antioxidants and gibberellin. *Nat. Commun.* **2022**, *13*, 529. [CrossRef]
3. Li, X.; Ruan, H.; Zhou, C.; Meng, X.; Chen, W. Controlling citrus Huanglongbing: Green sustainable development route is the future. *Front. Plant Sci.* **2021**, *12*, 760481. [CrossRef] [PubMed]
4. Liu, Y.; Xue, A.; Wang, S.; Hao, Y.; Cui, M.; Liu, L.; Luo, L. Metabolic response of *Citrus limon* to Asian citrus psyllid infestation revealed by EESI-MS and HPLC. *Anal. Biochem.* **2020**, *609*, 113973. [CrossRef]
5. Lee, J.A.; Halbert, S.E.; Dawson, W.O.; Robertson, C.J.; Keesling, J.E.; Singer, B.H. Asymptomatic spread of Huanglongbing and implications for disease control. *Proc. Natl. Acad. Sci. USA* **2015**, *112*, 7605–7610. [CrossRef]
6. Wang, K.; Liao, Y.; Meng, Y.; Jiao, X.; Huang, W.; Liu, T.C.-y. The early, rapid, and non-destructive detection of citrus Huanglongbing (HLB) based on microscopic confocal Raman. *Food Anal. Methods* **2019**, *12*, 2500–2508. [CrossRef]
7. Xue, A.; Liu, Y.; Li, H.; Cui, M.; Huang, X.; Wang, W.; Wu, D.; Guo, X.; Hao, Y.; Luo, L. Early detection of Huanglongbing with EESI-MS indicates a role of phenylpropanoid pathway in citrus. *Anal. Biochem.* **2022**, *639*, 114511. [CrossRef] [PubMed]
8. Shroff, R.; Vergara, F.; Muck, A.; Svatoš, A.; Gershenzon, J. Nonuniform distribution of glucosinolates in *Arabidopsis thaliana* leaves has important consequences for plant defense. *Proc. Natl. Acad. Sci. USA* **2008**, *105*, 6196–6201. [CrossRef]
9. Yan, S.; Bhawal, R.; Yin, Z.; Thannhauser, T.W.; Zhang, S. Recent advances in proteomics and metabolomics in plants. *Mol. Hortic.* **2022**, *2*, 17. [CrossRef]
10. Aharoni, A.; Goodacre, R.; Fernie, A.R. Plant and microbial sciences as key drivers in the development of metabolomics research. *Proc. Natl. Acad. Sci. USA* **2023**, *120*, e2217383120. [CrossRef] [PubMed]
11. Maeda, H.A.; Fernie, A.R. Evolutionary history of plant metabolism. *Annu. Rev. Plant Biol.* **2021**, *72*, 185–216. [CrossRef]
12. Perez de Souza, L.; Alseekh, S.; Scossa, F.; Fernie, A.R. Ultra-high-performance liquid chromatography high-resolution mass spectrometry variants for metabolomics research. *Nat. Methods* **2021**, *18*, 733–746. [CrossRef]
13. Alseekh, S.; Aharoni, A.; Brotman, Y.; Contrepois, K.; D'Auria, J.; Ewald, J.; Ewald, J.C.; Fraser, P.D.; Giavalisco, P.; Hall, R.D.; et al. Mass spectrometry-based metabolomics: A guide for annotation, quantification and best reporting practices. *Nat. Methods* **2021**, *18*, 747–756. [CrossRef] [PubMed]
14. Horn, P.J.; Chapman, K.D. Imaging plant metabolism in situ. *J. Exp. Bot.* **2023**, *75*, 1654–1670. [CrossRef]
15. Yin, Z.; Huang, W.; Fernie, A.R.; Yan, S. Mass spectrometry imaging techniques: A versatile toolbox for plant metabolomics. *Trends Plant Sci.* **2023**, *28*, 250–251. [CrossRef]
16. Yin, Z.; Dong, T.; Huang, W.; Du, M.; Chen, D.; Fernie, A.R.; Yi, G.; Yan, S. Spatially resolved metabolomics reveals variety-specific metabolic changes in banana pulp during postharvest senescence. *Food Chem. X* **2022**, *15*, 100371. [CrossRef] [PubMed]
17. Qin, R.; Li, P.; Du, M.; Ma, L.; Huang, Y.; Yin, Z.; Zhang, Y.; Chen, D.; Xu, H.; Wu, X. Spatiotemporal visualization of insecticides and fungicides within fruits and vegetables using gold nanoparticle-immersed paper imprinting mass spectrometry imaging. *Nanomaterials* **2021**, *11*, 1327. [CrossRef]
18. Yin, Z.; Huang, W.; Li, K.; Fernie, A.R.; Yan, S. Advances in mass spectrometry imaging for plant metabolomics—Expanding the analytical toolbox. *Plant J.* 2024; *early view*. [CrossRef]
19. Wang, T.; Cheng, X.; Xu, H.; Meng, Y.; Yin, Z.; Li, X.; Hang, W. Perspective on advances in laser-based high-resolution mass spectrometry imaging. *Anal. Chem.* **2020**, *92*, 543–553. [CrossRef]
20. Ma, S.; Leng, Y.; Li, X.; Meng, Y.; Yin, Z.; Hang, W. High spatial resolution mass spectrometry imaging for spatial metabolomics: Advances, challenges, and future perspectives. *TrAC Trends Anal. Chem.* **2023**, *159*, 116902. [CrossRef]
21. Chen, Y.-J.; Zeng, H.-S.; Jin, H.L.; Wang, H.B. Applications of mass spectrometry imaging in botanical research. *Adv. Biotechnol.* **2024**, *2*, 6. [CrossRef]
22. Dong, Y.; Aharoni, A. Image to insight: Exploring natural products through mass spectrometry imaging. *Nat. Prod. Rep.* **2022**, *39*, 1510–1530. [CrossRef] [PubMed]
23. Müller, W.H.; Verdin, A.; De Pauw, E.; Malherbe, C.; Eppe, G. Surface-assisted laser desorption/ionization mass spectrometry imaging: A review. *Mass Spectrom. Rev.* **2022**, *41*, 373–420. [CrossRef]
24. Lu, M.; Yang, X.; Yang, Y.; Qin, P.; Wu, X.; Cai, Z. Nanomaterials as assisted matrix of laser desorption/ionization time-of-flight mass spectrometry for the analysis of small molecules. *Nanomaterials* **2017**, *7*, 78. [CrossRef] [PubMed]
25. Pomastowski, P.; Buszewski, B. Complementarity of matrix- and nanostructure-assisted laser desorption/ionization approaches. *Nanomaterials* **2019**, *9*, 260. [CrossRef] [PubMed]
26. Dong, Y.; Li, B.; Malitsky, S.; Rogachev, I.; Aharoni, A.; Kaftan, F.; Svatoš, A.; Franceschi, P. Sample preparation for mass spectrometry imaging of plant tissues: A review. *Front. Plant Sci.* **2016**, *7*, 60. [CrossRef]
27. Wu, X.; Qin, R.; Wu, H.; Yao, G.; Zhang, Y.; Li, P.; Xu, Y.; Zhang, Z.; Yin, Z.; Xu, H. Nanoparticle-immersed paper imprinting mass spectrometry imaging reveals uptake and translocation mechanism of pesticides in plants. *Nano Res.* **2020**, *13*, 611–620. [CrossRef]
28. Xia, J.; Lou, G.; Zhang, L.; Huang, Y.; Yang, J.; Guo, J.; Qi, Z.; Li, Z.; Zhang, G.; Xu, S.; et al. Unveiling the spatial distribution and molecular mechanisms of terpenoid biosynthesis in *Salvia miltiorrhiza* and *S. grandifolia* using multi-omics and DESI–MSI. *Hortic. Res.* **2023**, *10*, uhad109. [CrossRef] [PubMed]

29. Hemalatha, R.G.; Pradeep, T. Understanding the molecular signatures in leaves and flowers by desorption electrospray ionization mass spectrometry (DESI MS) imaging. *J. Agric. Food Chem.* **2013**, *61*, 7477–7487. [CrossRef]
30. Hemalatha, R.G.; Ganayee, M.A.; Pradeep, T. Electrospun nanofiber mats as "smart surfaces" for desorption electrospray ionization mass spectrometry (DESI MS)-based analysis and imprint imaging. *Anal. Chem.* **2016**, *88*, 5710–5717. [CrossRef] [PubMed]
31. Wei, J.; Buriak, J.M.; Siuzdak, G. Desorption-ionization mass spectrometry on porous silicon. *Nature* **1999**, *399*, 243–246. [CrossRef]
32. Samarah, L.Z.; Tran, T.H.; Stacey, G.; Vertes, A. Mass spectrometry imaging of bio-oligomer polydispersity in plant tissues by laser desorption ionization from silicon nanopost arrays. *Angew. Chem. Int. Ed.* **2021**, *60*, 9071–9077. [CrossRef] [PubMed]
33. Liu, X.; Chen, Z.; Wang, T.; Jiang, X.; Qu, X.; Duan, W.; Xi, F.; He, Z.; Wu, J. Tissue imprinting on 2D nanoflakes-capped silicon nanowires for lipidomic mass spectrometry imaging and cancer diagnosis. *ACS Nano* **2022**, *16*, 6916–6928. [CrossRef] [PubMed]
34. Chen, D.; Du, M.; Huang, Y.; Xu, Y.; Chen, Y.; Ma, L.; Xie, Q.; Zhu, X.; Chen, Z.; Xu, H.; et al. Plasmonic polydopamine-modified $TiO_2$ nanotube substrates for surface-assisted laser desorption/ionization mass spectrometry imaging. *Nano Res.* **2023**, *16*, 3028–3039. [CrossRef]
35. Yang, J.; Zhang, W.; Zhang, H.; Zhong, M.; Cao, W.; Li, Z.; Huang, X.; Nie, Z.; Liu, J.; Li, P.; et al. Polydopamine-modified substrates for high-sensitivity laser desorption ionization mass spectrometry imaging. *ACS Appl. Mater. Interfaces* **2019**, *11*, 46140–46148. [CrossRef] [PubMed]
36. Yang, J.; Zhang, H.; Jia, J.; Zhang, X.; Ma, X.; Zhong, M.; Ouyang, Z. Antireflection surfaces for biological analysis using laser desorption ionization mass spectrometry. *Research* **2018**, *2018*, 5439729. [CrossRef] [PubMed]
37. Sakai, R.; Kondo, H.; Ishikawa, K.; Ohta, T.; Hiramatsu, M.; Tanaka, H.; Hori, M. Effects of high-quality carbon nanowalls ionization-assisting substrates on surface-assisted laser desorption/ionization mass spectrometry performance. *Nanomaterials* **2023**, *13*, 63. [CrossRef]
38. Kim, Y.-K.; Na, H.-K.; Kwack, S.-J.; Ryoo, S.-R.; Lee, Y.; Hong, S.; Hong, S.; Jeong, Y.; Min, D.-H. Synergistic effect of graphene oxide/MWCNT films in laser desorption/ionization mass spectrometry of small molecules and tissue imaging. *ACS Nano* **2011**, *5*, 4550–4561. [CrossRef]
39. Iakab, S.A.; Ràfols, P.; Tajes, M.; Correig-Blanchar, X.; García-Altares, M. Gold nanoparticle-assisted black silicon substrates for mass spectrometry imaging applications. *ACS Nano* **2020**, *14*, 6785–6794. [CrossRef]
40. Sun, S.; Liu, W.; Yang, J.; Wang, H.; Qian, K. Nanoparticle-assisted cation adduction and fragmentation of small metabolites. *Angew. Chem. Int. Ed.* **2021**, *60*, 11310–11317. [CrossRef]
41. Manjavacas, A.; Liu, J.G.; Kulkarni, V.; Nordlander, P. Plasmon-induced hot carriers in metallic nanoparticles. *ACS Nano* **2014**, *8*, 7630–7638. [CrossRef] [PubMed]
42. Dutkiewicz, E.P.; Su, C.-H.; Lee, H.-J.; Hsu, C.-C.; Yang, Y.-L. Visualizing vinca alkaloids in the petal of *Catharanthus roseus* using functionalized titanium oxide nanowire substrate for surface-assisted laser desorption/ionization imaging mass spectrometry. *Plant J.* **2021**, *105*, 1123–1133. [CrossRef] [PubMed]
43. Wang, X.-N.; Li, B. Monolithic gold nanoparticles/thiol-β-cyclodextrin-functionalized $TiO_2$ nanowires for enhanced SALDI MS detection and imaging of natural products. *Anal. Chem.* **2022**, *94*, 952–959. [CrossRef] [PubMed]
44. Chen, D.; Xu, Y.; Huang, Y.; Chen, Y.; Zhao, Y.; Yan, B.; Zhu, X.; Chen, Z.; Xu, H.; Yin, Z.; et al. Mapping molecular signatures in plant leaves, flowers, and fruits by a $TiO_2$ nanotube-based plasmonic chip for imprint mass spectrometry imaging. *ACS Agric. Sci. Technol.* **2023**, *3*, 119–130. [CrossRef]
45. Wu, Q.; Chu, J.L.; Rubakhin, S.S.; Gillette, M.U.; Sweedler, J.V. Dopamine-modified $TiO_2$ monolith-assisted LDI MS imaging for simultaneous localization of small metabolites and lipids in mouse brain tissue with enhanced detection selectivity and sensitivity. *Chem. Sci.* **2017**, *8*, 3926–3938. [CrossRef] [PubMed]
46. Sun, X.; Huang, L.; Zhang, R.; Xu, W.; Huang, J.; Gurav, D.D.; Vedarethinam, V.; Chen, R.; Lou, J.; Wang, Q.; et al. Metabolic fingerprinting on a plasmonic gold chip for mass spectrometry based in vitro diagnostics. *ACS Cent. Sci.* **2018**, *4*, 223–229. [CrossRef] [PubMed]
47. Alam, F.; Mohammadin, K.; Shafique, Z.; Amjad, S.T.; Asad, M.H.H.b. Citrus flavonoids as potential therapeutic agents: A review. *Phytother. Res.* **2022**, *36*, 1417–1441. [CrossRef] [PubMed]
48. Dong, Z.; Liu, X.; Srivastava, A.K.; Tan, Q.; Low, W.; Yan, X.; Wu, S.; Sun, X.; Hu, C. Foliar nutrition alleviate citrus plants from Asian citrus psyllid feeding by affecting leaf structure and secondary metabolism. *Sci. Hortic.* **2023**, *309*, 111667. [CrossRef]
49. Galili, G.; Amir, R.; Fernie, A.R. The regulation of essential amino acid synthesis and accumulation in plants. *Annu. Rev. Plant Biol.* **2016**, *67*, 153–178. [CrossRef] [PubMed]
50. Galili, G.; Amir, R. Fortifying plants with the essential amino acids lysine and methionine to improve nutritional quality. *Plant Biotechnol. J.* **2013**, *11*, 211–222. [CrossRef]
51. Gonda, I.; Lev, S.; Bar, E.; Sikron, N.; Portnoy, V.; Davidovich-Rikanati, R.; Burger, J.; Schaffer, A.A.; Tadmor, Y.; Giovannonni, J.J.; et al. Catabolism of L-methionine in the formation of sulfur and other volatiles in melon (*Ucumis melo* L.) fruit. *Plant J.* **2013**, *74*, 458–472. [CrossRef]
52. Roje, S. *S*-Adenosyl-L-methionine: Beyond the universal methyl group donor. *Phytochemistry* **2006**, *67*, 1686–1698. [CrossRef] [PubMed]

53. Hirai, M.Y.; Sugiyama, K.; Sawada, Y.; Tohge, T.; Obayashi, T.; Suzuki, A.; Araki, R.; Sakurai, N.; Suzuki, H.; Aoki, K.; et al. Omics-based identification of *Arabidopsis* Myb transcription factors regulating aliphatic glucosinolate biosynthesis. *Proc. Natl. Acad. Sci. USA* **2007**, *104*, 6478–6483. [CrossRef] [PubMed]
54. Asai, T.; Matsukawa, T.; Kajiyama, S. Metabolomic analysis of primary metabolites in citrus leaf during defense responses. *J. Biosci. Bioeng.* **2017**, *123*, 376–381. [CrossRef] [PubMed]
55. de Moraes Pontes, J.G.; Vendramini, P.H.; Fernandes, L.S.; de Souza, F.H.; Pilau, E.J.; Eberlin, M.N.; Magnani, R.F.; Wulff, N.A.; Fill, T.P. Mass spectrometry imaging as a potential technique for diagnostic of Huanglongbing disease using fast and simple sample preparation. *Sci. Rep.* **2020**, *10*, 13457. [CrossRef]
56. Hijaz, F.M.; Manthey, J.A.; Folimonova, S.Y.; Davis, C.L.; Jones, S.E.; Reyes-De-Corcuera, J.I. An HPLC-MS characterization of the changes in sweet orange leaf metabolite profile following infection by the bacterial pathogen *Candidatus* Liberibacter asiaticus. *PLoS ONE* **2013**, *8*, e79485. [CrossRef] [PubMed]
57. Jones, S.; Hijaz, F.; Davis, C.; Folimonova, S.; Manthey, J.; Reyes-De-Corcuera, J. GC-MS analysis of secondary metabolites in leaves from orange trees infected with HLB: A 9-month study. *Proc. Fla. State Hortic. Soc.* **2012**, *125*, 75–83.

*agronomy*

MDPI

*Article*

# Spring Abundance, Migration Patterns and Damaging Period of *Aleyrodes proletella* in the Czech Republic

**Kamil Holý [1] and Kateřina Kovaříková [2,*]**

[1] Group of Integrated Protection of Agricultural Crops against Pests, Crop Research Institute, Drnovska 507/73, 161 06 Prague, Czech Republic; holy@vurv.cz

[2] Group of Plant Secondary Metabolites in Crop Protection, Crop Research Institute, Drnovska 507/73, 161 06 Prague, Czech Republic

* Correspondence: kovarikova@vurv.cz; Tel.: +420-607-060-325

**Abstract:** The cabbage whitefly has become an important pest on brassica vegetables in Central Europe. It does not destroy the affected plants, but the product becomes unmarketable, causing considerable economic losses. The pest is also difficult to control due to its way of life and because it develops resistance to some of the active components of insecticides. In organic farming systems, insecticides are strictly restricted, but neither predators nor whitefly parasitoids are able to keep the pest at a tolerable level. It is, therefore, necessary to become familiar with the whitefly's life cycle and habits, including mass migration from winter hosts to vegetables. We inspected 44 rapeseed fields across the republic in the period 2014–2021 in order to find the connection between the presence of oilseed rape fields near vegetable growing areas (VGAs) and the abundance of the overwintering cabbage whiteflies. We also conducted regular weekly monitoring of whitefly occurrence in the main cultivation area of the Czech Republic (Polabí) with the aim of specifying critical data important for the successful control of this pest. We found that the cabbage whitefly incidences were many times higher in rapeseed fields close to VGAs compared to areas where the crops are not adjacent. The average number of whiteflies was 0.59 individuals per plant in VGA-1 (oilseed rape grown inside this area or up to 1 km far), 0.052 in VGA-2 (distance 3–10 km from vegetable fields) and 0.014 in VGA-3 (more than 20 km). In the extremely warm year 2016, the difference was up to sixty times. The first CW eggs laid on cruciferous vegetables were usually found around 20 May. The period of mass migration of CW adults to cruciferous vegetables was between 6 June and 2 August. At this time, vegetables are most vulnerable to damage. Successful control of the cabbage whitefly requires the use of fabric netting, combined with an insecticide as needed and trap plants as needed; the latter have to be destroyed before adult whiteflies hatch—typically in early July.

**Keywords:** *Aleyrodes proletella*; brassica crops; oilseed rape; ecology; migration; cabbage whitefly; Czech Republic

## 1. Introduction

The cabbage whitefly (CW) *Aleyrodes proletella* L. (Hemiptera: Aleyrodidae) is an oligophagous species causing economic losses to brassica vegetables. It is a native species of the Palearctic realm but has also spread to other continents, where it has the invasive species status. It is currently present on all the continents except Antarctica [1]; for example, its occurrence has been reported in Angola, Australia [2], India, the United States, Mexico, Brazil, China [3] and New Zealand [4]. It was regarded as a minor pest in Europe, so little research has focused on this species in the past [5], but its importance as a pest has increased considerably in recent decades [6–13]. Den Belder et al. [14] stated that *Aleyrodes proletella* is slowly spreading from Central to Western Europe and has recently become a serious pest in commercial brassica fields in the Netherlands.

The CW spreading is probably associated with an increase in winter rapeseed (*B. napus* L.) cultivation areas. Askoul et al. [15] examined the effects of various host plants on the reproductive parameters of *A. proletella*, such as fertility, survival rate and preoviposition period, and all of the parameters were significantly the best in winter rapeseed. In the UK, for example, the oilseed rape cultivation area increased only in the mid-1970s; brassicas as vegetables clearly dominated until then [16]. In the Czech Republic, the size of the rapeseed growing areas was under 50 thousand hectares until the end of the 1970s. After the 1990s, there was a jump shift when the cultivation area skyrocketed to 350 thousand ha. After 2000, the area decreased to 250 thousand ha, and there was another shift to about 340 thousand ha in 2007 in connection with EU directives on organic additives in diesel fuel. The rapeseed growing area was between 350 and 420 thousand hectares in 2010–2020 [17]. The oilseed rape area grew by 20% from the Czech Republic's ascension to the EU (1 May 2004) to 1 May 2016, and the total increase has even been 283.9% since 1989 [18].

In the Czech Republic, CW had no pest status in the past [19]. Until the beginning of the 21st century, no significant damage to crops existed in the Czech Republic. The first damage was recorded around 2008 in the southeastern part of the Czech Republic (Moravia region), where the climate is warmer. In the Polabí region, with a high concentration of vegetable fields, the damage occurred 1–2 years later (depending on location). Since 2012, *A. proletella* has been a regular pest on cruciferous vegetables in all lowland regions of the Czech Republic [20] and is also abundant in all oilseed rape fields.

Winter rapeseed provides a suitable shelter for the cabbage whitefly during the winter as well as a food source for the period when no vegetables grow in the fields. Its widespread cultivation results in so-called green bridges and, along with milder winters, has a critical influence on the current success rate of the cabbage whitefly [14,16]. CW does virtually no harm to rapeseed, but it creates numerous colonies there [13], from which the adults migrate en masse to vegetable fields in the course of June [16], where they pose a major problem. It has been noted that CW has two morphs with different specific flight capabilities [21]. While one of them can spread over a range of several hundred meters, the other migrates from origins up to several kilometres distant. The migrating morph (around 5 km) appears to be much more important in colonising vegetable fields than the local one (200–1000 m) [22]. Typically, the cabbage whitefly life cycle from egg (laid after wintering) to adult is completed within 3–6 weeks, depending on the temperature [13,23]. The number of generations that the whiteflies can achieve differs depending on the geographic location. While there tend to be 3–5 generations in the UK [24], there may be up to 10 generations under ideal conditions in Southern Europe [11].

Infested plants are characterised by the presence of CW or later by the appearance of honeydew; sap sucking has little effect on the crops, so no retarded or dead plants due to CW sucking have been observed, and no colour changes are visible on the leaves. After the registration of insecticides containing the active ingredient spirotetramat in 2008 and other effective insecticides such as spinosyns or cyantraniliprole [25], CW was a problem only in a small part of less intensively sprayed fields, but especially in organic vegetable production. A recent study by Müller et al. [26] detected field resistance to ketonenol insecticides in some CW populations in Belgium and Germany with cross-resistance to spirotetramat and spiromesifen, indicating an increased risk that insecticide protection against CW may become less reliable and the situation may worsen.

To draw an appropriate sensitive crop protection strategy, it is important to know the CW population trends. Critical data include knowledge of when whitefly populations peak on rapeseed and when they migrate to vegetables [5]. The objective of this investigation was to obtain a better understanding of cabbage whitefly migration and population development on sites selected based on their proximity to vegetable growing areas, validate the biological data for the Czech Republic and help complete missing information about the pest.

## 2. Materials and Methods

### *2.1. Spring Abundance of Whiteflies on Oilseed Rape*

The survey of the spring abundance of whiteflies was carried out in March or early April in 2014, 2016–2018 and 2021, up to a plant height of 40 cm. The assessment of the fields each year was carried out within a few days. The date was different each year depending on the weather, but the plants were always in the same phenophase.

Forty-four oilseed rape fields across the Czech Republic (Central Europe, 48°39′–50°59′ N, 12°19′–18°29′ E) were inspected for the presence of *A. proletella* (Figure 1). Nine selected sampling sites were located in (1) a vegetable growing area (VGA-1) with oilseed rape grown inside this area or up to 1 km away; twenty-two sites were (2) close to VGA at 3–10 km (VGA-2); and thirteen sites were considered controls with (3) no proximity of VGA at more than 20 km (VGA-3). At each site, ten oilseed rape plants with four repetitions were inspected, and numbers of adults, eggs and nymphs were recorded.

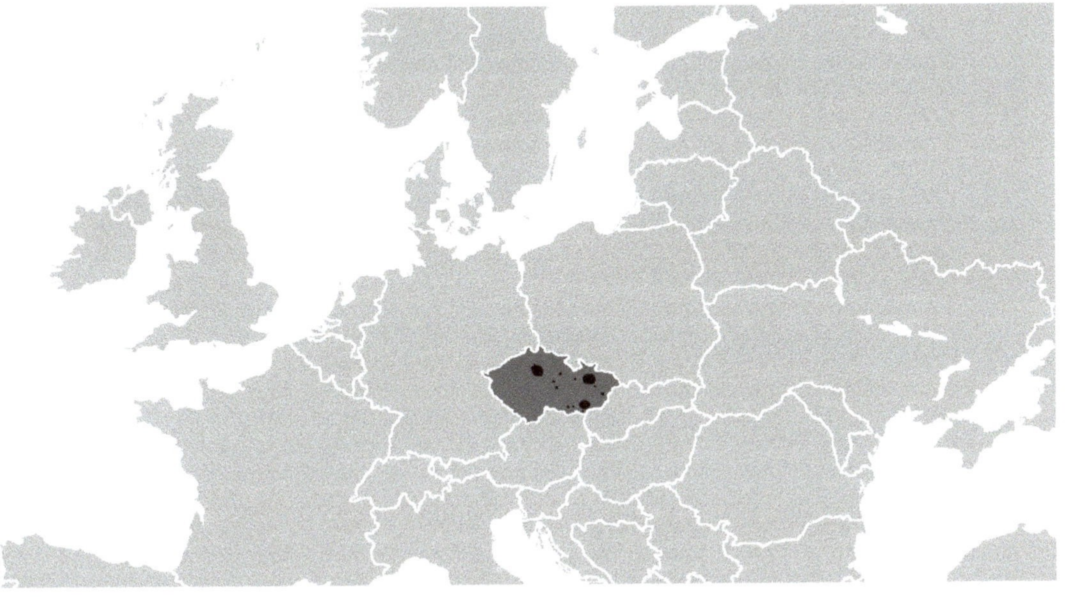

**Figure 1.** Schematic map of locations of investigated winter rapeseed fields in European context.

The location of the rapeseed fields changed a little from year to year due to crop rotation, but their distance was always within 2 km. The altitude of the sites ranged from 150 to 200 m a.s.l. for (1), 166 to 350 m a.s.l. for (2) and 175 to 530 m a.s.l. for (3). Mean temperatures in the Czech Republic for a given year and deviations from the long-term normal are summarised for selected months in Table 1.

**Table 1.** Mean temperatures and deviations from the long-term normal for the Czech Republic. Data taken from CHMI [27].

| Mean Temperatures and Deviations from the Long-Term Normal for the Czech Republic | January | February | March | April | Annual | Year |
|---|---|---|---|---|---|---|
| Temperature | 0.5 | 2.1 | 6.2 | 9.8 | 9.4 | 2014 |
| Deviation | 2.5 | 3 | 3.3 | 1.9 | 1.5 | |
| Temperature | −1.4 | 3 | 3.3 | 7.7 | 8.7 | 2016 |
| Deviation | 0.6 | 3.9 | 0.4 | −0.2 | 0.8 | |

**Table 1.** *Cont.*

| Mean Temperatures and Deviations from the Long-Term Normal for the Czech Republic | January | February | March | April | Annual | Year |
|---|---|---|---|---|---|---|
| Temperature | −5.6 | 1.1 | 5.9 | 6.9 | 8.6 | 2017 |
| Deviation | −3.6 | 2 | 3.4 | −1 | 1.1 | |
| Temperature | 1.8 | −3.5 | 0.8 | 12.7 | 9.6 | 2018 |
| Deviation | 3.8 | −2.6 | -2.1 | 4.8 | 1.7 | |
| Temperature | −1.1 | −0.8 | 2.6 | 5.4 | 8 | 2021 |
| Deviation | 0.3 | −0.4 | -0.6 | −3.1 | −0.3 | |

### 2.2. *Monitoring of Whiteflies, Their Natural Enemies and Migration Period of Whiteflies*

Regular monitoring of cruciferous vegetables was carried out from April to October 2018–2023 as a service for the growers and the Vegetable Union of Bohemia and Moravia (Czech Republic) to prepare a weekly report with the current occurrence of pests, their natural enemies and recommendations for crop protection. During the monitoring of vegetable pests, the first occurrence of adults, eggs and nymphs, as well as the first hatching of whitefly adults from the puparia on cruciferous vegetables, was recorded. The onset of mass migration from oilseed rape was estimated as the presence of more than five adult whiteflies per vegetable plant on at least 90% of the plants in the field.

The fields in the "Polabí" region were inspected once a week. Polabí is an area along the Elbe River where vegetable cultivation predominates. It has the highest concentration of vegetable fields in the Czech Republic. Three sites were monitored: the vicinity of Litoměřice (GPS: 50.5221728 N, 14.1204472 E), Obříství (GPS: 50.3071008 N, 14.4655750 E) and Semice (GPS: 50.1527447 N, 14.8755667 E) with numerous fields with different cruciferous vegetables (cabbage, kale, cauliflower, kohlrabi, etc.). A visual examination of plants was performed. In each site, three to five fields were inspected, depending on year and season. Therefore, a minimum of 40 plants per field was visually inspected in four different places of the field, and the abundance of pests and natural enemies was recorded.

From April to June, all life stages of CW were recorded; however, in this work, we mainly mention the first occurrences of whitefly stages in rapeseed and vegetables. After the mass migration had started, the number of adults was only estimated (increasing/decreasing from week to week) because after the beginning of mass migration, growers started spraying insecticides against CW adults regularly until harvest, and the abundance of adults thus depended more on the spray programme than on migration. All of the monitored fields were used for vegetable production, so no untreated control was available. Monitoring of CW larvae continued until the first adults developed on cruciferous vegetables. This is important because, from this period, cruciferous vegetables are an important source of CW for new vegetable plantings.

Predators of CW were recorded during visual examination of vegetable plants from the first occurrence of CW until the end of October. Hymenopteran parasitoids were examined only from puparia. Leaves with puparia of CW were collected from all three sites (more than 100 puparia on each site) at least twice per season; afterward, the leaves were placed in a plastic cup with mesh on the top and stored in laboratory conditions (23 ± 3 °C) until adults of CW or parasitoids hatched.

### 2.3. *Data Analysis*

The "spring abundance" data were statistically analysed with R version 4.3.1 [28]. The count data (numbers of adult whiteflies per plant) were fitted with a negative binomial family using the "glmmTMB" package model [29]. The proximity to VGA (1 = yes, 2 = between, 3 = no) and the year were set as the explanatory variables. Repetition was taken as a random effect. Residual plots from the package "DHARMa" [30] and AIC values were used to select the best model, which was chosen to calculate the variance analysis table (ANOVA function). Tukey's post hoc test ("emmeans" package) was applied for multiple comparisons [31]. The "multcomp" package was used to assign letters to indicate significant differences [32]. A total of 8400 observations were made.

## 3. Results

### 3.1. Spring Abundance of Whiteflies on Oilseed Rape

Observation of the first laid eggs on rapeseed was recorded at the beginning of March, specifically on 4 March 2014 in Mochov–VGA 1, when the plants were mostly 10–15 cm high. CW always lay fewer than 10 eggs per clutch initially. The first grey coloration of the eggs, indicating impending hatching, and more than 10 eggs per clutch were observed in mid-March, but usually in early April, when the plants reached a height of more than 20 cm. This period roughly corresponded to the beginning of apricot blossom.

The numbers of CW adults were significantly different on the sampled sites depending on their proximity to vegetable growing areas ($\chi^2{}_2 = 905.77$, $p < 0.001$). The effect of the year was also significant ($\chi^2{}_4 = 136.53$, $p < 0.001$). The highest number of CW individuals was found on oilseed rape plants in VGA-1 in all the sampling years. More details are shown in Table 2. No overwintering whitefly nymphs and puparia were found on the oilseed rape during our survey.

**Table 2.** Mean numbers of CW adults per plant (mean ± SE) per year over all experimental sites.

| Mean Number of *Aleyrodes proletella* Adults in Rape Fields per Plant in Early Spring during the Survey. Different Letters Mark Significant Differences in CW Numbers between Vegetable Growing Areas (VGAs) | | | |
|---|---|---|---|
| **Year** | **In VGA** | **Close to VGA** | **Far from VGA** |
| 2014 | $0.76 \pm 0.08$ [c] | $0.06 \pm 0.01$ [b] | $0.02 \pm 0.01$ [a] |
| 2016 | $1.28 \pm 0.12$ [c] | $0.05 \pm 0.01$ [b] | $0.02 \pm 0.01$ [a] |
| 2017 | $0.09 \pm 0.02$ [b] | $0.02 \pm 0.00$ [a] | $0.01 \pm 0.00$ [a] |
| 2018 | $0.49 \pm 0.06$ [c] | $0.07 \pm 0.01$ [b] | $0.01 \pm 0.00$ [a] |
| 2021 | $0.33 \pm 0.04$ [c] | $0.06 \pm 0.01$ [b] | $0.01 \pm 0.0$ [a] |

### 3.2. Monitoring of Whiteflies, Their Natural Enemies and Migration Period of Whiteflies

The first adults of the generation, which developed on oilseed rape and herbs, usually appeared and laid eggs in cruciferous vegetable around 20 May (Figure 2). Due to cold temperatures in the spring of 2021 and 2023, the occurrence was delayed until early June.

The number of migrating adults gradually increases, and as the oilseed rape dries, a mass migration occurs. The beginning of the mass migration to cruciferous vegetables was usually around 10 June; in 2021, it was postponed until the second half of June. Intense storms at the beginning of July 2021 killed most of the adults in the vegetables, so a significant increase in damage appeared later that year—from the end of July/beginning of August. Mass migration can be recognised by a sudden increase in the number of individuals on brassica plants. Adults can be observed during mass migration and also in other crops (e.g., celery, zucchini) in some years. Mass migration from oilseed rape usually lasted until the beginning of July; in 2021–2023, it was postponed until the end of July/beginning of August due to the late harvest. The rapeseed plants were still green at the beginning of July. The end of the mass migration to cruciferous vegetables is gradual, and it corresponds with the end of the rapeseed harvest in the surrounding fields. During July, individuals hatched from the rapeseed in uplands may also probably migrate to vegetables, up to tens of kilometres. The next generations overlap, and they develop on vegetables and partly on newly germinated rapeseed oil plants on the harvested fields. The main autumn migration from vegetables to new oilseed rape fields in the Czech Republic usually appears around mid-October, but in 2022 it happened at the beginning of September before an unusually long cold and rainy period from 17 September to 13 October appeared.

During our survey of the occurrence of natural enemies of CW in Czech fields since 2011, no significant predator or parasitoid was observed. The dominant syrphid species was *Episyrphus balteatus*. Coccinellidae and other predators were usually low in numbers. *Encarsia tricolor* was recorded on unsprayed vegetables only in 2011.

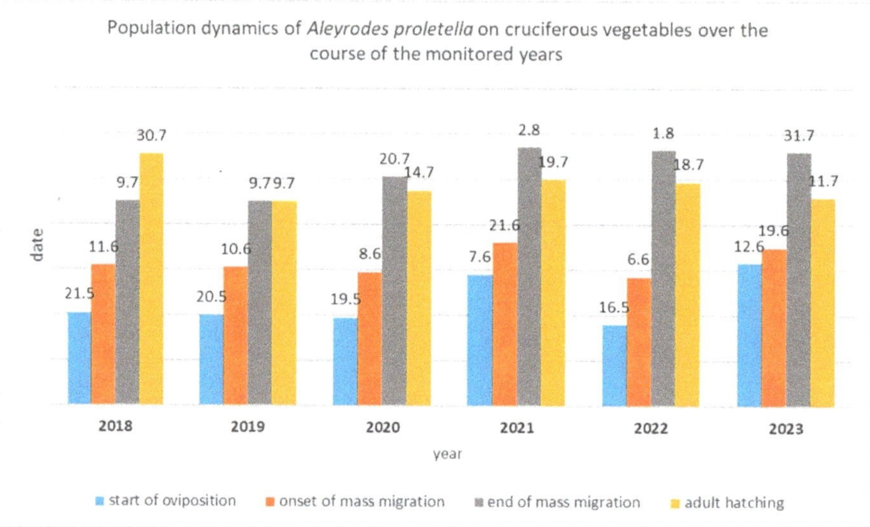


**Figure 2.** Population dynamics of *Aleyrodes proletella* on cruciferous vegetables over the course of the monitored years.

## 4. Discussion

Richter and Hirthe [13] reported that overwintering *A. proletella* adults in Germany began ovipositing on oilseed rape, notably at the end of April, and dispersed to new host plants in June. In our case, the first eggs laid on rapeseed were found at the beginning of March 2014, which was a little earlier, but this might have been an effect of different weather. More eggs are usually laid at the end of March or the beginning of April. Also, in contrast to the observations of Iheagwam [21] in southern England, we found no fourth-instar larvae on overwintering rapeseed in March/April. On the contrary, the author states that during winter, there are often many of them, which is due to different winter temperatures in both regions. Because there are no cruciferous vegetables left in the field from autumn to spring in the Czech Republic, we assume that the first generation of CW develops mainly on winter oilseed rape and wild plants (March–May). The second CW generation then develops on rapeseed and partly on wild herbs and vegetables (May–July). The number of generations is generally dependent on the geographic region (temperature) and may also overlap [24]. Therefore, it is not easy to determine the exact order.

Whitefly migration occurs for several reasons–plant death, drying up, or overcrowding–more than six adults per approx. 6.5 cm$^2$ [33]. Iheagwam [21] also stated that the high density of larvae slows down the development. In our case, the trigger was probably the ripening of rapeseed when the plants were still green but no longer provided favourable conditions for the development of the next CW generation. This was also stated by Richter and Hirthe [13]. Our findings are similar to the results of Richter and Hirthe [13]. Between the years 2014 and 2021, the average number of whiteflies per 10 plants varied between 1 and 13 individuals on winter rapeseed. Let us say that the average number was five individuals per 10 plants in normal years, whilst 2016 and 2017 were exceptionally warm and cold, respectively. This means that CW numbers in the spring did not grow substantially from 2009 to 2011.

Migration and damage rates depend on the weather. During a cold and rainy period in June and July, the mass migration is often interrupted for one or two weeks. Also, intensive rain can reduce the CW population in vegetables. Cruciferous vegetable damage usually occurs from the end of June or the beginning of July. Autumn migration from vegetables to overwintering sites takes place during September/October. In some years, mass migration can actually be observed during warm October days, when adults can be found even in city centres (e.g., in the centre of Prague), far away from the fields. Prague city centre is at least 7 km from the nearest field.

We assume that less than 1% of CW adults are killed by predators (mainly syrphid larvae). Syrphid larvae prefer to feed on cabbage aphids (*Brevicoryne brassicae*); only after eating all the aphids do they start feeding on CW adults. Other predators were usually low in numbers, probably due to frequent spraying with broad-spectrum pyrethroids and better migration capability. Unlike syrphid larvae, they can fly to other nearby crops after aphid predation. In neighbouring Germany, mainly *Encarsia tricolor* parasitises CW nymphs [34,35]. This species was recorded on unsprayed vegetables in the Czech Republic only in 201120. Since then, no additional CW parasitoids (adults or parasitic whitefly nymphs) have been observed, even in fields without pesticide application. Interestingly, the mass occurrence of CW did not lead to an increase in the activity of predators or parasitoids, and this rich food source remained unused.

The first cruciferous vegetables are planted in March, and, depending on the weather, they are covered with netting until mid/late April. Field edges are especially vulnerable to whitefly infestation [5]. The mesh is very effective protection of young plants against colonisation by overwintering CW females. To some extent, it is also possible to use systemic insecticides. With contact insecticides, it is necessary to wet the underside of the leaves well using droplegs [36]. Trap plants can be used to reduce the pest population [37]. On the other hand, trap plants can also be a source of CW, so it is necessary to know when the first adults of the generation who are developing cruciferous vegetables usually appear. According to our observation, it is usually in the first half of July. Remains of cruciferous vegetables on harvested fields can be a habitat for CW and other pests if they are not removed. For this reason, it is necessary to remove post-harvest residues immediately after harvest from the beginning of July and carefully plough them into the soil without leaving plant remains on the surface because adults would hatch from puparia even on dead leaves. The landing behaviour of whiteflies can also be reduced by using reflective and contrast-minimising foils [38]; this method should also be included in integrated and organic farming systems.

The present findings can contribute to whitefly management, especially in organic farming, where the use of pesticides is limited. Suitable means of CW control appear to be (1) a netting system at the right time (before the mass migration), which can be combined with the application of insecticides if necessary, (2) the use of trap plants that can be destroyed by ploughing in shortly before new adults hatch. Several studies [35,39], including the present one, have shown that successful whitefly control cannot be based on naturally occurring predators and parasitoids because their contribution is not sufficient. Laurenz and Mayhöfer [35] tested the implementation of banker plants with *Aleyrodes lonicerae* as an alternative host for natural enemies. Due to the high labour costs, we do not foresee outdoor use of this method.

## 5. Conclusions

Oilseed rape is a source of CW for surrounding fields with cruciferous vegetables. The first adults migrate from oilseed rape to cruciferous vegetables at the end of May, and a mass migration usually begins between the beginning of June and the beginning of July. Migration to overwintering sites takes place in September–October, and a mass migration usually takes place in October. No significant reduction in CW population by parasitoids or predators was observed in any of the monitored areas.

**Author Contributions:** Conceptualisation, K.H.; methodology, K.H.; formal analysis, K.K.; investigation, K.H. and K.K.; data curation, K.K.; writing—original draft preparation, K.K. and K.H.; writing—review and editing, K.K. and K.H.; supervision, K.H.; project administration, K.H. All authors have read and agreed to the published version of the manuscript.

**Funding:** This research was funded by Institutional Support of the MINISTRY OF AGRICULTURE of the Czech Republic, grant number MZe RO0423 and MZe The National Action Plan to Reduce the Use of Pesticides in the CR 2024.

**Data Availability Statement:** The raw data supporting the conclusions of this article will be made available by the corresponding author upon request.

**Conflicts of Interest:** The authors declare no conflicts of interest.

## References


1. Evans, G.A. The Whiteflies (Hemiptera: Aleyrodidae) of the World and Their Host Plants and Natural Enemies. Version: 070606. 2007. Available online: https://keys.lucidcentral.org/keys/v3/whitefly/PDF_PwPETC/world-whitefly-catalog-Evans.pdf (accessed on 22 January 2024).
2. De Barro, P.J.; Carver, M. Cabbage Whitefly, *Aleyrodes proletella* (L.) (Hemiptera: Aleyrodidae), Newly Discovered in Australia. *Aust. J. Entomol.* **1997**, *36*, 255–256. [CrossRef]
3. Chen, M.M.; Guo, R.; Zhang, J.L.; Wan, G.H.; Yang, J.; Zhang, G.F. Rapid Identification of *Aleyrodes proletella* (Hemiptera: Aleyrodidae), a New Invasive Whitefly Species in Mainland China, Based on SS-COI Marker. *Acta Entomol. Sin.* **2015**, *58*, 579–586.
4. Dale, P.S.; Hayes, J.C.; Johannesson, J. New Records of Plant Pests in New Zealand. *N. Zeal. J. Agric. Res.* **1976**, *19*, 265–269. [CrossRef]
5. Collier, R.; Collins, S. *Biology of the Cabbage Whitefly, Aleyrodes proletella*; Final Report; Agriculture and Horticulture Development Board: Coventry, UK, 2015.
6. Martin, J.H.; Mifsud, D.; Rapisarda, C. The Whiteflies (Hemiptera: Aleyrodidae) of Europe and the Mediterranean Basin. *Bull. Entomol. Res.* **2000**, *90*, 407–448. [CrossRef] [PubMed]
7. Loomans, A.J.M.; Staneva, I.; Huang, Y.; Bukovinskine-Kiss, G.; van Lenteren, J.C. When Native Non-target Species Go Indoors: A New Challenge to Biocontrol of Whiteflies in European Greenhouses. *IOBC/WPRS Bull.* **2002**, *25*, 139–142.
8. Muñiz, M.; Nebreda, M. Differential Variation in Host Preference of *Aleyrodes proletella* (L.) on Some Cauliflower Cultivars. *IOBC/WPRS Bull.* **2003**, *26*, 49–52.
9. Trdan, S.; Modic, Š.; Bobnar, A. The Influence of Cabbage Whitefly (*Aleyrodes proletella* L., Aleyrodidae) Abundance on the Yield of Brussels Sprouts. *IOBC/WPRS Bull.* **2003**, *26*, 265–270.
10. van Rijn, P.; den Belder, E.; Elderson, J.; Vlaswinkel, M.; van Alebeek, F. Perspectives for Functional Agro Biodiversity in Brussels Sprouts. *IOBC/WPRS Bull.* **2008**, *34*, 121–124.
11. Pajović, I. Seasonal Dynamics of Most Detrimental Pest Insects Species on Cabbage Plants in Montenegro. *Agric. For.* **2011**, *51*, 25–42.
12. Saucke, H.; Schultz, B.; Wedemeyer, R.; Liebig, N.; Zimmermann, O.; Katz, P. Biotechnische Regulierung der Kohlmottenschildlaus in Kohlgemüse—Sachstand und Perspektiven. *Gesunde Pflanzen* **2011**, *63*, 183–189. [CrossRef]
13. Richter, E.; Hirthe, G. Hibernation and Migration of *Aleyrodes proletella* in Germany. *IOBC/WPRS Bull.* **2014**, *107*, 143–149.
14. den Belder, E.; Landure, J.; Elderson, J.; Vlaswinkel, M.; Willemse, J.; van Alebeek, F.; van Rijn, P.; van Gurp, H. Green Bridges over the Winter: Consequences for *Brassica* Pests. *IOBC-WPRS Bull.* **2008**, *34*, 29–32.
15. Askoul, K.; Richter, E.; Vidal, S.; Lusebrink, I. Life History Parameters of *Aleyrodes proletella* (Hemiptera: Aleyrodidae) on Different Host Plants. *J. Econ. Entomol.* **2018**, *112*, 457–464. [CrossRef]
16. Daniel, C.; Collier, R.; Thomas, J.; Hommes, M. Effects of Landscape and Region on Pests and Pathogens in *Brassica* vegetables and Oilseed Rape. *IOBC-WPRS Bull.* **2016**, *118*, 99–105.
17. ČSÚ. Grafy—Vývoj Ploch Osevů v Letech 1960–2020. Graf 3. 2020. Available online: https://www.czso.cz/csu/czso/soupis-ploch-osevu-k-31-5-2020 (accessed on 28 March 2022).
18. Hůserek, M. *Zhodnocení Trendu Vývoje Struktury Plodin v ČR*; Diplomová práce; Mendelova Univerzita v Brně, Agronomická fakulta: Brno, Czech Republic, 2017.
19. Miller, F. *Zemědělská Entomologie*; Nakladatelství ČSAV: Praha, Czech Republic, 1956; p. 1056s.
20. Holý, K. Účinnost vybraných přípravků na molici vlaštovičníkovou a mšici zelnou. *Zahradnictví* **2013**, *12*, 82–84.
21. Iheagwam, E.U. Effects of Temperature on Development of the Immature Stages of the Cabbage Whitefly, *Aleyrodes proletella* (Homoptera: Aleyrodidae). *Ent. Exp. Appl.* **1978**, *23*, 91–95. [CrossRef]
22. Ludwig, M.; Ludwig, H.; Conrad, C.; Dahms, T.; Meyhöfer, R. Cabbage Whiteflies Colonise Brassica Vegetables Primarily from Distant, Upwind Source Habitats. *Entomol. Exp. Appl.* **2019**, *167*, 713–721. [CrossRef]
23. Alonso, D.; Gómez, A.A.; Nombela, G.; Muñiz, M. Temperature-dependent Development of *Aleyrodes proletella* (Homoptera: Aleyrodidae) on Two Cultivars of Broccoli under Constant Temperatures. *Environ. Entomol.* **2009**, *38*, 11–17. [CrossRef]
24. Butler, C.G. On the Ecology of *Aleurodes brassicae* Walk. (Hemiptera). *Trans. R. Entomol. Soc. Lond.* **1938**, *87*, 291–311. [CrossRef]

25. Kovaříková, K.; Holý, K.; Skuhrovec, J.; Saska, P. The Efficacy of Insecticides Against Eggs and Nymphs of *Aleyrodes proletella* (Hemiptera: Aleyrodidae) under Laboratory Conditions. *Crop. Prot.* **2017**, *98*, 40–45. [CrossRef]
26. Müller, V.; Maiwald, F.; Lange, G.; Nauen, R. Mapping and Characterization of Target-Site Resistance to Cyclic Ketoenol Insecticides in Cabbage Whiteflies, *Aleyrodes proletella* (Hemiptera: Aleyrodidae). *Insects* **2024**, *15*, 178. [CrossRef] [PubMed]
27. CHMI. Územní Teploty. Available online: https://www.chmi.cz/historicka-data/pocasi/uzemni-teploty (accessed on 3 July 2024).
28. R Core Team. *R: A Language and Environment for Statistical Computing*; R Foundation for Statistical Computing: Vienna, Austria, 2024.
29. Brooks, M.E.; Kristensen, K.; van Benthem, K.J.; Magnusson, A.; Berg, C.W.; Nielsen, A.; Skaug, H.J.; Maechler, M.; Bolker, B.M. glmmTMB Balances Speed and Flexibility Among Packages for Zero-inflated Generalized Linear Mixed Modeling. *R J.* **2017**, *9*, 378–400. [CrossRef]
30. Hartig, F. DHARMa: Residual Diagnostics for Hierarchical (Multi-Level/Mixed) Regression Models. R Package Version 0.4.6. 2022. Available online: https://CRAN.R-project.org/package=DHARMa (accessed on 13 December 2023).
31. Lenth, R. Emmeans: Estimated Marginal Means, aka Least-Squares Means. R Package Version 1.8.8. 2023. Available online: https://CRAN.R-project.org/package=emmeans (accessed on 11 December 2023).
32. Hothorn, T.; Bretz, F.; Westfall, P. Simultaneous Inference in General Parametric Models. *Biom. J.* **2008**, *50*, 346–363. [CrossRef] [PubMed]
33. Butler, C.G. A Further Contribution to the Ecology of *Aleurodes brassicae* Walk. (Hemiptera). *Proc. R. Entomol. Soc. Lond.* **1938**, *13*, 161–172. [CrossRef]
34. Laurenz, S.; Schmidt, S.; Balkenhol, B.; Meyhöfer, R. Natural Enemies Associated with the Cabbage Whitefly *Aleyrodes proletella* in Germany. *J. Plant Dis. Prot.* **2019**, *126*, 47–54. [CrossRef]
35. Laurenz, S.; Meyhöfer, R. Conservation of Non-Pest Whiteflies and Natural Enemies of the Cabbage Whitefly *Aleyrodes proletella* on Perennial Plants for Use in Non-Crop Habitats. *Insects* **2021**, *12*, 774. [CrossRef]
36. Richter, E. Populationsdynamik und Bekämpfung der Kohlmottenschildlaus *Aleyrodes proletella* im Gemüsebau. In Proceedings of the 57. Deutsche Pflanzenschutztagung, Berlin, Germany, 6–9 September 2010; p. 216. [CrossRef]
37. Hondelmann, P.; Paul, C.; Meyhöfer, R. Combining Crop Resistance and Trap Plants for Cabbage Whitefly Control. *IOBC/WPRS Bull.* **2016**, *118*, 15–21.
38. Niemann, J.U.; Menssen, M.; Poehling, H.M. Manipulation of Landing Behaviour of Two Whitefly Species by Reflective Foils. *J. Plant. Dis. Prot.* **2021**, *128*, 97–108. [CrossRef]
39. Springate, S. The Cabbage Whitefly *Aleyrodes proletella* and its Natural Enemies on Wild Cabbage *Brassica oleracea* on the Kent Coast. *Trans. Kent Field Club* **2017**, *20*, 42–58.

agronomy

MDPI

*Article*

# Enhanced Tomato Pest Detection via Leaf Imagery with a New Loss Function

**Lufeng Mo [1,2,†], Rongchang Xie [1,†], Fujun Ye [3,*], Guoying Wang [1,*], Peng Wu [1] and Xiaomei Yi [1]**

1 College of Mathematics and Computer Science, Zhejiang A&F University, Hangzhou 311300, China; molufeng@zafu.edu.cn (L.M.); 2021611011064@stu.zafu.edu.cn (R.X.); wp@zafu.edu.cn (P.W.); yxm@zafu.edu.cn (X.Y.)

2 Information and Education Technology Center, Zhejiang A&F University, Hangzhou 311300, China

3 Network and Data Center, Communication University of Zhejiang, Hangzhou 310018, China

* Correspondence: yefuj@cuz.edu.cn (F.Y.); wgy@zafu.edu.cn (G.W.)

† These authors contributed equally to this work.

**Abstract:** Pests have caused significant losses to agriculture, greatly increasing the detection of pests in the planting process and the cost of pest management in the early stages. At this time, advances in computer vision and deep learning for the detection of pests appearing in the crop open the door to the application of target detection algorithms that can greatly improve the efficiency of tomato pest detection and play an important technical role in the realization of the intelligent planting of tomatoes. However, in the natural environment, tomato leaf pests are small in size, large in similarity, and large in environmental variability, and this type of situation can lead to greater detection difficulty. Aiming at the above problems, a network target detection model based on deep learning, YOLONDD, is proposed in this paper. Designing a new loss function, NMIoU (Normalized Wasserstein Distance with Mean Pairwise Distance Intersection over Union), which improves the ability of anomaly processing, improves the model's ability to detect and identify objects of different scales, and improves the robustness to scale changes; Adding a Dynamic head (DyHead) with an attention mechanism will improve the detection ability of targets at different scales, reduce the number of computations and parameters, improve the accuracy of target detection, enhance the overall performance of the model, and accelerate the training process. Adding decoupled head to Head can effectively reduce the number of parameters and computational complexity and enhance the model's generalization ability and robustness. The experimental results show that the average accuracy of YOLONDD can reach 90.1%, which is 3.33% higher than the original YOLOv5 algorithm and is better than SSD, Faster R-CNN, YOLOv7, YOLOv8, RetinaNet, and other target detection networks, and it can be more efficiently and accurately utilized in tomato leaf pest detection.

**Keywords:** attention mechanism; pest images; small targets; target detection

## 1. Introduction

With a growing global population and accelerated urbanization, the sustainability of agricultural production has become a global challenge. Tomato is one of the most important cash crops, while tomato yield is affected by insect pests of tomato leaves, among which leafminer flies, thrips, tabacco budworm, and spider leaf mite [1–3] have a greater impact on tomato yield. However, traditional pest detection methods [4–6] rely on manual visual inspection, which is not only time-consuming and labor-intensive but also subject to subjective judgment, making it difficult to achieve wide-scale and efficient monitoring.

In recent years, deep learning has achieved significant results in pest identification [7–9], which can better overcome the challenges of traditional machine learning. Research on deep learning in pest detection can be categorized into two approaches: pest detection and classification and pest-induced leaf infestation feature detection. The research on pest detection and classification mainly focuses on improving the classical deep learning methods, such as P. Venk et al. [10], which achieved good results on pest datasets of three peanut crops by integrating VIT, PCA, and MFO; Pattnaik G et al. [11], which feature extraction of pests by HOG and LBP, and the extracted feature maps are fed into SVM [12] classifiers for training; In contrast, leaf damage caused by insect pests can be detected in two ways: by quantifying the extent of insect damage to the leaf and by detecting the location of the insect-damaged leaf. For example, Liang et al. [13] developed polynomial and logistic regression models for leaf extraction to estimate leaf damage; Da Silva et al. [14] used image segmentation to preserve the leaf region, augmented the dataset with a synthesis technique, and trained the network with a model for detecting pest-induced damage to leaves; Fang et al. [15], Zhu R et al. [16], Zhu L et al. [17], and others used the improved YOLO series of models to identify pest-induced leaf damage and achieved good detection results.

The detection process of pests is found to have low resolution, and small pest detection is much more difficult and less accurate than large pest detection. Therefore, the detection of small target goals is challenging in agricultural production environments. Currently, mainstream detection methods are added to the mainstream target detection model through multi-scale feature fusion, anchor frame optimization, and loss function optimization. For example, Ye et al. [18] improved YOLOv8 by proposing slice-assisted fine-tuning and slice-assisted hyper-inference (SAHI), designing a generalized efficient layer aggregation network (GELAN), introducing the MS structure, introducing the BiFormer attention mechanism, and using the MPDIoU loss function in order to solve the small targets in tea pests, and achieved a better detection effect in tea pests. Tian Y et al., [19] in order to solve insect pests in agriculture by improving the network in the feature extraction and feature fusion parts, the method is indeed feasible and has better results in the detection of small targets. However, the existing target detection algorithms still have limitations for small target pests. The occlusion and light caused by leaves, etc., bring difficulties in model detection, which affects the feature extraction ability of the network model.

In natural environments, tomato pests are small in size, large in similarity, and large in environmental variability, and these types of situations can make detection more difficult. Lippi M et al. [20], Mamdouh N et al. [21]. Yang S et al. [22]. performed algorithmic improvements based on the YOLO series for target detection of hazelnut pests, olive fruit fly, and maize pests. Their study proved that the YOLO series of algorithms has a better detection effect. Although the YOLO series of primary target detection models offers better advantages in terms of detection speed and model size, they are more suitable as benchmark models for pest detection. However, the Mosaic data augmentation used in YOLOv5 causes the originally smaller targets to become even smaller, resulting in poorer generalization ability of the model. Moreover, YOLOv5 is not very stable at detecting small targets and requires a large amount of training data to achieve high accuracy.

In response to the above issues, this paper proposes a deep learning-based target detection model, YOLONDD, for tomato leaf pest detection. The model is based on the YOLOv5 structure. Firstly, we design a new loss function NMIoU by borrowing the idea of loss function NWD and loss function MPDIoU to enhance the ability of the model to detect and recognize objects at different scales and to improve the robustness of the model. Meanwhile, we added the Dynamic head [23] module in order to improve the multi-scale feature extraction capability of the network model and to solve the challenge of too small a target in tomato leaf pest detection. Finally, we found that the number of parameters in the network model increased, and the training time was too long. In order to reduce the number of parameters and computational complexity of the network model, we introduce the Decoupled head [24] module in YOLOv5, which enhances the generalization ability and robustness of the model. The main contributions of this study include:

- We propose an improved tomato leaf pest detection model, YOLONDD, to solve the problems of uneven pest scales and low detection accuracy in the detection process.
- To enhance the ability of the model to detect and recognize objects at different scales, we design a new loss function NMIoU, which is adaptive to changes in different scales.

This loss function enhances the model's detection and recognition of small targets in the feature map.

- In order to better extract the feature information of small targets, we add the Dynamic head module, which gradually extracts the information of the feature map through the scale-aware module, spatial-aware module, and task-aware module. This method can adaptively adjust the size of the sensing field to adapt to the scale change in different targets and improve the detection ability of different scale targets.
- In order to reduce the number of parameters and training time of the model, we introduce the decoupling head module, the decoupling head helps the model to extract the target location and category information, learn the feature map information, and fuse it through different network branches, so that the model's generalization ability and robustness are enhanced.

In addition, in order to verify the robustness of the proposed method, experiments were conducted using tomato pest images as well as infestation images in a real environment. The experimental results show that the method proposed in this paper is able to meet the requirements in terms of accuracy and other aspects of the dataset and has a better detection effect on tomato pest images.

## 2. Materials and Methods

### 2.1. Main Ideas

In this paper, a deep learning-based detection model, YOLONDD, is proposed, and its structure is shown in Figure 1.

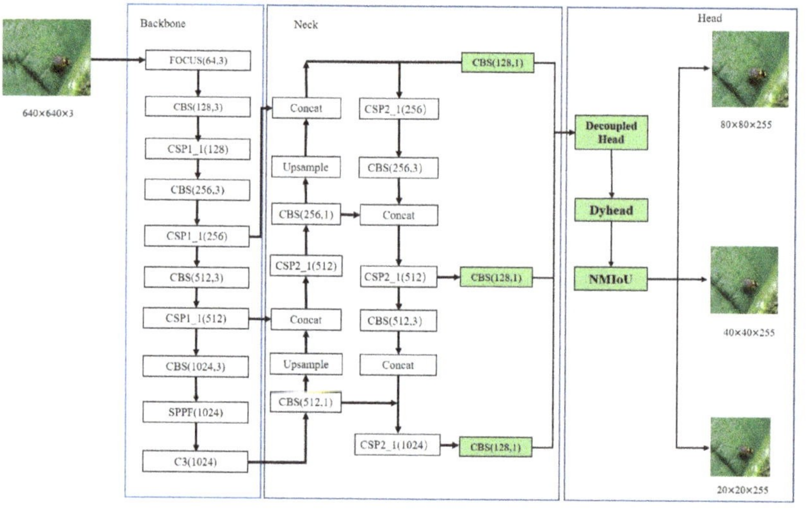


**Figure 1.** Network structure diagram of YOLONDD, where the green rectangles are the improvement aspects of the model.

According to Figure 1, the main ideas of YOLONDD include three aspects: designing a new loss function NMIoU, adding a DyHead with an attention mechanism, and adding a decoupled head to Head.

Using the NWD [25] (Normalized Wasserstein Distance) metric in combination with MPDIoU [26] (Mean Pairwise Distance Intersection over Union), the NMIoU loss function is proposed, which can improve the sensitivity to the target frame and the small-scale detection ability, and the balance coefficient can be adjusted to achieve the weight of Loss.

DyHead is a target detection framework based on an attention mechanism. It can adaptively adjust the receptive field size to adapt to the scale change in different targets and improve the detection ability of targets at different scales. Meanwhile, DyHead utilizes a shared feature extraction network to reduce the number of computations and parameters.

The coupled head used in YOLOv5 has a simple design idea, which requires a large number of parameters and computational resources and is prone to overfitting. The decoupled head can extract the target location and category information separately, learn them through different network branches separately, and finally fuse them. It can effectively reduce the number of parameters and computational complexity and enhance the generalization ability and robustness of the model.

2.1.1. Designing a New Loss Function NMIoU

Small target detection is a very challenging problem in target detection in pest data where the pests are only a few pixels in size in the image. As the appearance information of small target pests is more difficult to detect, the current state-of-the-art detectors cannot get better results on small targets. We can also see from our experiments that loss functions such as IoU [27], CIoU [28], and DIoU are very sensitive to the positional deviation of small targets and greatly reduce the detection performance when they are used in anchor-based detectors. In order to solve this problem, we propose a loss function based on the combination of the NWD metric and the MPDIoU loss function-NMIoU, where the MPDIoU loss function is schematically shown in Figure 2.

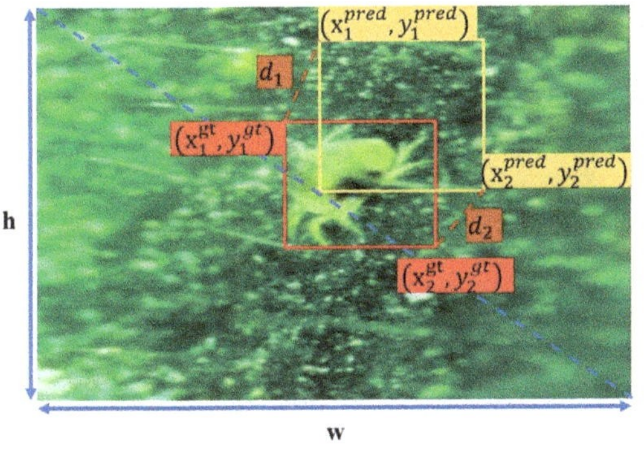


**Figure 2.** Schema of the MPDIoU loss function. Red rectangle means the ground truth and yellow one is the predicted result, where x and y represent the coordinates.

The loss function used in YOLOv5 is CIoU, which takes into account the distance between the target frame and the Anchor, the overlap, the scale, and the penalty term to make the target frame regression more stable. However, CIoU suffers from high computational complexity, does not take into account the case where the correct frame has the same aspect ratio as the predicted frame, is highly sensitive to the target frame, and is not applicable to large-scale differences.

The CIoU loss function cannot be optimized when the predicted box has the same aspect ratio as the actual labeled box, but the width and height values are completely different. MPDIoU can solve the above problem well; it is a comparative measure of bounding box similarity based on the distance of the minimum point, which takes into account all the relevant factors considered in the existing loss functions, such as the overlapping or non-overlapping area, the distance of the centroids, and the deviation of width and height. MPDIoU is determined by the four vertices of the bounding box in the upper left corner $(x_1, y_1)$, upper right corner $(x_2, y_1)$, lower left corner $(x_1, y_2)$, and lower right corner $(x_2, y_2)$. The general bounding box is determined by the center coordinates $(c_x, c_y)$ and the width $w$ and height $h$. For this reason, we can calculate the four vertex coordinates as shown in the formula. The four vertex coordinates are obtained [29], and the distance between the two sets of vertices is calculated, and the minimum of these distances is chosen $MPD$. Then, the ratio of the overlapping region of the predicted and real boxes to their concatenated region is calculated. The final MPDIoU loss function can be defined as Equation.

$$x_1 = C_{xpred} - \frac{w_1}{2} \quad (1)$$

$$y_1 = C_{ypred} - \frac{h_1}{2} \quad (2)$$

$$x_2 = C_{xgt} - \frac{w_2}{2} \quad (3)$$

$$y_2 = C_{xgt} - \frac{h_2}{2} \quad (4)$$

$$DIoU = \frac{B_{pred} \cap B_{gt}}{B_{pred} \cup B_{gt}} \quad (5)$$

$$L_{MPDIOU} = 1 - \frac{MPD}{DIoU} \quad (6)$$

In Equation, $x_1$, $y_1$, $x_2$, $y_2$ are the coordinates of the four vertices of the bounding box, $C$ is the coordinate of the center of the bounding box, $w$ is the width of the bounding box, and $h$ is the length of the bounding box.

The NWD metric is introduced to address the problem of high sensitivity to target boxes and its inapplicability to large-scale differences. The traditional IoU and its variants are discarded for the boxes with low confidence in the bounding box because, in these bounding boxes, the foreground and background pixels are concentrated on the center and the boundary of the bounding box, respectively, which leads to the low useful information in the box, and to this point, it affects the low confidence. NWD solves this problem well by assigning different weights to the pixels in the bounding box, with the center having the highest weight and decreasing from the center to the boundary. For this purpose, the bounding box is modeled as a two-dimensional Gaussian distribution, and then the distance of the distribution is calculated by the Wasserstein distance. The two-dimensional Gaussian distribution of the bounding box is defined by the centroid $(x, y)$, width $w$, and height $h$. For the two-bounding box $\mathcal{N}_a$, $\mathcal{N}_b$ Gaussian distributions, the Wasserstein distance can be obtained by calculating the difference between their means and covariances, and finally, the value domain of the Wasserstein distance is restricted to be between 0 and 1 by normalization, and the normalization factor $C$ is the maximum value of the two maximum values of the diagonal lengths of the bounding box.

$$W^2_{2(\mathcal{N}_a,\mathcal{N}_b)} = \left\| \left( \left[cx_a, cy_a, \frac{w_a}{2}, \frac{h_a}{2}\right]^T, \left[cx_b, cy_b, \frac{w_b}{2}, \frac{h_b}{2}\right]^T \right) \right\|_2^2 \quad (7)$$

$$NWD(\mathcal{N}_a, \mathcal{N}_b) = \exp\left( -\frac{\sqrt{W_2^2(\mathcal{N}_a, \mathcal{N}_b)}}{C} \right) \quad (8)$$

In addition, NWD integration into YOLO is improved based on the label assignment, NMS, and regression loss function of IoU. In the NWD-based label assignment strategy, for training RPNs, positive labels will be assigned to two types of anchors: anchors with the highest NWD values and NWD values greater than $\theta n$ and anchors with NWD values higher than the positive threshold $\theta p$ for any real frame. Therefore, negative labels will be assigned to anchors if their NWD values are lower than the negative threshold $\theta n$ for all real frames. In addition, anchors with neither positive nor negative labels assigned do not participate in the training process. where $\theta p$ and $\theta n$ are the original detectors. NMS is based on NWD. First, it sorts all the prediction frames based on their scores. The highest-scoring prediction frame M is selected, and all other prediction frames with significant overlap with M are suppressed. This process is recursively applied to the remaining boxes. Based on the regression loss of NWD, IoU-Loss is not able to provide gradients for the optimization network in some cases, and for this reason, the NWD metric is designed as a loss function:

$$\mathrm{NWD} = 1 - \mathrm{NWD}(\mathcal{N}_a, \mathcal{N}_b) \tag{9}$$

As mentioned above, the NMIoU loss function is obtained by combining MPDIoU and NWD, and we just need to give reasonable balance coefficients to the NMIoU loss function to regulate the loss weights of MPDIoU and NWD. The formula for the NMIoU loss function is as follows:

$$\mathrm{NMIoU} = \lambda \mathrm{MPDIoU} + (1 - \lambda)\mathrm{NWD} \tag{10}$$

In Equation, $\lambda$ is the equilibrium coefficient that regulates the loss weights of MPDIoU and NWD.

#### 2.1.2. Adding a DyHead with an Attention Mechanism

The detection head is a crucial component of the target detection model, whose role is to process the features of the last layer of the network and generate the results of target detection. In the target detection task, the detection head assumes several important roles. First, the detection head is responsible for target localization. The location information of the target frame is predicted through regression. Second, the detection head is also responsible for target classification. By predicting the category to which the target belongs through a classifier, the model is able to identify and label different objects in the image through the classification function of the detection head. In addition, the detection head is also responsible for generating the target score. The target score is used to measure the confidence or importance of each target in the detection results. This helps in filtering out the targets with high confidence levels, thereby improving the accuracy and reliability of the detection results. The role of the detection head in the target detection task is critical and indispensable. Through the functions of target localization, target classification, and target score generation, the detection head is able to achieve accurate localization and classification of targets in an image, providing important target detection results for real-world application scenarios.

For this reason, this paper introduced the detection head DyHead in YOLOv5, which contains an attention mechanism. Dyhead consists of a scale-aware attention module, a spatial-aware attention module, and a task-aware attention module. When the scale-aware attention module processes the original feature map $F_1$, it performs a global average pooling operation on the feature map, which aggregates features on different scales to obtain a multi-scale global representation. Subsequently, the above multi-scale feature maps are integrated using $1 \times 1$ convolution to fuse the features on different scales. Relu and Hard Sigmoid functions are used to enhance the nonlinear representation of the model for the multi-scale feature maps. Finally, the computed weights are multiplied with the original feature map to obtain the new feature map $F_2$. When the Perceptual Attention module processes $F_2$, the Index function extracts the positional information of the feature map, and subsequently, the extracted positional information is processed by a $3 \times 3$ deformable convolution and used to adjust the offsets of the convolution kernel, the sparse sampling, and the aggregation of the feature elements by the Sigmoid function. The offset function uses the obtained offsets to increase the weight share of the target shallow contour and edge information in the network. Finally, the acquired position information is appended to a new feature map $F_3$. $F_3$ is fed into the task-aware attention module, which reduces the spatial dimensionality of the feature map through an average pooling operation, followed by two fully connected layers to learn the inter-channel relationships. After the fully connected layers, the ReLU activation function is introduced to enhance the nonlinear representation of the features, and the scaling of the features is adjusted by a normalization operation.

The normalization uses the scaling factor and offset factor in Batch Normalization for task-awareness, and the output of the task-aware attention module is used to obtain the desired feature map $F_4$ through a composite operation, as shown in Figure 3.

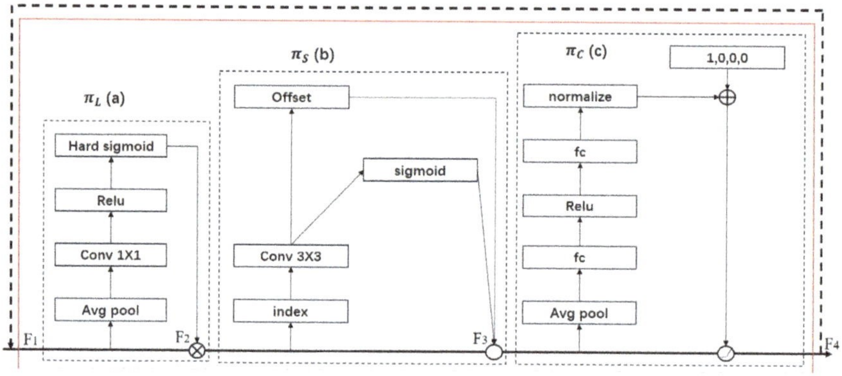


**Figure 3.** DyHead structure. (**a**) Scale-aware attention module. (**b**) Spatial-aware attention module. (**c**) Task-aware attention module.

It uses the attention mechanism to unify the different target detection heads. As can be seen in Figure 4, the initial features are noisy due to domain differences, and for this reason, they cannot focus on the target well; firstly, after the original features are processed by the scale-aware attention module, the features become more sensitive to the targets at different scales; secondly, when the feature maps processed by the scale-aware attention module are processed by the spatial location-aware attention module, the features become more sparse, focusing on the foreground targets at different locations; Finally, after processing by the task-aware attention module, the features will form different activations based on different downstream tasks through the attention mechanism between feature levels for scale perception, between spatial locations for spatial perception, and within the output channel for task perception, as shown in Figure 4.

$$W(\mathcal{F}) = \pi_C(\pi_S(\pi_L(\mathcal{F})\cdot\mathcal{F})\cdot\mathcal{F})\cdot\mathcal{F} \tag{11}$$

$$\pi_L(\mathcal{F})\cdot\mathcal{F} = \sigma\left(f\left(\frac{1}{SC}\sum_{S,C}\mathcal{F}\right)\right)\cdot\mathcal{F} \tag{12}$$

$$\pi_S(\mathcal{F})\cdot\mathcal{F} = \frac{1}{L}\sum_{l=1}^{L}\sum_{k=1}^{K} W_{l,k}\cdot\mathcal{F}(l; p_k + \Delta_{k;}C)\cdot\Delta m_k \tag{13}$$

$$\pi_C(\mathcal{F})\cdot\mathcal{F} = \max\left(\alpha^1(\mathcal{F})\cdot\mathcal{F}_c + \beta^1(\mathcal{F}), \alpha^2(\mathcal{F})\cdot\mathcal{F}_c + \beta^2(\mathcal{F})\right) \tag{14}$$

In Equation, $L$ is the number of layers, $f(\cdot)$ is a linear function with $1 \times 1$ convolutional approximation, $\sigma(x)$ is the hard_sigmoid activation function, $k$ is the number of sparsely sampled positions, $W_{l,k}$ is the weight of the convolutional kernel, $\Delta_k$ is the positional offset, $p_k$ is the spatial offset, $\Delta m_k$ is a self-learnable importance metric factor with respect to the position $p_k$, $\mathcal{F}_c$ is the slice of the feature tensor $\mathcal{F}$ on a particular channel $C$, $\alpha^1(\mathcal{F})$ and $\alpha^2(\mathcal{F})$ the weights obtained by network learning, and $\beta^1(\mathcal{F})$ and $\beta^2(\mathcal{F})$ are bias terms.

DyHead is introduced into YOLOv5, which is applied to a one-stage detector. The scale-aware attention module, the spatial location-aware attention module, and the task-aware attention module are combined as a group, and then the number of their cycles is selected for image processing, as shown in Figure 5.

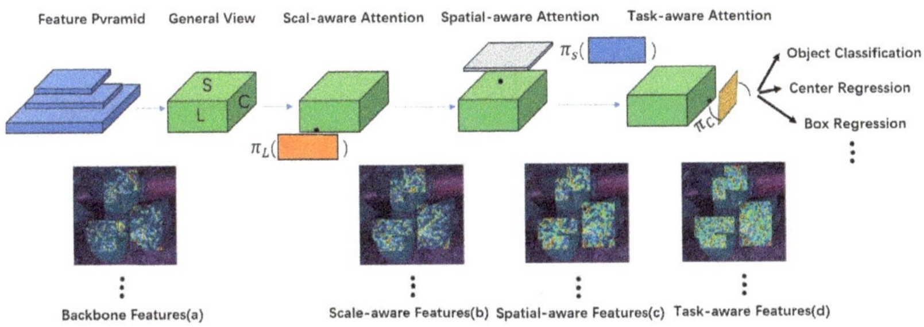


**Figure 4.** DyHead image processing process. (**a**) Original feature map. (**b**) The scale-aware attention module processes the a-map. (**c**) The spatial-aware attention module processes the b-map as a feature map. (**d**) The task-aware attention module processes the c-map as a feature map. Where $\pi_L$ is a scale-aware attention module, $\pi_S$ is a spatial-aware attention module, and $\pi_C$ is a task-aware attention module. The feature map's blue and cyan colors indicate regions with low detection values, while yellow, orange, and red indicate regions with higher detection values. They are in a progressive order.

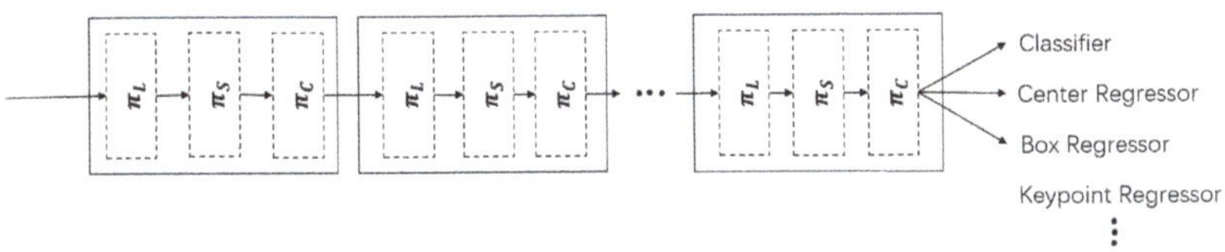


**Figure 5.** DyHead applied a one-stage detector. $\pi_L$ is a scale-aware attention module, $\pi_S$ is a spatial-aware attention module, and $\pi_C$ is a task-aware attention module.

Normalization in DyHead uses the configuration of Group Normalization [30] (GN). GN aims to address the problem of performance degradation of Batch Normalization [31] (BN) on small batches of data or when the batch size varies, especially in non-convolutional layers of RNNs and CNNs. The core idea of GN is to perform normalization on each training batch to divide the channels in each feature map of the network into groups and normalize the channels within each group. Doing so reduces the model's dependence on the batch size while maintaining the efficiency and effectiveness of the normalization operation. Specifically, the channels of the feature map are divided into G groups, each containing C/G channels. If the size of the feature map is [N, H, W, C], where N is the batch size, H and W are the height and width of the feature map, and C is the number of channels, then each group will contain C/G channels; the mean and variance are calculated for the channels within each group, and the channels in each group are normalized using these statistics. The normalization formula is as follows:

$$\hat{x}_g = \frac{x_g - \mu_g}{\sqrt{\sigma_g^2 + \epsilon}} \tag{15}$$

$$y_g = \gamma \cdot \hat{x}_g + \beta_g \tag{16}$$

In Equation, $\hat{x}_g$ is the input to the g layer of the network, $\mu_B$ is the mean of feature i for all samples in the current batch g, $\sigma_B^2$ is the variance of feature $g$ for all samples in the current batch $g$, and $\epsilon$ is a very small constant that prevents dividing by zero and ensures numerical stability. $\gamma$ and $\beta$ are the learnable scaling and offsetting parameters.

The final normalized feature map is fed into the next layer of the network. It does this by calculating the mean and variance of all the samples in each batch and then using these statistics to normalize the inputs for the current batch. This results in a more stable distribution of inputs to the network and reduces the internal covariance bias, thus allowing the use of larger learning rates, faster training, and less sensitivity to initialization weights.

2.1.3. Adding Decoupled Head to Head

There is no decoupled head in YOLOv5, only a coupling header. The coupling head typically requires the feature map output from the convolutional layers to be fed directly into several fully connected or convolutional layers in order to generate outputs for the target locations and categories. Such a design brings a lot of parameters and computational resources to the model, and the model is also prone to overfitting. In contrast, a more efficient decoupled head structure is designed in YOLOv6 with the help of a hybrid channel strategy, which reduces the number of 3 × 3 convolutional layers in the decoupled head of YOLOX to only one. The width of the head is jointly scaled by the width multipliers of the Backbone and Neck. The delay is reduced while maintaining accuracy, mitigating the additional delay overhead associated with the 3 × 3 convolution in the decoupled head. This is shown in Figure 6.

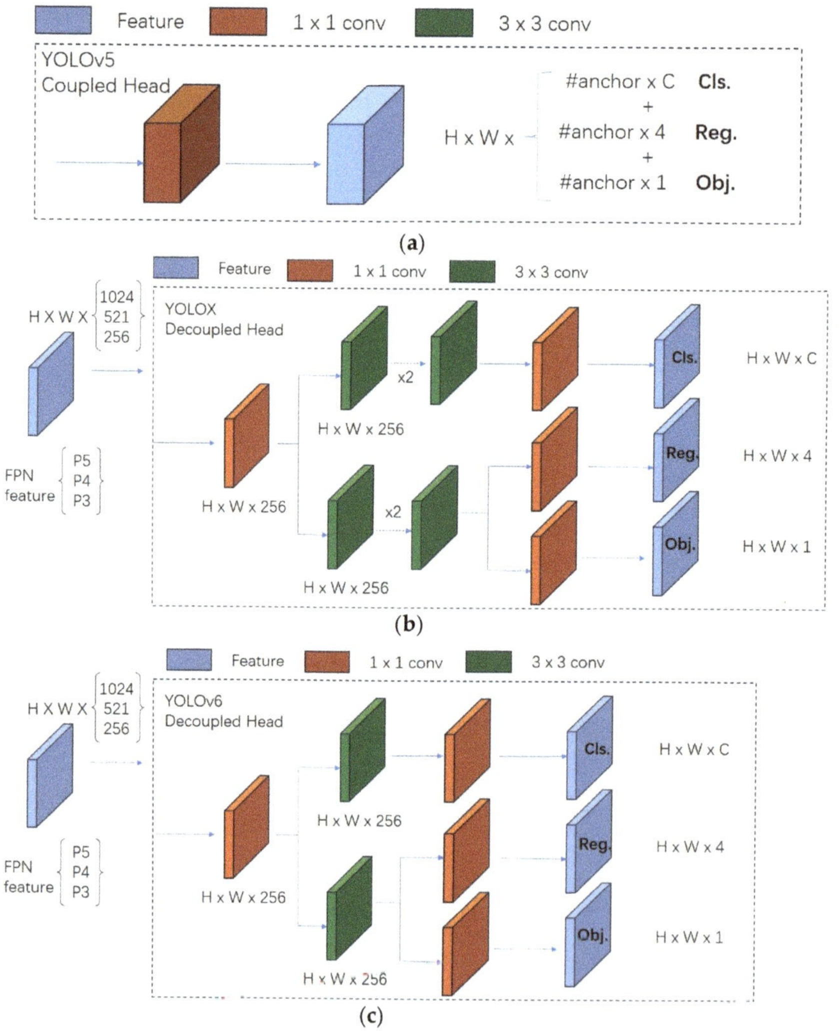


**Figure 6.** Decoupled Head. (**a**) YOLOv5-coupled head. (**b**) YOLOX-decoupled head. (**c**) YOLOv6 decoupled head.

In this paper, with the help of YOLOv6's decoupled head idea, the decoupled head is introduced for YOLOv5. The feature maps of P3, P4, and P5 in YOLOv5 are inputted, and the feature maps are processed by a 1 × 1 convolution and divided into two branches, and then the processed features are all processed by a 3 × 3 convolution, and the final output of the feature maps is processed by another 1 × 1 convolution, and the classification maps are obtained, respectively, a coordinate position map and a target frame confidence map.

## 2.2. Datasets

### 2.2.1. Data Acquisition and Preprocessing

The original 1426 images of the pest dataset used in the experiments in this paper were obtained from Kaggle [32]. Among them, the Leafminer flies were not included in the original dataset, which was collected by ourselves from the Internet, totaling 192 images. The Kaggle dataset was collected from inside the greenhouse, and in order to be closer to tomato leaf pests in the natural environment, we collected tomato leaf pests through the cultivation help grower website, where tomato pests were collected from tomato leaf pests photographed by the technicians in the natural environment, and we added the collected images to the public dataset. The tomato leaf pests on the website are all from the tomato leaf pests photographed by technicians in the natural environment, and we added the collected images to the public dataset, including 89 images of leafminer flies, 63 images of thrips, 68 images of tabacco budworm, and 510 images of spider leaf mite.

Secondly, data augmentation techniques such as rotation, mirroring, cropping, and scaling are used to increase the sample capacity and improve generalization. Some of the data are shown in Figure 7.

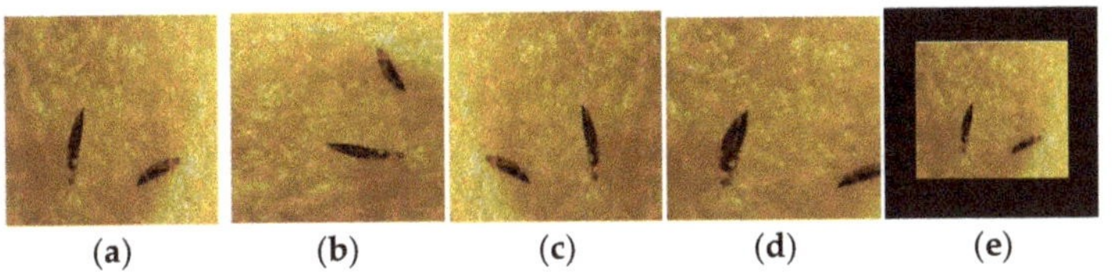


**Figure 7.** Data augmentation. (**a**) Original image. (**b**) Rotation. (**c**) Mirroring. (**d**) Cropping. (**e**) Scaling.

Finally, the pest images in the dataset were labeled by the annotation software LabelImg 1.8.6 (EASY EAI, China) [33] to generate XML files. Although the YOLO series has provisions for dataset labeling files, the dataset was uniformly stored in PASCAL VOC8 [34] data format for better comparison experiments of various methods as well as experimental efficiency.

### 2.2.2. Pest Image Datasets

The images in this dataset were collected based on the pathological features, eggs, larvae, and adults caused by the pests on the leaves. Also, according to the type of pests in the images, the tomato pest images in this dataset are divided into leafminer flies, thrips, Tabacco Budworm, and spider leaf mite datasets, and the images are divided into a training set, a testing set, and a validation set according to 8:1:1, and some of the data are shown in Figure 8.

Among them, 1426 images of pests containing 192, 180, 167, and 887 images of Leafminer flies, Thrips, Tabacco Budworm, and spider leaf mite tomato leaf pests, respectively. This dataset was expanded to 3479 images by image enhancement, as detailed in Table 1.

**Figure 8.** Images of infested tomato leaves and their labels. (**a**) Leafminer flies. (**b**) Thrips. (**c**) Tabacco Budworm. (**d**) Spider leaf mite. Red boxes are labeled boxes.

**Table 1.** Size of pest datasets.

| Pest | Raw Data | Enhanced Data | Training Set | Testing Set | Validation Set |
|---|---|---|---|---|---|
| Leafminer flies | 192 | 768 | 614 | 77 | 77 |
| Thrips | 180 | 720 | 576 | 72 | 72 |
| Tabacco Budworm | 167 | 668 | 534 | 67 | 67 |
| Spider leaf mite | 887 | 1323 | 1058 | 132 | 132 |
| Total | 1426 | 3479 | 2783 | 348 | 348 |

### 2.3. *Experiment Scheme*

This section carries out the experimental design in order to test the performance of the YOLONDD model proposed in this paper. Firstly, software and hardware equipment configuration is followed by dataset production and preprocessing required for this experiment, as well as the hyperparameter setting of the experimental network. Finally, the robustness test and ablation experiment.

#### 2.3.1. Hardware and Software Configuration

In this paper, the PyTorch deep learning framework is used to train and test the network model, and the specific experimental configuration is shown in Table 2.

**Table 2.** Experimental software and hardware configuration.

| Item | Detail |
|---|---|
| CPU | Ryzen 5 5600X 6-Core Processor @3.70 GHz (AMD, CA, USA) |
| GPU | RTX3060Ti (8 G) (NVIDIA, CA, USA) |
| RAM | 16 GB |
| Operating system | Windows 11 64-bit (Microsoft, WA, USA) |
| CUDA | CUDA12.2 |
| Python | Python 3.7 |

#### 2.3.2. Determination of Training Parameters

The original YOLOv5 model, under the premise of the initial learning rate of 0.0001 and a Batch-size of 8, performed well on the PASCAL VOC2012 and COCO datasets. On this basis, according to the commonly used empirical values of network training hyperparameters, the hyperparameters of the YOLONDD network are finally determined after repeated tests, as shown in Table 3.

**Table 3.** The optimized hyperparameters.

| Epoch | Batch | Lr | Input-Shape |
|---|---|---|---|
| 100 | 16 | 0.0001 | 256 × 256/640 × 640 |

2.3.3. Comparison Experiments

(1) In order to determine the base loss function of this paper, a variety of different loss functions were added to the tomato leaf pest images for comparison experiments, and AP was selected as an index to test the detection performance of this paper's Model.
(2) In order to determine the equilibrium coefficient between NWD and MPDIoU, values are taken in the interval [0, 1], comparison experiments are conducted, and AP is selected as an indicator to test the detection performance of the model in this paper.
(3) In order to test the normalization functions GN and BN in DyHead as well as the number of cycles of DyHead on the tomato leaf pest images, comparative experiments were conducted, and AP and training time were selected as indicators to test the detection performance of the model in this paper.
(4) In order to test the performance of the model YOLONDD proposed in this paper in the tomato pest image detection task, comparative experiments are conducted with YOLONDD with traditional target detection models such as Faster R-CNN [35], SSD [36], YOLOv7 [37], RetinaNet, YOLOv5, and YOLOv8. The experiment divides the total dataset into a training set, testing set, and validation set according to the ratio of 80%, 10%, and 10%, which are used to train the model and conduct the test, and selects AP and F1 as the indexes to test the detection performance of this paper's model, and FPS, training time, and single-image prediction time are selected as the indexes to verify the detection efficiency of the YOLONDD proposed in this paper.

In addition, detection comparison experiments are conducted on the constructed tomato pest dataset to test the generalization ability of the model and verify the robustness of YOLONDD proposed in this paper.

2.3.4. Ablation Experiments

To verify the effectiveness of the designed new loss function MNIoU, the introduction of the detection head DyHead with an attention mechanism, and the introduction of the decoupled head, seven sets of ablation experiments are performed on the total dataset.

(1) YOLON: On the basis of the original YOLOv5 network, the original loss function CIOU is replaced by the loss function NMIOU designed in this paper.
(2) $YOLOD_1$: On the basis of the original YOLOv5 network, introducing the DyHead.
(3) $YOLOD_2$: On the basis of the original YOLOv5 network, adding a decoupled head.
(4) $YOLOND_1$: On the basis of YOLON, the DyHead detection head is introduced into the network model.
(5) $YOLOND_2$: On the basis of YOLON, a decoupled head is introduced in the network model.
(6) $YOLOD_1D_2$: On the basis of $YOLOD_1$, a decoupled head is introduced in the network model.
(7) YOLONDD: Based on $YOLOND_1$, a decoupled head is introduced in the network model, which is the method proposed in this paper.

## 2.4. Evaluation Indicators

In this paper, Average Precision (AP), Frames Per Second (FPS), and F1 score (F1) are used as important evaluation metrics for the detection of tomato leaf pests to analyze the network detection performance.

(1) Average Precision (AP)

AP is the area enclosed by the PR curve and the coordinate axis, and is calculated as follows:

$$\text{Precision} = \frac{\text{TP}}{\text{TP} + \text{FP}} \tag{17}$$

$$\text{Recall} = \frac{\text{TP}}{\text{TP} + \text{FP}} \tag{18}$$

$$\text{AP} = \int_0^1 \text{p}(\text{r})\text{dr} \tag{19}$$

In Equation, TP is the number of predicted bounding boxes that are correctly categorized and have the correct coordinates of the bounding box, FP is the number of predicted bounding boxes that are incorrectly categorized, FN is the number of bounding boxes that are not predicted, p (Precision) is the precision rate, and r (Recall) is the recall rate.

(2) Frames Per Second (FPS)

FPS is the number of frames per second transmitted, which indicates the number of images that can be processed in a second or the time it takes to process an image to evaluate the detection speed, the shorter the time, the faster the speed. The calculation formula is as follows:

$$FPS = 1/Latency \tag{20}$$

(3) F1 score (F1)

The F1 score is a kind of reconciled average of model precision and recall, calculated as follows:

$$F1 = \frac{\text{2PrecisionRecall}}{\text{Precision} + \text{Recall}} \tag{21}$$

## 3. Results

### 3.1. Loss Function Analysis

In order to explore the advantages and disadvantages of different loss functions in detecting the dataset, this paper conducts controlled experiments on CIoU, EIoU, DIoU, GIoU, WIoU, SIoU, and MPDIoU to select the optimal loss function. The results are shown in Table 4.

**Table 4.** Comparison of different loss functions.

| Loss Function | mAP/% |
|---|---|
| CIoU | 87.2 |
| EIoU | 87.9 |
| DIoU | 88.9 |
| GIoU | 89.2 |
| SIoU | 89.0 |
| MPDIoU | 89.4 |

As can be seen from Table 4, MPDIoU is 2.52%, 1.71%, 0.56%, 0.22%, and 0.45% higher than CIoU, EIoU, DIoU, GoU, and SIoU. Therefore, MPDIoU is adopted as our base loss function.

### 3.2. NMIoU Balance Coefficient Analysis

In order to explore the advantages and disadvantages of different Losses of NDW and Losses of MPDIoU-specific gravity for the detection of the dataset, we conducted controlled experiments on different Loss-specific gravity to select the optimal Loss-specific gravity. The results are shown in Table 5.

**Table 5.** Comparison of the NMIoU balance factor.

| Ratio | mAP/% |
|---|---|
| CIoU | 87.2 |
| NMIoU ($\lambda = 0$) | 88.4 |
| NMIoU ($\lambda = 0.1$) | 88.6 |
| NMIoU ($\lambda = 0.2$) | 88.4 |
| NMIoU ($\lambda = 0.3$) | 88.4 |
| NMIoU ($\lambda = 0.4$) | 88.6 |
| NMIoU ($\lambda = 0.5$) | 87.9 |
| NMIoU ($\lambda = 0.6$) | 89.1 |
| NMIoU ($\lambda = 0.7$) | 88.3 |
| NMIoU ($\lambda = 0.8$) | 88.5 |
| NMIoU ($\lambda = 0.9$) | 89.0 |
| NMIoU ($\lambda = 1$) | 88.9 |

From Table 5, it can be seen that when the balance coefficient $\lambda$ of MPDIoU and NWD in YOLOv5 is 0.6, NMIoU is higher than the others. Therefore, $\lambda = 0.6$ is used as the balance coefficient of the loss function NMIoU.

### 3.3. Comparison of DyHead Normalization Functions GN and BN

In order to explore the advantages and disadvantages of different DyHead normalization functions for the detection of the dataset, this paper added the BN normalization function, and conducted controlled experiments on GN and BN to select the optimal normalization function and the number of cycles. The results are shown in Table 6.

**Table 6.** Comparison of DyHead normalization functions GN and BN.

| Normal Function | Cycle Times | mAP/% | Training Time/h |
|---|---|---|---|
| Auto(YOLOv5) | 0 | 87.2 | 2.098 |
| GN | 2 | 88.9 | 3.693 |
| | 4 | 88.7 | 4.201 |
| | 6 | 87.8 | 5.090 |
| BN | 2 | 88.5 | 3.215 |
| | 4 | 87.9 | 4.066 |
| | 6 | 88.2 | 4.763 |

As can be seen from Table 6, DyHead achieves better results in the tomato leaf pest dataset when DyHead in YOLOv5 uses the GN normalization function and achieves an accuracy of 88.90% at a cycle time of 2, which is higher compared with YOLOv5, GN4, GN6, BN2, BN4, and BN6. GN2 has a time of 3.693 h, which is improved relative to the others. Relative to YOLOv5 itself, YOLOv5 uses a DyHead detection head, which needs to be processed cyclically through the scale-aware attention module, the spatial location-aware attention module, and the task-aware attention module, resulting in an increase in time, which is an unavoidable time increase. Relative to the short training time of BN = 2, the difference of the main normalization module GN needs to group the channels when calculating the mean and variance, which will introduce some additional computational overhead. Furthermore, GN introduces the concept of groups, which will increase the total number of parameters of the model, thus affecting the training speed. However, from Table 6, it can be seen that the mAP of GN = 2 has the highest accuracy. Therefore, the GN normalization function is adopted and looped twice as the optimal parameter pairing for DyHead.

### 3.4. Detection Performance of the YOLONDD Model

The detection results of YOLONDD and other models are shown in Table 7, and the thermal effects of them are shown in Table 8.

**Table 7.** Effect of detection using different models.

| Model | Leafminer Flies | Thrips | Tabacco Budworm | Spider Leaf Mite |
|---|---|---|---|---|
| Original image |  |  |  | 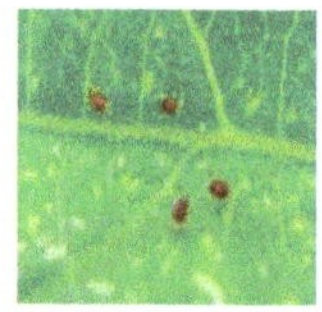 |
| Labeled image |  |  |  | 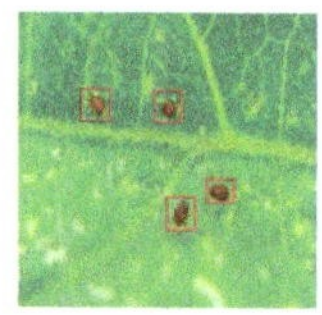 |
| Faster R-CNN |  |  |  |  |
| SSD |  |  |  | 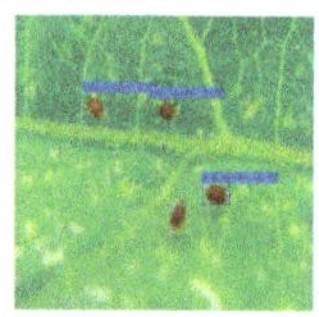 |
| RetinaNet |  |  |  |  |
| YOLOv7 |  |  |  | 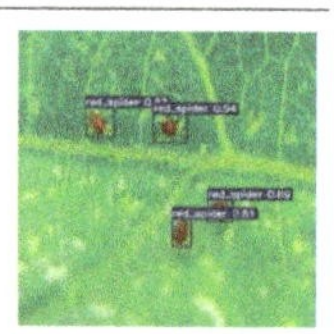 |
| YOLOv5 |  |  |  |  |

**Table 7.** *Cont.*

| Model | Leafminer Flies | Thrips | Tabacco Budworm | Spider Leaf Mite |
|---|---|---|---|---|
| YOLOv8 |  |  |  | 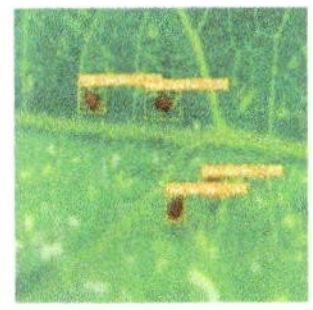 |
| YOLONDD |  |  |  | 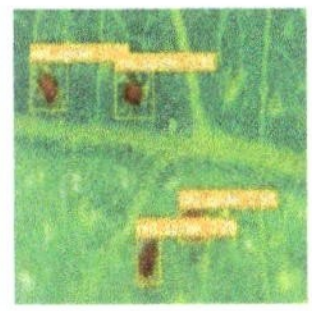 |

**Table 8.** Thermal effects of different models. The red box is the labeled box. The blue and cyan colors in the detection box indicate areas with low detection values, and the yellow, orange, and red colors indicate areas with high detection values. Their order is progressive.

| Model | Leafminer Flies | Thrips | Tabacco Budworm | Spider Leaf Mite |
|---|---|---|---|---|
| Labeled images |  |  |  | 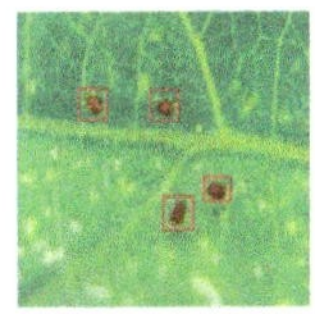 |
| Faster R-CNN | 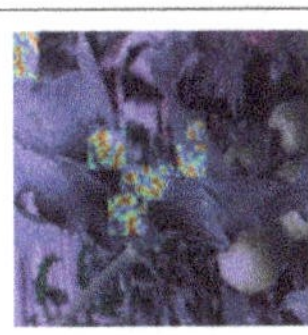 |  | 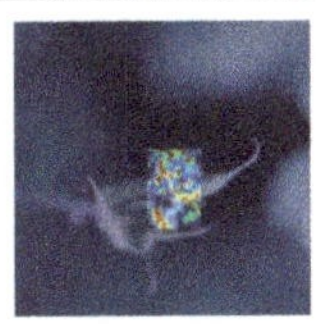 |  |
| SSD | 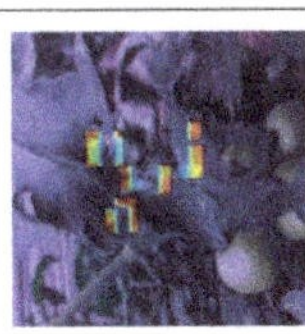 |  | 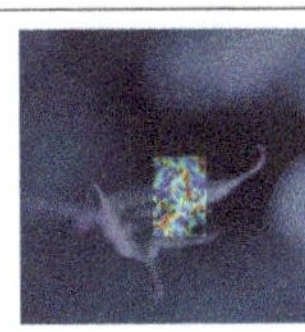 |  |
| RetinaNet |  |  | 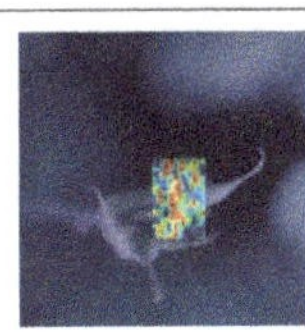 |  |

**Table 8.** *Cont.*

| Model | Leafminer Flies | Thrips | Tabacco Budworm | Spider Leaf Mite |
|---|---|---|---|---|
| YOLOv7 |  |  | 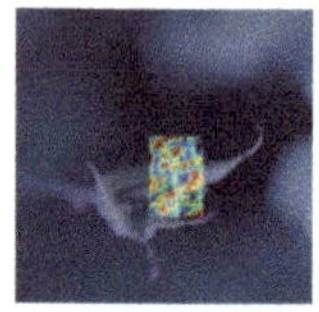 |  |
| YOLOv5 |  |  | 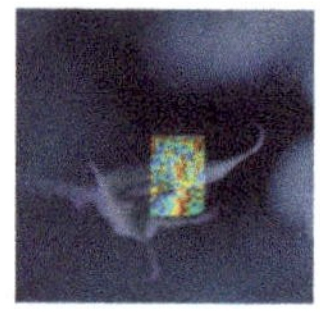 |  |
| YOLOv8 |  |  | 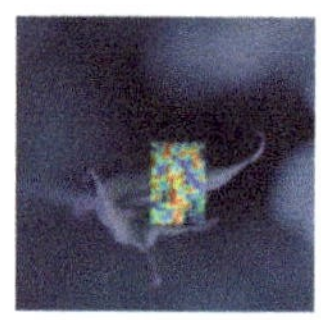 |  |
| YOLONDD |  |  |  |  |

From Table 7, it can be seen that YOLONDD has a higher detection accuracy for tomato leaf pests compared with the other six models, and there is less leakage of pests for small targets.

From Table 8, it can be seen that YOLONDD performs better in tomato leaf pests compared with the other six models, and the red and yellow high-value areas are more concentrated.

The performance of tomato leaf pest image detection using YOLONDD and the other six models was compared using mAP, FPS, training time, F1, and single image prediction time as metrics, and the results are shown in Tables 9–11.

**Table 9.** mAP of different models for tomato leaf pest detection.

| Model | Leafminer Flies | Thrips | Tobacco Budworm | Spider Leaf Mite | mAP/% |
|---|---|---|---|---|---|
| Faster R-CNN | 60 | 71 | 95 | 84 | 77.5 |
| SSD | 59.93 | 85.19 | 96.56 | 85.19 | 77.58 |
| YOLOv5 | 79.8 | 84.5 | 95.8 | 88.5 | 87.2 |
| RetinaNet | 65 | 79 | 98 | 87 | 82.35 |
| YOLOv7 | 81.8 | 90.3 | 96.8 | 89.6 | 89.6 |
| YOLOv8 | 62.9 | 76.3 | 93 | 88.3 | 80.1 |
| YOLONDD | 83.6 | 90.6 | 96.3 | 89.9 | 90.1 |

As can be seen from the table, the mAP of YOLONDD is 16.14%, 16.25%, 3.33%, 9.41%, 0.56%, and 12.48% higher than SSD, Faster R-CNN, YOLOv5, RetinaNet, YOLOv7, and YOLOv8, respectively. The experimental data proves that the proposed NMIoU loss function improves target frame sensitivity and small-scale detection ability.

**Table 10.** Efficiency of different models for tomato leaf pest detection.

| Model | Training Time/h | Single Image Prediction/ms | FPS |
|---|---|---|---|
| Faster R-CNN | 7.45 | 72.21 | 13.87 |
| SSD | 1.89 | 11.24 | 88.95 |
| YOLOv5 | 2.098 | 11.80 | 84.75 |
| RetinaNet | 6.18 | 31.99 | 31.25 |
| YOLOv7 | 4.41 | 19.80 | 50.50 |
| YOLOv8 | 0.68 | 15.5 | 64.52 |
| YOLONDD | 3.347 | 18.90 | 52.91 |

**Table 11.** F1 scores of different models for tomato leaf pest detection.

| Model | Leafminer Flies | Thrips | Tobacco Budworm | Spider Leaf Mite | mF1/% |
|---|---|---|---|---|---|
| Faster R-CNN | 50 | 59 | 84 | 65 | 64.50 |
| SSD | 54 | 46 | 90 | 81 | 67.75 |
| YOLOv5 | 74.75 | 82.19 | 93.42 | 86.28 | 84.16 |
| RetinaNet | 67 | 81 | 95 | 84 | 81.75 |
| YOLOv7 | 78.79 | 86.88 | 93.02 | 85.63 | 86.08 |
| YOLOv8 | 65.04 | 72.81 | 87.55 | 83.69 | 77.27 |
| YOLONDD | 81.39 | 87.15 | 98.94 | 86.94 | 88.60 |

From the table, we can see that the training time of YOLONDD is 3.347 h, which is faster than the other models. The single-image prediction time of YOLONDD is 18.90 ms, which is the fastest. The FPS of YOLONDD with Batch size = 1 is 52.91 ms, which is faster than Faster R-CNN, RetinaNet, and YOLOv7. YOLOv8 is faster than this model because it uses a more efficient architecture, a better feature extractor, and a more effective neck and head structure to improve training efficiency. SSD uses a more lightweight network structure, and SSD performs target detection on multiple scales of feature maps, which allows the model to capture targets at different resolutions and will improve the efficiency of training.

Compared with the YOLOv5 model, the training time of YOLONDD is 59.53%, and the single-image prediction time is increased by 7.1 ms, which is due to the fact that the YOLOv5 model itself has a simpler model structure and a smaller number of parameters, which reduces the model training time as well as the single-image prediction time. The slow training speed of YOLONDD is due to the use of the DyHead detection head, which extracts the feature maps several times for optimization, thus reducing the training speed. However, it can be seen from Tables 7–9 that its detection accuracy is not as good as that of the present model, YOLONDD. Overall, in the present model, by replacing the loss function NMIoU with CIoU, adding the detection head with the attention mechanism DyHead as well, and changing the coupling head of YOLOv5 to a decoupled head, the detection time is increased compared with the YOLOv5 model, but the model detection accuracy is also improved.

As can be seen from the table, the mF1 of YOLONDD is 88.60%, which is faster than the other models. Especially in Leafminer flies, there is an 8.88% improvement relative to the original YOLOv5. It shows that the YOLONDD can extract the features of pests well and improve the detection ability of images.

### 3.5. Results of Ablation Experiments

Ablation experiments were carried out based on the ablation experiment scheme described in Section 2.3.4. The results of the experiments are shown in Table 12.

**Table 12.** Performance of different ablation methods.

| Method | mAP/% |
|---|---|
| YOLOv5 | 87.2 |
| YOLON | 89.1 |
| YOLOD$_1$ | 88.9 |
| YOLOD$_2$ | 89.0 |
| YOLOND$_1$ | 88.0 |
| YOLOND$_2$ | 88.9 |
| YOLOD$_1$D$_2$ | 87.7 |
| YOLONDD | 90.1 |

From the results of the ablation experiments in Table 12, it can be seen that the mean average precision (mAP) of YOLON, YOLOD$_1$, and YOLOD$_2$, when paired individually, are all improved compared with that of YOLOv5, which illustrates the fact that the replacement of the loss function CIoU by the loss function NMIoU, the introduction of the detection head with an attentional mechanism, DyHead, and the introduction of a decoupled head, can improve the model's detection precision to some extent.

It can be seen that the accuracy decreases relative to YOLON, YOLOD$_1$, and YOLOD$_2$, indicating that DyHead and decoupled heads have different focuses in feature extraction and representation. When they are combined, feature incoherence occurs, resulting in a model that does not effectively utilize all the useful information. The loss function NMIoU is combined with DyHead, while the NMIoU loss function focuses on optimizing the positioning accuracy of the bounding box.

Combining the three, it can be seen that the accuracy is at its highest, and the decoupled head processes the feature maps of P3, P4, and P5 in batches, which can result in classification maps, coordinate position maps, and target box confidence maps. Whereas DyHead enhances the feature representation through the attention mechanism, the NMIoU loss function focuses on optimizing the positioning accuracy of the bounding box. The decoupled head can just bring out the focused advantages of DyHead and NMIoU.

The training detection efficiency of each ablation experiment model was also compared, and the results are shown in Table 13.

**Table 13.** Efficiency of different ablation methods.

| Method | Training Time/h | Single Image Prediction Time/ms |
|---|---|---|
| YOLOv5 | 2.098 | 11.8 |
| YOLON | 2.204 | 12.6 |
| YOLOD$_1$ | 3.693 | 23.1 |
| YOLOD$_2$ | 2.594 | 13.5 |
| YOLOND$_1$ | 3.10 | 23.3 |
| YOLOND$_2$ | 2.209 | 13.5 |
| YOLOD$_1$D$_2$ | 3.189 | 26.6 |
| YOLONDD | 3.347 | 18.9 |

As can be seen from Table 13, the training time and prediction time of the YOLON, YOLOD$_1$, and YOLOD$_2$ models increased relative to the traditional YOLOv5, indicating that the inclusion of the detection head DyHead with attention mechanism, loss function, and the introduction of the decoupled head alone increase the computational complexity of the model, which in turn increases the training time of the model. With YOLOD$_1$, the training time of the model is reduced by adding the decoupled head and NMIoU separately. Because YOLOv5 adopts the structure of a coupled head, adding DyHead for this purpose will increase the complexity of the model, and DyHead will perform cyclic feature extraction on the image twice, which will enhance the number of parameters in the model even more and increase the training time. However, the decoupled head will process the feature maps of P3, P4, and P5 in batches, which will save a lot of time. Adding the NMIoU loss function can simplify the process of calculating the target position, and the structure of NMIoU can improve the convergence speed. Therefore, adding a decoupled head and NMIoU loss function can relatively reduce the training time of the model.

## 4. Discussion

In traditional pest management, the identification of pests relies on the farmers' empirical judgment, which is the disadvantage of the approach that could be more efficient and more balanced on subjective judgment. According to the characteristics of pests in natural environments, factors such as the occlusion of pests, complex backgrounds, colors, and target size can affect the model's effectiveness for detection and identification. Therefore, there are challenges in detecting small target pests accurately and quickly. Currently, there are several studies on crop pest detection and recognition, but there are fewer studies on small-target crop pests. Hu et al. [38] improved YOLOv5 for tomato pest detection, and the model's mAP could reach 98.10%. Still, because the authors' dataset was annotated with the entire diseased leaf, they could not verify that the model performed well in complex scenarios and with small targets. Wang et al. [39] improved YOLOv3 for early detection of tomato pests and diseases in complex backgrounds, and the model achieved better detection results, but the model has leakage in small targets. Although all these studies have excellent performance in pest and disease detection, they need more research on small targets with complex backgrounds. For this reason, this paper proposes a method for pest detection of small targets in complex contexts. The validation results on the self-constructed dataset show that YOLONDD is better than other models in detecting small targets under complex backgrounds, which can effectively detect pests and improve the accuracy of pest detection.

Although YOLONDD has a mAP of 90.1% in the dataset, which can satisfy the current detection task, there is still a leakage of detection, indicating that there is still room for improvement in the model's ability to detect small-target pests in complex backgrounds. Combined with the heat map, it can be seen that YOLONDD pays attention to the areas where pests or pest infestation parts exist, which indicates that some pests or pest infestation parts cannot be detected. The main reason for this may be that although the newly designed NMIou loss function reduces some of the missed detections due to leaf occlusion, some of the frames with low scores are still filtered when the vertex coordinates are calculated for the overlapping frames, which is a problem that needs to be paid attention to in the future.

Currently, some tomato leaf pests are used as research objects in this paper, but many pests still have practical applications, and more pest datasets are needed to improve the model's applicability. The pest collection method must also be further standardized, considering factors such as shading and light conditions to enhance the model's generalization ability.

## 5. Conclusions

Fast and accurate identification of tomato pests can help tomato farmers detect pests in their tomato crops and improve tomato yields, as well as agro-horticultural to improve plant health and visual perception, and can help specialist experts easily determine the type of pests.

A network target detection model based on deep learning, YOLONDD, was proposed in this paper. A new loss function NMIoU is designed in the model to improve the ability of anomaly processing, improve the model's ability to detect and recognize objects at different scales, and improve the robustness to scale changes; the detection head DyHead containing an attention mechanism is introduced to improve the ability to detect targets at different scales. Adding a DyHead with an attention mechanism improves the detection ability of targets at different scales, reduces the number of computations and parameters, improves the accuracy of target detection, improves the overall performance of the model, and accelerates the training process. Adding a decoupled head to Head can effectively reduce the number of parameters and the computational complexity, and enhance the model's generalization ability and robustness.

The model proposed in this paper, YOLONDD, was shown to detect tomato leaf pests more effectively in images than other test models. The mAP of the whole dataset of tomato pest images reached 90.1%. All of these are better than the target detection models such as Faster R-CNN, SSD, YOLOv5, RetinaNet, YOLOv7, YOLOv8, etc., and realize more efficient and accurate detection on tomato leaf pest images.

In this paper, we achieved a more accurate detection effect on the results of pest images, but there are still some issues to face. The future directions of related research include the following aspects: how to detect and eliminate the shades in the tomato leaf images; how to accurately detect pests if the colors of pests and leaves are similar; and how to recognize pests if there are some obstacles on parts of pests, such as other leaves or stalks, so as to reduce the influence of shadows, colors, and occlusions on the detection of tomato pests in complex backgrounds of real environments.

**Author Contributions:** Conceptualization, R.X. and G.W.; methodology, R.X., G.W., F.Y., L.M., P.W. and X.Y.; software, R.X.; validation, R.X.; formal analysis, R.X., P.W. and X.Y.; investigation, R.X., G.W., F.Y., L.M., P.W. and X.Y.; data curation, R.X.; writing—original draft. R.X.; resources, G.W., F.Y. and L.M.; writing—review and editing, G.W.; funding acquisition, L.M. and P.W. All authors have read and agreed to the published version of the manuscript.

**Funding:** This study was supported by the Key Research and Development Program of Zhejiang Province (Grant number: 2021C02005) and Zhejiang Provincial Commonweal Projects (Grant number: LGG21F020001).

**Data Availability Statement:** The code for our proposed model YOLONDD and dataset used in the experiments can be found on GitHub: https://github.com/zafucslab/YOLONDD (accessed on 6 May 2024).

**Conflicts of Interest:** The authors declare no conflicts of interests.

## References

1. Crispi, G.M.; Valente, D.S.M.; de Queiroz, D.M.; Momin, A.; Fernandes-Filho, E.I.; Picanço, M.C. Using Deep Neural Networks to Evaluate Leafminer Fly Attacks on Tomato Plants. *Agriengineering* **2023**, *5*, 273–286. [CrossRef]
2. Xu, Y.; Gao, Z.; Zhai, Y.; Wang, Q.; Gao, Z.; Xu, Z.; Zhou, Y. A CNNA-Based Lightweight Multi-Scale Tomato Pest and Disease Classification Method. *Sustainability* **2023**, *15*, 8813. [CrossRef]
3. Asiry, K.A.; Huda, M.N.; Mousa, M.A.A. Abundance and Population Dynamics of the Key Insect Pests and Agronomic Traits of Tomato (*Solanum lycopersicon* L.) Varieties under Different Planting Densities as a Sustainable Pest Control Method. *Horticulturae* **2022**, *8*, 976. [CrossRef]
4. Zeng, X.; Huang, C.; Zhan, L. Image recognition method of agricultural pests based on multisensor image fusion technology. *Adv. Multimed.* **2022**, *2022*, 6359130. [CrossRef]
5. Harris, C.G.; Andika, I.P.; Trisyono, Y.A. A Comparison of HOG-SVM and SIFT-SVM Techniques for Identifying Brown Planthoppers in Rice Fields. In Proceedings of the 2022 IEEE 2nd Conference on Information Technology and Data Science (CITDS), Debrecen, Hungary, 16–18 May 2022; pp. 107–112.
6. Kasinathan, T.; Singaraju, D.; Uyyala, S.R. Insect classification and detection in field crops using modern machine learning techniques. *Inf. Process. Agric.* **2021**, *8*, 446–457. [CrossRef]
7. Zhang, W.; Huang, H. AgriPest-YOLO: A rapid light-trap agricultural pest detection method based on deep learning. *Front. Plant Sci.* **2022**, *13*, 1079384. [CrossRef] [PubMed]
8. Türkoğlu, M.; Hanbay, D. Plant disease and pest detection using deep learning-based features. *Turk. J. Electr. Eng. Comput. Sci.* **2019**, *27*, 1636–1651. [CrossRef]
9. Kuzuhara, H.; Takimoto, H.; Sato, Y.; Kanagawa, A. Insect pest detection and identification method based on deep learning for realizing a pest control system. In Proceedings of the 2020 59th Annual Conference of the Society of Instrument and Control Engineers of Japan (SICE), Chiang Mai, Thailand, 23–26 September 2020; pp. 709–714.
10. Venkatasaichandrakanth, P.; Iyapparaja, M. Pest Detection and Classification in Peanut Crops Using CNN, MFO, and EViTA Algorithms. *IEEE Access* **2023**, *11*, 54045–54057. [CrossRef]
11. Pattnaik, G.; Parvathi, K. Automatic detection and classification of tomato pests using support vector machine based on HOG and LBP feature extraction technique. In *Progress in Advanced Computing and Intelligent Engineering: Proceedings of ICACIE 2019*; Springer: Singapore, 2021; Volume 2, pp. 49–55.
12. Cortes, C.; Vapnik, V. Support-vector networks. *Mach. Learn.* **1995**, *20*, 273–297. [CrossRef]

13. Liang, W.; Kirk, K.R.; Greene, J.K. Estimation of soybean leaf area, edge, and defoliation using color image analysis. *Comput. Electron. Agric.* **2018**, *150*, 41–51. [CrossRef]
14. da Silva Vieira, G.; Rocha, B.M.; Fonseca, A.U.; de Sousa, N.M.; Ferreira, J.C.; Cabacinha, C.D.; Soares, F. Automatic detection of insect predation through the segmentation of damaged leaves. *Smart Agric. Technol.* **2022**, *2*, 100056. [CrossRef]
15. Fang, W.; Guan, F.; Yu, H.; Bi, C.; Guo, Y.; Cui, Y.; Su, L.; Zhang, Z.; Xie, J. Identification of wormholes in soybean leaves based on multi-feature structure and attention mechanism. *J. Plant Dis. Prot.* **2023**, *130*, 401–412. [CrossRef]
16. Zhu, R.; Hao, F.; Ma, D. Research on Polygon Pest-Infected Leaf Region Detection Based on YOLOv8. *Agriculture* **2023**, *13*, 2253. [CrossRef]
17. Zhu, L.; Li, X.; Sun, H.; Han, Y. Research on CBF-YOLO detection model for common soybean pests in complex environment. *Comput. Electron. Agric.* **2024**, *216*, 108515. [CrossRef]
18. Ye, R.; Gao, Q.; Qian, Y.; Sun, J.; Li, T. Improved YOLOv8 and SAHI Model for the Collaborative Detection of Small Targets at the Micro Scale: A Case Study of Pest Detection in Tea. *Agronomy* **2024**, *14*, 1034. [CrossRef]
19. Tian, Y.; Wang, S.; Li, E.; Yang, G.; Liang, Z.; Tan, M. MD-YOLO: Multi-scale Dense YOLO for small target pest detection. *Comput. Electron. Agric.* **2023**, *213*, 108233. [CrossRef]
20. Lippi, M.; Bonucci, N.; Carpio, R.F.; Contarini, M.; Speranza, S.; Gasparri, A. A yolo-based pest detection system for precision agriculture. In Proceedings of the 2021 29th Mediterranean Conference on Control and Automation (MED), Puglia, Italy, 22–25 June 2021; pp. 342–347.
21. Mamdouh, N.; Khattab, A. YOLO-based deep learning framework for olive fruit fly detection and counting. *IEEE Access* **2021**, *9*, 84252–84262. [CrossRef]
22. Yang, S.; Xing, Z.; Wang, H.; Dong, X.; Gao, X.; Liu, Z.; Zhang, X.; Li, S.; Zhao, Y. Maize-YOLO: A new high-precision and real-time method for maize pest detection. *Insects* **2023**, *14*, 278. [CrossRef] [PubMed]
23. Dai, X.; Chen, Y.; Xiao, B.; Chen, D.; Liu, M.; Yuan, L.; Zhang, L. Dynamic head: Unifying object detection heads with attentions. In Proceedings of the IEEE/CVF Conference on Computer Vision and Pattern Recognition, Virtual, 19–25 June 2021; pp. 7373–7382.
24. Li, C.; Li, L.; Jiang, H.; Weng, K.; Geng, Y.; Li, L.; Ke, Z.; Li, Q.; Cheng, M.; Nie, W.; et al. YOLOv6: A single-stage object detection framework for industrial applications. *arXiv* **2022**, arXiv:2209.02976.
25. Wang, J.; Xu, C.; Yang, W.; Yu, L. A normalized Gaussian Wasserstein distance for tiny object detection. *arXiv* **2021**, arXiv:2110.13389.
26. Siliang, M.; Yong, X. Mpdiou: A loss for efficient and accurate bounding box regression. *arXiv* **2023**, arXiv:2307.07662.
27. Jiang, B.; Luo, R.; Mao, J.; Xiao, T.; Jiang, Y. Acquisition of localization confidence for accurate object detection. In Proceedings of the European Conference on Computer Vision (ECCV), Munich, Germany, 8–14 September 2018; pp. 784–799.
28. Zheng, Z.; Wang, P.; Liu, W.; Li, J.; Ye, R.; Ren, D. Distance-IoU loss: Faster and better learning for bounding box regression. *Proc. AAAI Conf. Artif. Intell.* **2020**, *34*, 12993–13000. [CrossRef]
29. Mo, L.; Shi, L.; Wang, G.; Yi, X.; Wu, P.; Wu, X. MISF: A Method for Measurement of Standing Tree Size via Multi-Vision Image Segmentation and Coordinate Fusion. *Forests* **2023**, *14*, 1054. [CrossRef]
30. Wu, Y.; He, K. Group normalization. In Proceedings of the European Conference on Computer Vision (ECCV), Munich, Germany, 8–14 September 2018; pp. 3–19.
31. Ioffe, S.; Szegedy, C. Batch normalization: Accelerating deep network training by reducing internal covariate shift. In Proceedings of the International Conference on Machine Learning, Lille, France, 7–9 July 2015; pp. 448–456.
32. Available online: https://www.kaggle.com/datasets/kaustubhb999/tomatoleaf (accessed on 21 June 2022).
33. Available online: https://github.com/HumanSignal/labelImg (accessed on 1 July 2022).
34. Li, Q.; Xie, B.; You, J.; Bian, W.; Tao, D. Correlated logistic model with elastic net regularization for multilabel image classification. *IEEE Trans. Image Process.* **2016**, *25*, 3801–3813. [CrossRef] [PubMed]
35. Ren, S.; He, K.; Girshick, R.; Sun, J. Faster R-CNN: Towards real-time object detection with region proposal networks. *Adv. Neural Inf. Process. Syst.* **2015**, 28. [CrossRef]
36. Liu, W.; Anguelov, D.; Erhan, D.; Szegedy, C.; Reed, S.; Fu, C.Y.; Berg, A.C. Ssd: Single shot multibox detector. In Proceedings of the Computer Vision–ECCV 2016: 14th European Conference, Amsterdam, The Netherlands, 11–14 October 2016; Proceedings, Part I 14. Springer International Publishing: Berlin/Heidelberg, Germany, 2016; pp. 21–37.
37. Wang, C.Y.; Bochkovskiy, A.; Liao HY, M. YOLOv7: Trainable bag-of-freebies sets new state-of-the-art for real-time object detectors. In Proceedings of the IEEE/CVF Conference on Computer Vision and Pattern Recognition, Vancouver, BC, Canada, 24 June 2023; pp. 7464–7475.
38. Hu, W.; Hong, W.; Wang, H.; Liu, M.; Liu, S. A Study on Tomato Disease and Pest Detection Method. *Appl. Sci.* **2023**, *13*, 10063. [CrossRef]
39. Wang, X.; Liu, J.; Zhu, X. Early real-time detection algorithm of tomato diseases and pests in the natural environment. *Plant Methods* **2021**, *17*, 43. [CrossRef]

*agronomy*

MDPI

*Article*

# Developing an Effective Push–Pull System for Managing Outbreaks of the Invasive Pest *Bactrocera dorsalis* (Diptera: Tephritidae) in *Nephelium lappaceum* Orchards

**Jian Wen [1,2,†], Zhe Shan [1,†], Yan Zou [1], Xianwu Lin [1], Zhifu Cui [3], Rihui Yan [1,2] and Fengqin Cao [1,2,*]**

[1] Key Laboratory of Insect Molecular Biology, and Green Prevention and Control of Tropical Plant Diseases and Pests, Hainan University, Haikou 570228, China; 996095@hainanu.edu.cn (J.W.); 13833900795@163.com (Z.S.); zouyan0825@126.com (Y.Z.); linxianw@hainanu.edu.cn (X.L.); ryanl@hainanu.edu.cn (R.Y.)

[2] School of Tropical Agriculture and Forestry, Hainan University, Haikou 570228, China

[3] Hainan State Farms of Academy of Sciences Group Co., Ltd., Haikou 570226, China; zhifu555@163.com

* Correspondence: 990957@hainanu.edu.cn

† These authors contributed equally to this work.

**Abstract:** Outbreaks of the oriental fruit fly, *Bactrocera dorsalis* (Hendel), present significant challenges to global fruit production, necessitating effective control measures that minimize environmental risks and pesticide resistance. This study aimed to develop and evaluate the effectiveness of four distinct push–pull control strategies for managing *B. dorsalis* outbreaks in a *Nephelium lappaceum* orchard. These strategies involved the inclusion of low-concentration abamectin, spraying repellent with a drone or manually, using methyl eugenol (ME) or food bait and employing either two types of attractants and repellents or a single type. The findings indicated that incorporating the low-concentration abamectin into the push–pull system, utilizing ME as an attractant instead of food lures and manually applying abamectin and attractants were all effective in reducing the *B. dorsalis* population size and minimizing fruit damage. While increasing the diversity of repellents and attractants enhanced the long-term effectiveness of the system, it did not result in a significant decrease in *B. dorsalis* population size or fruit damage rate compared to using a single repellent or attractant. In conclusion, the push–pull strategy emerged as a viable method for managing *B. dorsalis* outbreaks, offering potential benefits in reducing environmental risks and pesticide resistance. However, the study underscored the importance of the context-specific construction of push–pull strategies to optimize their effectiveness in orchard settings.

**Keywords:** integrated pest management; invasive pests; tropical fruits; pest behavior

## 1. Introduction

The oriental fruit fly (*Bactrocera dorsalis*) is a highly destructive pest that poses a significant threat to global fruit production [1–4]. Originating from tropical and subtropical regions of Asia, this pest has a broad adaptability to various host plants [5] and exhibits robust reproductive capacities [6,7]. As a result, it has significantly expanded its geographical range from China, Southeast Asia and India to the Hawaiian Islands [8]. Due to its widespread distribution and economic impact, it is considered a globally recognized quarantine pest [1,4].

There are various methods available for controlling *B. dorsalis*, typically employed preemptively before outbreaks occur [9,10]. These methods include biological control [11,12], pheromone traps [10,13] and the sterile insect technique [14,15]. While chemical control is widely recognized as an effective measure for managing pest outbreaks [16], it also poses risks of environmental contamination and can contribute to the development of insecticide resistance [17], thereby hindering sustainable long-term pest management efforts. Consequently, there is an urgent need to develop effective integrated *B. dorsalis* management methods that can reduce pesticide resistance development and systemic resistance in insects [18], thus, reducing damage in fruit production.

The push–pull strategy is an integrated behavioral control method that is widely applied in agricultural and medical insect control [19–23]. Its core concept is to repel pests from protected resources while simultaneously attracting them to an alternative attractive source, resulting in pest elimination [20], and to manipulate the abundance and distribution of natural enemies [24]. There are two implementation approaches for the push–pull strategy [20]. The first involves planting non-crop plants with repellent properties and combining them with trap crops [25]. The second approach utilizes the combination of repellents and attractants to lure and trap pests [26]. Usually, non-toxic components are used in the push–pull strategy, aiming to reduce pesticide usage, while combining different stimuli and coordinating pest distributions to enhance efficiency [20].

The push–pull strategy has demonstrated significant effectiveness in controlling crop pests such as fall armyworm *Spodoptera frugiperda* (J.E. Smith) [27], cotton bollworm *Helicoverpa armigera* (Hübner) [28] and Colorado potato beetle *Leptinotarsa decemlineata* (Say) [29]. It has also shown success in the integrated management of medical insects [30] and *B. dorsalis* [31]. In conclusion, the push–pull strategy holds promise as an integrated pest control method for managing *B. dorsalis* [31]. It has the potential to reduce pesticide resistance and mitigate environmental pollution [20]. However, there is currently no literature available regarding the effectiveness of the push–pull strategy during pest outbreaks or how to construct a push–pull system suitable for outbreak control.

The objective of this study was to develop an effective push–pull strategy to manage *B. dorsalis* in a commercial rambutan orchard. The approach was to test four variations of a push–pull strategy in comparison to the use of a standard insecticide. The four experiments (Experiments 1–4) were systematically conducted in a single orchard, with a 30-day respite between each experiment. According to the different push–pull strategy designs (see below), the "push" was established by using allicin and/or d-limonene as the repellents. Allicin has been proven to repel *B. dorsalis* and other fruit fly species, acting as a repellent [32–34]. Additionally, d-limonene, a natural compound found in plants of the Rutaceae family, can also serve as a deterrent for *B. dorsalis* [35]. The "pull" was established by using methyl eugenol (ME bait) and/or food bait.

The first push–pull strategy compared the effectiveness of a push (allicin)–pull (ME) system with and without the use of low-concentration abamectin in controlling the outbreak of *B. dorsalis*. Abamectin is considered environmentally friendly, displaying minimal residue levels in water and easy biodegradability by soil microorganisms [36]. It is also known to have low toxicity towards avian species and humans [37,38]. However, certain studies show that it is classified as a highly toxic pesticide for mice and rats and has negative effects on soil microbial communities in the short term [39]. To preserve the eco-friendliness of these push–pull strategies, a low concentration of abamectin was used in all the experiments. The second approach compared the effectiveness of a manually sprayed repellent (allicin) and abamectin in the push–pull system (with ME as the attractant) with drone spraying in controlling *B. dorsalis*. The third approach compared the control efficacy of push–pull systems using food bait as an attractant to those using ME as an attractant, with allicin as the repellent in both systems. Finally, the fourth approach evaluated the impact of increasing the variety of attractants (ME + food bait) and repellents (allicin + d-limonene) on the effectiveness of the push–pull system.

The series of research holds substantial implications for implementing context-specific push–pull control strategies to manage *B. dorsalis* outbreaks and reduce pesticide usage.

## 2. Materials and Methods

### 2.1. Study Site

These experiments were conducted in a commercial rambutan, *Nephelium lappaceum* (Sapindales: Sapindaceae), orchard located in Baotang Li and Miao Autonomous County, Hainan Province, China (18°36′39.0″ N, 109°43′12.9″ E). The orchard spanned an area of approximately 60,000 m$^2$ and comprised approximately 450 *N. lappaceum* trees, all of which were over 5 years old. Situated in a hilly area at an elevation ranging from 120 to 150 m, the orchard experienced severe infestations by *B. dorsalis* during the fruit maturation period from May to October. Based on our preliminary investigations, the *B. dorsalis* infestation began in early May, with trap counts ranging from 30 flies/trap/day. By early June, this count escalated to 300–500 flies/trap/day, and this peak infestation period persisted until late September when fruit harvesting was completed. Importantly, the orchard had not utilized chemical pesticides for approximately 8 months (from October 2022 to June 2023).

In the orchard, a total of 9 large plots (Figure 1a) were selected, with each plot covering an area of approximately 2500 m$^2$ and containing approximately 35 trees. The distance between these plots exceeded 100 m. To minimize variability between plots, treatments in each experiment (Experiments 1–4) were randomly allocated to different plots, with each treatment having 3 replicates (plots). Furthermore, measures were taken to position the plots as far away from the edges of the orchard as feasible.

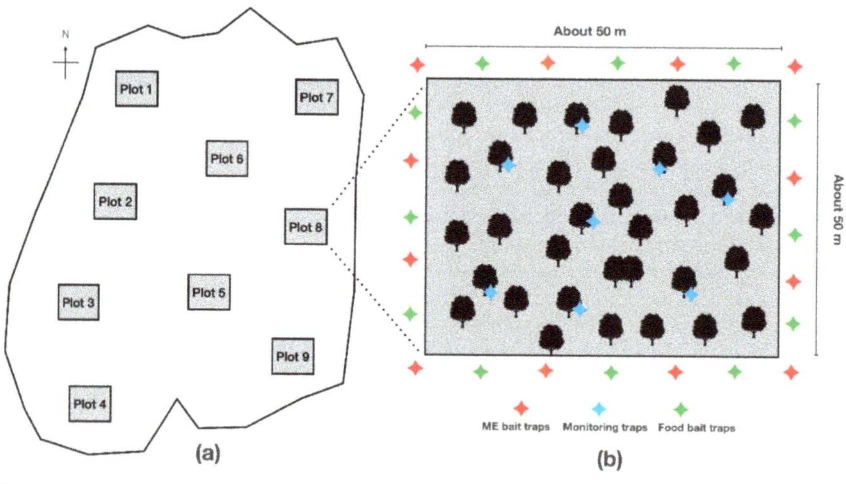


**Figure 1.** The orchard was divided into nine plots (**a**). In Experiments 1, 2, 3 and 4, each plot was randomly assigned a different push–pull strategy or a control treatment using bromophos + β-cypermethrin without employing the push–pull strategy. (**b**) The schematic representation of the trap distribution in the push–pull system. In Experiments 1 and 2, methyl eugenol (ME) bait traps (red) were used as attractants in the push–pull systems, while food bait traps (green) were not used. In Experiment 3, the ME traps were removed from the push–pull strategy with only food bait. In Experiment 4, both ME traps and food bait traps were deployed for the push–pull strategy using two types of attractants and repellents. For strategies using a single type of repellent and attractant, the food bait traps were not utilized.

## 2.2. Works Taken before All Experiments

Prior to each of the four experiments (on 1 June, 8 July, 15 August and 22 September 2023), the initial population of *B. dorsalis* was monitored across the 9 plots. The ME trap method described by Biasazin et al. [40] was employed. Unlike their use of McPhail traps, we utilized a white bottle and covered it with a black lip (h = 20 cm, d = 10 cm, see Figure S1a in the Supplementary Material) and treated it with 1 mL of methyl eugenol (ME, Pherobio Technology Co., Ltd., Beijing, China). Eight traps were suspended at a height of 1.8 m above the ground, positioned approximately 30 m apart within the plot (see Figure 1b). After a one-day period, the numbers of captured *B. dorsalis* were counted.

Subsequently, four independent experiments were conducted in a single orchard from 2 June to 29 September 2023, with a 30-day interval, while comparing the reduction rates of *B. dorsalis* populations and the rates of fruit damage among them (calculation of these rates was provided below). The investigations revealed that treatments involving early spraying did not yield a significant impact on the *B. dorsalis* population after a period of 30 days. The average initial number among the four experiments was 502.5, 466.2, 507.5 and 483.8 flies, respectively, and no significant differences were found (Kruskal–Wallis test: $\chi^2 = 2.43$, df = 3, $p = 0.225$; performed using GraphPad Prism software version 8.0 for Windows (San Diego, CA, USA, www.graphpad.com (accessed on 4 September 2023))) as the population returned to pre-spraying levels. Concurrently, the 30-day break aligned with the harvest cycle of mature *N. lappaceum* fruits, effectively curtailing the interference of residual fruit bodies on the potency of ensuing treatments. Moreover, within this temporal scope, the tropical climate of Hainan Island provided a relatively stable thermal range of 28 to 32 °C, mitigating the potential distortion of experimental outcomes attributable to thermal fluctuations. Additionally, the methods summary for these four experiments is shown in Table 1.

**Table 1.** The methods summary for these four push–pull strategies involved the use of repellents and attractants to represent the push and pull components, respectively.

| Push (Repellent Used) | Pull (Attractant Used) |
|---|---|
| Experiment 1: Push–Pull System with and without Low-Concentration Pesticide | |
| Allicin | Methyl eugenol |
| Allicin + abamectin | Methyl eugenol |
| Experiment 2: Drone-Based System vs. Manual-Based System | |
| Drone-applied allicin + abamectin | Methyl eugenol |
| Manually applied allicin + abamectin | Methyl eugenol |
| Experiment 3: Food Bait vs. ME Bait | |
| Manually applied allicin + abamectin | Food bait |
| Manually applied allicin + abamectin | Methyl eugenol |
| Experiment 4: Two Types of Repellents and Attractants vs. a Single Type | |
| Allicin + d-limonene + abamectin | Food bait + methyl eugenol |
| Allicin | Methyl eugenol |

### 2.3. *Experiment 1: Push–Pull System with and without Low-Concentration Pesticide*

Based on our pre-experimental monitoring, we found a high population density of *B. dorsalis* in the orchard, with 502.5 flies per trap per day at the beginning of June 2023. The traditional push–pull system is typically most effective as a preventive measure against pest outbreaks when pest populations are relatively low [20,38,41,42]. However, when pest populations become too high, they can overwhelm the control system, leading to saturation and reduced efficacy. To mitigate the issues associated, we incorporated low-concentration abamectin into the push–pull strategy.

The experiments were conducted from 2 June to 9 June 2023. Out of nine plots, three were randomly selected for the implementation of push–pull systems without the use of abamectin in each plot. Within each selected plot, manual spraying of the repellent (allicin) was performed, while attractants (ME traps) were placed around the plot's perimeter.

For the spraying application, allicin (80%, Xinwolong Biochemical Co., Ltd., Nanyang, China) was dissolved in water at a concentration of 3.2 ppm (recommended concentration: 3–5 ppm). A volume of 360 mL per hectare was applied (recommended rate: 300–900 mL/ha). This solution was evenly sprayed on the tree canopy and lower levels between 10:00 a.m. and 11:00 a.m. The spraying process utilized a high-pressure agricultural sprayer (YL-160Lsnx2, Fujiwara, Taizhou, China) (refer to Figure S1b in the Supplementary Material).

Three additional plots were randomly selected, maintaining the same treatment procedure but including a low concentration of abamectin in the allicin spray. Initially, a mixture was prepared by combining allicin (3.2 ppm) and abamectin (0.2 ppm, Guanyongqiaodi Agricultural Technology Co., Ltd., Zhoukou, China) at a 1:1 ratio. Subsequently, 720 mL of the resulting blend was applied per hectare (each component at 360 mL per hectare). The abamectin concentration recommended by the manufacturer for application on fruit trees ranged from 2 to 5 ppm, with an application rate of 800 to 1200 mL per hectare. In this experiment, a lower concentration of abamectin was used than the recommended level to maintain the non-toxic nature of the push–pull system. The resulting solution was manually sprayed within the orchard, following the previously outlined procedure.

In the three remaining plots, a standard insecticidal control mixture commonly used in China was applied. This mixture comprised 22% β-cypermethrin (2 ppm, Henglida Technology Co., Ltd., Jinan, China) and 50% bromophos (2 ppm, Xianda Technology Co., Ltd., Zhoukou, China) pesticides in a 1:1 ratio. Each component was applied at a rate of 360 mL and 300 g per hectare, respectively (recommended concentrations for both were 2–5 ppm and the application rates were 300–1200 mL and 300–1000 g per hectare, respectively). The application was carried out manually, following established methodologies. It is worth noting that as this experiment was conducted in a commercial orchard, there was no blank control group, as non-treatment was not permitted by the farmers due to economic reasons.

We deployed 12 attraction ME traps around the perimeter (Figure 1b). These traps were hung at a height of 1.8 m and positioned 3–5 m away from the edge of each plot to serve as the "push" attractant. On the third and seventh day of the experiment, 8 monitoring traps were suspended inside each plot (Figure 1b), following the same procedure as described earlier.

For assessing the fruit damage rate, we employed the five-point sampling method, which entailed selecting five *N. lappaceum* trees from each plot and collecting five mature fruits from each cardinal direction (east, south, west and north; height = 1–2 m) per tree, totaling twenty fruits per tree. These fruits were peeled and meticulously examined to detect *B. dorsalis* larvae and calculate the damage rate. Since most larvae did not hatch within the first three days and did not cause visible damage symptoms (visible damage symptoms typically appeared after five days of infection), we assessed the fruit damage rate only on the seventh day. This technique was also applied in Experiments 2, 3 and 4.

### 2.4. Experiment 2: Drone-Based System vs. Manual-Based System

In these experiments, we incorporated drone or manual spraying techniques to apply repellent and abamectin. The experiments took place across identical nine experimental plots spanning from 9 July to 16 July 2023, precisely one month after the conclusion of Experiment 1. Due to finding out in Experiment 1 that including abamectin in the spraying system could reduce the fruit fly population and fruit damage rate (see Section 3 and Figure 2) more efficiently than without, in Experiments 2, 3 and 4, all the spraying systems included abamectin.

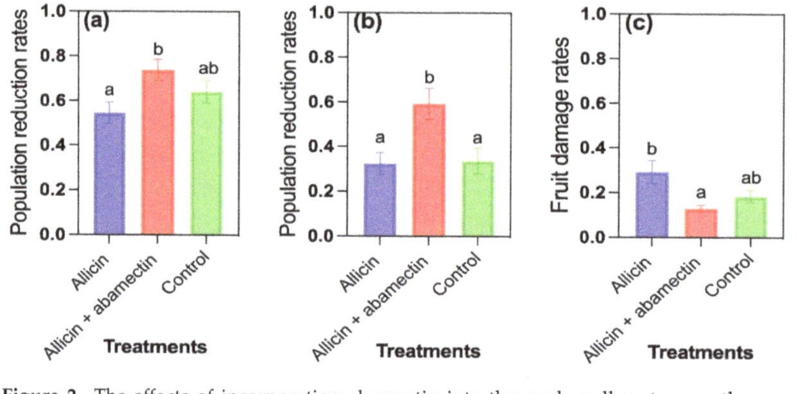


**Figure 2.** The effects of incorporating abamectin into the push–pull system on the population reduction rate of *B. dorsalis* after 3 days (**a**) and 7 days (**b**), as well as the fruit damage rate (**c**), were tested by using the Kruskal–Wallis tests. "Allicin" indicates the system with allicin as the push repellent, while "allicin + abamectin" refers to the push repellent incorporating abamectin. Both systems used ME as the attractant. "Control" represents the pesticide treatment (bromophos + β-cypermethrin) without using the push–pull strategy. The absence of significant differences between treatments is indicated by the same letters above the distinct bars ($p < 0.05$).

For the drone spraying method, we employed a DJIT30 drone (Dajiang Innovation Technology Inc., Shenzhen, China; see Figure S1c) in three randomly selected plots. The drone featured a flow rate of 0.379 L/min, producing fog droplets ranging from 130 to 250 μm in size. The spray mixture, comprising 80% allicin (repellent) and 5% abamectin (3.2 ppm and 0.2 ppm, respectively), was prepared following the procedures outlined in Experiment 1. Spraying operations were conducted by the drone at a height of 1.8 m above the tree canopy, with application occurring between 10:00 a.m. and 11:00 a.m.

For the manual-based system, we utilized the same materials (allicin, abamectin and ME) and preparation methods as in the drone spraying method. However, we applied the repellents and abamectin manually within the experimental plots (three plots), following the manual spraying treatment outlined in Experiment 1. The remaining three plots served as the standard insecticidal control, with the same preparation and application protocols as outlined in Experiment 1.

### 2.5. Experiment 3: Food Bait vs. ME Bait

The study aimed to ascertain the most suitable bait type for establishing an effective pull–push control system for *B. dorsalis*. Food baits typically consist of hydrolyzed proteins or sugars, which appeal to both male and female flies [43–45]. In contrast, ME bait contains components of sex pheromones, specifically targeting male flies [10,13]. Therefore, these baits operate through distinct mechanisms [1,46].

The experiments took place from 16 August to 23 August 2023. In this trial set, three random plots received the same treatments as outlined in the manual-based system described in Experiment 2. Additionally, three other plots were subjected to a combination of food bait and ME bait, forming a food and ME bait system. The food bait used was a commercial product (Q/BI001-2022, Bioglobal Agricultural Science Co., Ltd., Shenzhen, China), primarily composed of fermented sucrose. These food baits were mixed with abamectin and water at a ratio of 1:0.1:20, resulting in a solution of 3000 mL of food bait, 300 mL of abamectin (0.2 ppm) and 60 L of water per hectare. This mixture was then added to the food bait traps (each containing 30 mL). The food baits were suspended according to the depiction in Figure S1d and arranged following the layout illustrated in Figure 1b. The remaining three plots functioned as the standard insecticidal control, following the previously described protocols.

### 2.6. Experiment 4: Two Types of Repellents and Attractants vs. a Single Type

In the fourth experiment, conducted from 23 September to 29 September 2023, we introduced 95% d-limonene (Macklin Biochemical Co., Ltd., Shanghai, China) as an additional repellent and food bait (as described previously) as an additional attractant. The aim was to evaluate whether incorporating multiple types of repellents and attractants could enhance the efficacy of the push–pull system.

For the push–pull systems incorporating two types of repellents and attractants, we prepared a concoction comprising allicin (repellent 1, 3.2 ppm), d-limonene (repellent 2) and 5% abamectin (0.2 ppm) mixed in a ratio of 1:0.5:1. This mixture was then diluted at a ratio of 1:2500, resulting in 360 mL per hectare, and manually applied to the *N. lappaceum* trees. Additionally, 12 food bait traps (as attractant 1) and ME bait traps (as attractant 2) were strategically positioned around the plot's perimeter, following the layout depicted in Figure 1b.

For treatments utilizing a single type of repellent (allicin) or attractant (ME), the "push" involved manually applying allicin combined with abamectin, while the "pull" utilized ME bait. The remaining three plots functioned as the standard insecticidal control.

### 2.7. Data Analysis

For each plot, the reduction rate of *B. dorsalis* population was calculated by subtracting the final average number of flies caught in the 8 ME monitoring traps from the initial average number, then dividing the result by the initial average number. The fruit damage rate was determined by counting the number of damaged fruits and dividing it by the total number of fruits collected from five trees. To assess the normality and homogeneity of variance, the Shapiro–Wilk test and Levene's test were employed, respectively. Due to the non-normal distribution of the data, the variance analysis of *B. dorsalis* population reduction rates and fruit damage rates among different treatments within the same experiment (Experiment 1, 2, 3 or 4) and days (third or seventh day) was conducted using the Kruskal–Wallis test [47]. Multiple comparisons were carried out using the two-stage linear step-up procedure of Benjamini, Krieger and Yekutieli. All statistical analyses were performed using GraphPad Prism software version 8.0 for Windows (San Diego, CA, USA, www.graphpad.com (accessed on 4 September 2023)).

## 3. Results

### 3.1. Push–Pull System with and without Low-Toxicity Pesticide

The incorporation of abamectin into the push–pull system significantly influenced both the population reduction rates and fruit damage rates of *B. dorsalis*. The statistical analysis revealed significant effects at both 3 days ($\chi^2 = 6.15$, df = 2, $p = 0.046$) and 7 days ($\chi^2 = 8.67$, df = 2, $p = 0.013$) on the population reduction rates. After a 3-day period, the push–pull system with abamectin achieved a 74.9% reduction rate in the *B. dorsalis* population, which was significantly higher than the 54.6% reduction observed in the push–pull system without abamectin (Figure 2a). After 7 days, the system containing abamectin had a notable reduction rate of approximately 60%, significantly outperforming both the system without abamectin and the control (Figure 2b). Additionally, the inclusion of abamectin resulted in a significant decrease in the fruit damage rate to 13.0% after 7 days, markedly lower than that observed in the system without abamectin (Figure 2c).

### 3.2. Drone-Based System vs. Manual-Based System

The method of manual spraying exhibited significant effects on the population reduction rate (3d: $\chi^2 = 10.59$, df = 2, $p = 0.005$; 7d: $\chi^2 = 8.84$, df = 2, $p = 0.012$) of *B. dorsalis* and fruit damage rates ($\chi^2 = 6.03$, df = 2, $p = 0.042$) compared to drone spraying. After 3 days, the reduction rate of *B. dorsalis* populations in the drone spraying treatment was only 35.4%, significantly lower than both manual spraying (65.5%, Figure 3a) and the control (62.4%, Figure 3a). Following 7 days, the reduction rate of *B. dorsalis* populations in the manual spraying treatment was 57.4%, whereas the drone spraying treatment had declined to 28.0%, significantly lower than the manual spraying treatment (Figure 3b). In terms of fruit damage rate, although the variance analysis revealed differences, multiple comparisons indicated no significant distinctions among the fruit damage rates of drone spraying, manual spraying and control treatments (Figure 3c).

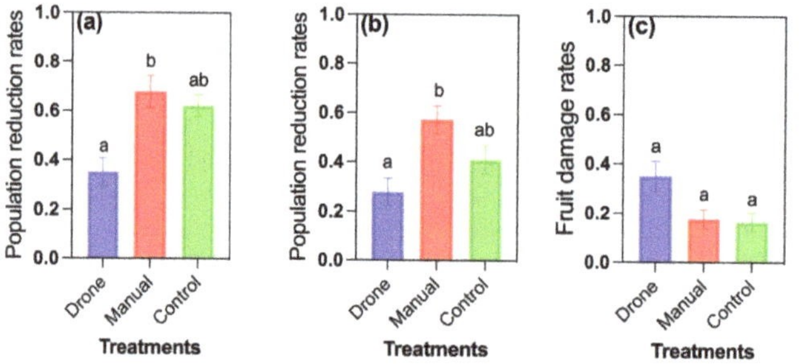


**Figure 3.** The impact of a drone-based push–pull system and a manual-based push–pull system on the population reduction rate of *B. dorsalis* after 3 days (**a**) and 7 days (**b**), as well as the fruit damage rate (**c**), was tested by using Kruskal–Wallis tests. "Drone" refers to the push–pull system utilizing drones, while "Manual" denotes the push–pull system operated manually. "Control" represents the pesticide treatment (bromophos + β-cypermethrin) without using the push–pull strategy. The absence of significant differences between treatments is indicated by the same letters above the distinct bars ($p < 0.05$).

### 3.3. Food Bait vs. ME Bait

The choice of bait (ME bait or food bait) employed to establish the push–pull system significantly affected the population reduction rate (3d: $\chi^2 = 6.41$, df = 2, $p = 0.040$; 7d: $\chi^2 = 8.14$, df = 2, $p = 0.018$) of *B. dorsalis* and fruit damage rate ($\chi^2 = 8.72$, df = 2, $p = 0.006$). After 3 days, the push–pull system established with food bait resulted in a population reduction rate of 45.5% for *B. dorsalis*, significantly lower than the reduction rates of 70.9% in the ME bait system and 63.4% in the control (Figure 4a). After 7 days, the reduction rate of *B. dorsalis* populations in the push–pull system using food bait remained significantly lower than that of the ME bait system (Figure 4b). In terms of the fruit damage rate, both the ME bait system and the control demonstrated notably lower rates compared to the push–pull system using food bait (Figure 4c).

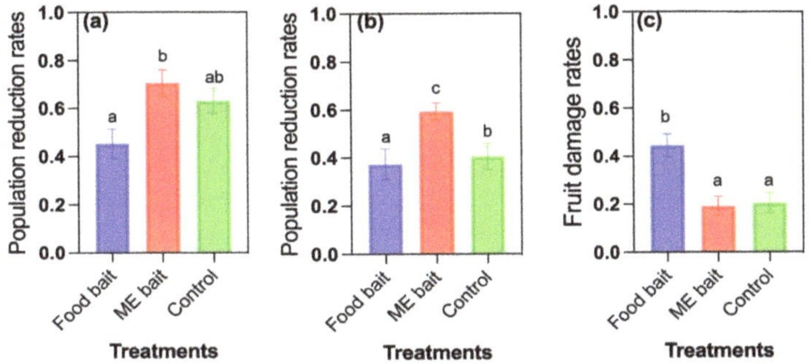


**Figure 4.** The influence of a food bait push–pull system and a ME bait push–pull system on the population reduction rate of *B. dorsalis* after 3 days (**a**) and 7 days (**b**), as well as the fruit damage rate (**c**), was tested by using Kruskal–Wallis tests. "Food bait" refers to the push–pull system employing food bait, while "ME bait" indicates the push–pull system using ME bait. "Control" represents the pesticide treatment (bromophos + β-cypermethrin) without using the push–pull strategy. The absence of significant differences between treatments is denoted by the same letters above the distinct bars ($p < 0.05$).

### 3.4. Two Types of Repellents and Attractants vs. A Single Type

Comparing push–pull systems established with two types of repellents and attractants to the push–pull system established with a single type, no significant impact on the reduction rate of *B. dorsalis* populations was observed after 3 days ($\chi^2 = 4.47$, df = 2, $p = 0.107$; Figure 5a). However, after 7 days, the push–pull systems constructed with two types exhibited higher reduction rates than the single-type system ($\chi^2 = 13.93$, df = 2, $p = 0.001$; Figure 5b), although the differences were not statistically significant. Nevertheless, all three treatments displayed similar fruit damage rates of approximately 20%, with no significant distinctions observed among them ($\chi^2 = 0.08$, df = 2, $p = 0.968$; Figure 5c).

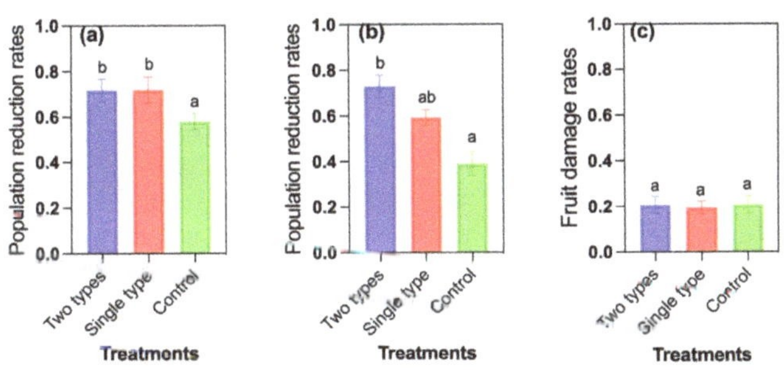


**Figure 5.** The effects of a push–pull system with two types of repellents and attractants and a push–pull system with a single type on the population reduction rate of *B. dorsalis* after 3 days (**a**) and

7 days (**b**), as well as the fruit damage rate (**c**), were tested by using Kruskal–Wallis tests. "Two types" refer to the push–pull system utilizing two types of repellents and attractants, while "Single type" indicates the push–pull system using a single type of repellents and attractants. "Control" represents the pesticide treatment (bromophos + β-cypermethrin) without using the push–pull strategy. The absence of significant differences between treatments is indicated by the same letters above the distinct bars ($p < 0.05$).

## 4. Discussion

The current study, conducted within a rambutan orchard, aimed to compare and evaluate various push–pull strategies for mitigating the outbreak of *B. dorsalis* and minimizing damage to fruit crops. The results indicated that integrating the low-concentration pesticide abamectin into the push–pull system, along with using methyl eugenol (ME) as an attractant instead of food bait and employing manual application, effectively reduced the population of *B. dorsalis* and fruit damage. Additionally, the introduction of a variety of attractants and repellents did not enhance the control's effectiveness compared to systems containing only single attractants and repellents.

The outcomes of this investigation clearly demonstrated an improvement in suppressing the *B. dorsalis* population with the integration of abamectin into the push–pull system. In scenarios without abamectin, the reduction rate of *B. dorsalis* population did not exceed 60%, whereas with abamectin, it surpassed 70%, accompanied by a significant reduction in fruit damage. While prior literature often favored non-toxic practices in push–pull strategies to align with their underlying philosophy [20,32], this investigation provided evidence that incorporating low-concentration compounds like abamectin during pest outbreaks offers a practical and effective approach. This strategy deviated from conventional push–pull usage, which typically focuses on pre-outbreak prevention. However, this study highlighted the need for further research to elucidate the mechanisms underlying abamectin-induced population regulation and to explore potential synergistic effects. To achieve this, subsequent experiments should employ a factorial design with main factors of interest: abamectin (with and without), attractant (ME and food) and repellent (allicin and d-limonene).

In comparing methodologies within the push–pull system, manual spraying emerged as more effective than drone-assisted spraying. While the latter offered benefits such as labor reduction and the ability to reach elevated areas [48,49], its application was less effective in orchards characterized by complex terrains and dense foliage, particularly when addressing the upper canopy layers, where insecticides' penetration is obstructed [50]. Migratory pests such as *B. dorsalis*, which might retreat to the lower canopy, thereby circumvent drone-delivered treatments, presenting an ongoing risk. As a result, drone spraying manifested a suboptimal control efficiency. In contrast, manual spraying, although labor-intensive, allowed for a more targeted dispersion of insecticides throughout the various strata of the canopy, effectively pressing the residual pest population to vacate the orchard, thus, bolstering the overall efficacy of pest control.

The choice of suitable repellents and attractants is crucial when designing a push–pull system. Our study demonstrated the superiority of using methyl eugenol (ME) in constructing the push–pull system for controlling *B. dorsalis*, which was consistent with the findings of Ugwu [51]. One possible reason for this is that the odor of food baits is more easily affected by environmental factors such as vegetation type and tree canopy density compared to ME [52]. Therefore, ME should be preferred in constructing a push–pull system for managing *B. dorsalis*.

While the integration of d-limonene and food bait did not significantly enhance the immediate efficacy of the push–pull system in mitigating *B. dorsalis* infestations, their inclusion was observed to extend control effectiveness for up to 7 days. Notably, even after seven days, the system's ability to control the pest population remained robust, exceeding a 70% threshold. The underlying factors for this sustained effectiveness are yet to be fully elucidated; however, it may be attributed in part to the short-range killing effect of d-limonene on dipteran species, as demonstrated by Showler et al. [53]. Nevertheless, increasing the variety of attractants or repellents inevitably raises labor and material costs. Therefore, if cost efficiency is a concern, employing a single type of attractant or repellent can still yield satisfactory results. However, if a more long-lasting pest control effect is desired, it may be worth considering increasing the variety of attractants and repellents.

In this experiment, the implementation of a push–pull control system proved to be more effective in reducing both the *B. dorsalis* population and fruit damage compared to relying solely on chemical pesticides. This effectiveness may have stemmed from the push–pull system's ability to decrease *B. dorsalis* resistance, as noted by Hassanali et al. [54], through the integration of various control measures from different perspectives. This integration reduced the prolonged exposure of pests to specific control methods, thereby decreasing the frequency of contact and adaptation time to a single control method, ultimately, slowing down the development of resistance [20].

Although we used abamectin in this experiment, the concentration we employed (0.2 ppm) was significantly lower than the recommended concentration. Additionally, individuals not killed by low-concentration abamectin, such as those with higher resistance, are still regulated by the push–pull system, being lured into traps and subsequently killed, thereby reducing the opportunity for resistant individuals to reproduce. Consequently, this approach would aid in slowing the development of pesticide resistance in *B. dorsalis*. From an environmental pollution perspective, using abamectin alone at high concentrations for controlling *B. dorsalis* requires substantial amounts and yields poor control efficacy. Preliminary experiments showed that using abamectin alone at 5 ppm (recommended concentration) resulted in only a 61.44% reduction in the pest population after 7 days. Achieving a higher reduction rate (e.g., exceeding 70%) required combining abamectin with other pesticides like β-cypermethrin (unpublished data). In contrast, utilizing the push–pull system (see Figure 5a) with a low concentration of abamectin (0.2 ppm) achieved over a 70% reduction in pest population. Thus, our push–pull system reduced the need for abamectin and other pesticides, contributing to a reduction in pesticide resistance development and pesticide pollution.

## 5. Conclusions

These series of comparative experiments provided compelling evidence that integrating low-concentration abamectin into the push–pull system, along with using methyl eugenol (ME) as an attractant instead of food bait, combined with manual application, constituted an optimal push–pull strategy for reducing *B. dorsalis* populations and preserving fruit quality during pest outbreaks. Moreover, this integrated approach demonstrated potential in mitigating the development of *B. dorsalis* resistance and reducing pesticide usage, thereby minimizing environmental pollution. However, the push–pull system incurred higher labor and material costs and presented challenges in design and implementation compared to conventional control methods, necessitating professional training and guidance for farmers. Additionally, this optimal system may be more advantageous in mountainous terrains, while the use of drones may be a more ideal option in large, flat areas, emphasizing the need for context-dependent push–pull strategies [20,42].

**Supplementary Materials:** The following supporting information can be downloaded at: https://www.mdpi.com/article/10.3390/agronomy14050890/s1.

**Author Contributions:** Conceptualization, J.W., X.L., Z.C. and R.Y.; Software, J.W.; Investigation, J.W., Z.S. and Y.Z.; Data curation, J.W.; Writing—original draft, J.W., Z.S. and F.C.; Funding acquisition, Z.C. and F.C. All authors have read and agreed to the published version of the manuscript.

**Funding:** The present study was supported by two funding sources: the project "Introduction, Integration, and Demonstration of Green Control Technologies for Longan" (grant ID: ZDYF2019083) and the "Hainan University Research Start-up Fund" (grant ID: RZ2300002725).

**Data Availability Statement:** The data presented in this study are openly available in FigShare at https://doi.org/10.6084/m9.figshare.25676319.v1.

**Acknowledgments:** We would like to express our sincere gratitude and appreciation to the Qixianling Global Fruit Exhibition Center for granting us the opportunity to conduct our experimental research within their orchard. We are sincerely grateful for their generosity and cooperation throughout the duration of this study.

**Conflicts of Interest:** Author Zhifu Cui was employed by the company Hainan State Farms of Academy of Sciences Group Co., Ltd. The remaining authors declare that the research was conducted in the absence of any commercial or financial relationships that could be construed as a potential conflict of interest.

## References

1. Liu, H.; Zhang, D.; Xu, Y.; Wang, L.; Cheng, D.; Qi, Y.; Zeng, L.; Lu, Y. Invasion, expansion, and control of *Bactrocera dorsalis* (Hendel) in China. *J. Integr. Agric.* **2019**, *18*, 771–787. [CrossRef]
2. Boinahadji, A.K.; Coly, E.V.; Diedhiou, C.A.; Sembene, P.M. Oviposition preference and offspring performance of the oriental fruit fly *Bactrocera dorsalis* (Diptera, Tephritidae) on eight host plants. *Int. J. Adv. Res.* **2020**, *8*, 931–937. [CrossRef] [PubMed]
3. Pangihutan, J.C.; Dono, D.; Hidayat, Y. The potency of minerals to reduce oriental fruit fly infestation in chili fruits. *PeerJ* **2022**, *10*, e13198. [CrossRef] [PubMed]
4. Weems, H.V.; Heppner, J.B.; Nation, J.L.; Steck, G.J. Oriental Fruit Fly, Bactrocera dorsalis (Hendel) (Insecta: Diptera: Tephritidae). Available online: https://edis.ifas.ufl.edu/publication/IN240 (accessed on 10 September 2023).
5. USDA. A Review of Recorded Host Plants of Oriental Fruit Fly, Bactrocera dorsalis (Hendel) (Diptera: Tephritidae), Version 3.1. 2016. Available online: https://www.ars.usda.gov/research/publications/publication/?seqNo115=347371 (accessed on 7 October 2023).
6. Choi, K.S.; Samayoa, A.C.; Hwang, S.Y.; Huang, Y.B.; Ahn, J.J. Thermal effect on the fecundity and longevity of *Bactrocera dorsalis* adults and their improved oviposition model. *PLoS ONE* **2020**, *15*, e0235910. [CrossRef] [PubMed]
7. Yu, C.; Zhao, R.; Zhou, W.; Pan, Y.; Tian, H.; Yin, Z.; Chen, W. Fruit fly in a challenging environment: Impact of short-term temperature stress on the survival, development, reproduction, and trehalose metabolism of *Bactrocera dorsalis* (Diptera: Tephritidae). *Insects* **2022**, *13*, 753. [CrossRef] [PubMed]
8. Stephens, A.E.A.; Kriticos, D.J.; Leriche, A. The current and future potential geographical distribution of the oriental fruit fly, *Bactrocera dorsalis* (Diptera: Tephritidae). *Bull. Entomol. Res.* **2007**, *97*, 369–378. [CrossRef]
9. Jaffar, S.; Lu, Y. Toxicity of some essential oils constituents against oriental fruit fly, *Bactrocera dorsalis* (Hendel) (Diptera: Tephritidae). *Insects* **2022**, *13*, 954. [CrossRef]
10. Sim, S.B.; Curbelo, K.M.; Manoukis, N.C.; Cha, D.H. Evaluating *Bactrocera dorsalis* (Hendel) (Diptera: Tephritidae) response to methyl eugenol: Comparison of three common bioassay methods. *J. Econ. Entomol.* **2022**, *115*, 556–564. [CrossRef]
11. Stibick, J.N.L. Natural Enemies of True Fruit Flies (Tephritidae). 2004. Available online: https://www.ippc.int/static/media/uploads/resources/natural_enemies_of_true_fruit_flies.pdf (accessed on 7 October 2023).
12. Garcia, F.R.M.; Ovruski, S.M.; Suárez, L.; Cancino, J.; Liburd, O.E. Biological control of Tephritid fruit flies in the Americas and Hawaii: A review of the use of parasitoids and predators. *Insects* **2020**, *11*, 662. [CrossRef]
13. Manoukis, N.C.; Cha, D.H.; Collignon, R.M.; Shelly, T.E. Terminalia larval host fruit reduces the response of *Bactrocera dorsalis* (Diptera: Tephritidae) adults to the male lure methyl eugenol. *J. Econ. Entomol.* **2018**, *111*, 1644–1649. [CrossRef]
14. Orankanok, W.; Chinvinijkul, S.; Thanaphum, S.; Sitilob, P.; Enkerlin, W.R. Area-wide integrated control of oriental fruit fly *Bactrocera dorsalis* and guava fruit fly *Bactrocera correcta* in Thailand. In *Area-Wide Control of Insect Pests*; Vreysen, M.J.B., Robinson, A.S., Hendrichs, J., Eds.; Springer: Dordrecht, The Netherlands, 2007; pp. 517–526. [CrossRef]
15. Shelly, T.; McInnis, D. Sterile insect technique and control of tephritid fruit flies: Do species with complex courtship require higher overflooding ratios? *Ann. Entomol. Soc. Am.* **2016**, *109*, 1–11. [CrossRef]
16. Vargas, R.I.; Piñero, J.C.; Leblanc, L. An overview of pest species of *Bactrocera* fruit flies (Diptera: Tephritidae) and the integration of biopesticides with other biological approaches for their management with a focus on the Pacific Region. *Insects* **2015**, *6*, 297–318. [CrossRef] [PubMed]
17. Jin, T.; Liang, G.; Zeng, L.; Lu, Y. Detoxification enzymes activities in different *Bactrocera dorsalis* (Hendel) populations and their relationship with the resistant levels. *J. Environ. Entomol.* **2014**, *36*, 58–67.
18. Zayed, M.S.; Taha, E.K.A.; Hassan, M.M.; Elnabawy, E.S.M. Enhance systemic resistance significantly reduces the silverleaf whitefly population and increases the yield of wweet pepper, *Capsicum annuum* L. var. annuum. *Sustainability* **2022**, *14*, 6583. [CrossRef]
19. Nalyanya, G., Moore, C.B.; Schal, C. Integration of repellents, attractants, and insecticides in a "push-pull" strategy for managing German cockroach (Dictyoptera: Blattellidae) populations. *J. Med. Entomol.* **2000**, *37*, 427–434. [CrossRef] [PubMed]
20. Cook, S.M.; Khan, Z.R.; Pickett, J.A. The use of push-pull strategies in integrated pest management. *Annu. Rev. Entomol.* **2007**, *52*, 375–400. [CrossRef] [PubMed]

21. Ahmad, S.; Jaworski, C.C.; Ullah, F.; Jamil, M.; Badshah, H.; Ullah, F.; Luo, Y. Efficacy of lure mixtures in baited traps to attract different fruit fly species in guava and vegetable fields. *Front. Insect Sci.* **2023**, *2*, 984348. [CrossRef] [PubMed]
22. Khan, Z.R.; Midega, C.A.O.; Amudavi, D.M.; Hassanali, A.; Pickett, J.A. On-farm evaluation of the 'push-pull' technology for the control of stemborers and striga weed on maize in western Kenya. *Field Crops Res.* **2008**, *106*, 224–233. [CrossRef]
23. Lang, J.; Chidawanyika, F.; Khan, Z.R.; Schuman, M.C. Ecological chemistry of pest control in push-pull intercropping systems: What we know, and where to go? *Chimia* **2022**, *76*, 906. [CrossRef]
24. Mala, M.; Baishnab, M.; Mollah, M.M.I. Push-pull strategy: An integrated approach to manage insect-pest and weed infestation in cereal cropping systems. *J. Biosci. Agric. Res.* **2020**, *25*, 2122–2127. [CrossRef]
25. Guea, O.G.M.; Castrejón-Ayala, F.; Robledo, N.; Jiménez-Pérez, A.; Sánchez-Rivera, G. Plant selection for the establishment of push-pull strategies for Zea mays-*Spodoptera frugiperda* pathosystem in Morelos, Mexico. *Insects* **2020**, *11*, 349. [CrossRef] [PubMed]
26. Byers, J.A.; Levi-Zada, A. Modelling push-pull management of pest insects using repellents and attractive traps in fruit tree orchards. *Pest Manag. Sci.* **2022**, *78*, 3630–3637. [CrossRef] [PubMed]
27. Khan, Z.R.; Pittchar, J.O.; Midega, C.A.O.; Pickett, J.A. Push-pull farming system controls fall armyworm: Lessons from Africa. *Outlooks Pest Manag.* **2018**, *29*, 220–224. [CrossRef]
28. Pyke, B.; Rice, M.; Sabine, B.; Zalucki, M.P. The push-pull strategy-behavioral control of Heliothis. *Aust. Cotton Grow.* **1987**, *9*, 7–9.
29. Martel, J.W.; Alford, A.R.; Dickens, J.C. Synthetic host volatiles increase efficacy of trap cropping for management of Colorado potato beetle, *Leptinotarsa decemlineata* (Say). *Agric. For. Entomol.* **2005**, *7*, 79–86. [CrossRef]
30. Zhang, R.L.; Chen, D.; Liu, J.; Zhuang, G.F.; Zhang, Z. The use of push-pull strategy in medical pests integrated management. *Chin. J. Vector Biol. Control* **2016**, *6*, 624–628.
31. Verghese, A.; Mouly, R.; Shivananda, T.; Soumya, C.; Rashmi, M. A push-pull strategy for the management of the Oriental fruit fly, *Bactrocera dorsalis* (Hendel) in mango. *Pest Manag. Hortic. Ecosyst.* **2021**, *26*, 269–271.
32. Adefegha, S.A.; Ogunsuyi, O.B.; Oboh, G. Purple onion in combination with garlic exerts better ameliorative effects on selected biomarkers in high-sucrose diet-fed fruit fly (*Drosophila melanogaster*). *Comp. Clin. Pathol.* **2020**, *3*, 713–720. [CrossRef]
33. Kayode, O.T.; Rotimi, D.; Emmanuel, F.; Iyobhebhe, M.; Kayode, A.A.A.; Ojo, A.O. Contraceptive and biochemical effect of juice extract of *Allium cepa*, *Allium sativum*, and their combination in Canton fruit flies. *J. Food Biochem.* **2021**, *18*, e13821. [CrossRef]
34. Keita, N.D.; Sarr, A.G.R.J.; Diome, T.; Cissokho, B. Study of the effects of biological extracts of "heals everything" *Gymnanthemum amygdalinum* and garlic *Allium sativum* on fruit flies *Bactrocera dorsalis* and their stage L3 larvae. *J. Entomol. Zool. Stud.* **2024**, *12*, 38–43. [CrossRef]
35. Wen, T.; Sang, M.; Wang, M.; Han, L.; Gong, Z.; Tang, X.; Long, X.; Xiong, H.; Peng, H. Rapid detection of d-limonene emanating from citrus infestation by *Bactrocera dorsalis* (Hendel) using a developed gas-sensing system based on QCM sensors coated with ethyl cellulose. *Sens. Actuators B Chem.* **2021**, *328*, 129048. [CrossRef]
36. De Souza, R.B.; Guimarães, J.R. Effects of avermectins on the environment based on its toxicity to plants and soil invertebrates-a review. *Water Air Soil Pollut.* **2022**, *233*, 259. [CrossRef] [PubMed]
37. Huang, Y.; Hong, Y.; Huang, Z.; Zhang, J.; Huang, Q. Avermectin induces the oxidative stress, genotoxicity, and immunological responses in the Chinese mitten crab, *Eriocheir sinensis*. *PLoS ONE* **2019**, *14*, e0225171. [CrossRef] [PubMed]
38. Pirasath, S.; Nageswaran, B.; Karunasena, R.P.V.; Gevakaran, M. Acute abamectin toxicity: A case report. *Toxicol. Commun.* **2021**, *5*, 66–68. [CrossRef]
39. Qiu, D.; Xu, N.; Zhang, Q.; Zhou, W.; Wang, Y.; Zhang, Z.; Yu, Y.; Lu, T.; Sun, L.; Zhou, N.Y.; et al. Negative effects of abamectin on soil microbial communities in the short term. *Front. Microbiol.* **2022**, *5*, 1053153. [CrossRef] [PubMed]
40. Biasazin, T.D.; Wondimu, T.W.; Herrera, S.L.; Larsson, M.; Mafra-Neto, A.; Gessese, Y.W.; Dekker, T. Dispersal and competitive release affect the management of native and invasive Tephritid fruit flies in large and smallholder farms in Ethiopia. *Sci. Rep.* **2021**, *11*, 2690. [CrossRef] [PubMed]
41. Silva, M.A.; Bezerra-Silva, G.C.D.; Mastrangelo, T. The host marking pheromone application on the management of fruit flies-A review. *Braz. Arch. Biol. Technol.* **2012**, *55*, 835–842. [CrossRef]
42. Alkema, J.T.; Dicke, M.; Wertheim, B. Context-dependence and the development of push-pull approaches for integrated management of *Drosophila suzukii*. *Insects* **2019**, *10*, 454. [CrossRef]
43. Bharathi, T.E.; Sathiyanandam, V.K.R.; David, P.M.M. Attractiveness of some food baits to the melon fruit fly, *Bactrocera cucurbitae* (Coquillett) (Diptera: Tephritidae). *Int. J. Trop Insect Sci.* **2004**, *24*, 125–134. [CrossRef]
44. Ekesi, S.; Mohamed, S.; Tanga, C.M. Comparison of food-based attractants for *Bactrocera invadens* (Diptera: Tephritidae) and evaluation of Mazoferm-Spinosad bait spray for field suppression in mango. *J. Econ. Entomol.* **2014**, *107*, 299–309. [CrossRef]
45. Hasnain, M.; Saeed, S.; Naeem-Ullah, U.; Ullah, S. Development of synthetic food baits for mass trapping of *Bactrocera zonata* S. (Diptera: Tephritidae). *J. King Saud Univ. Sci.* **2022**, *34*, 101667. [CrossRef]
46. Ono, H.; Hee, A.K.; Jiang, H. Recent advancements in studies on chemosensory mechanisms underlying detection of semiochemicals in Dacini fruit flies of economic importance (Diptera: Tephritidae). *Insects* **2021**, *12*, 106. [CrossRef] [PubMed]
47. Kruskal, W.H.; Wallis, W.A. Use of ranks in one-criterion variance analysis. *JASA* **1952**, *47*, 583–621. [CrossRef]
48. Abbas, A.; Zhang, Z.; Zheng, H.; Alami, M.M.; Alrefaei, A.F.; Abbas, Q.; Naqvi, S.A.H.; Rao, M.J.; Mosa, W.F.A.; Hussain, A.; et al. Drones in plant disease assessment, efficient monitoring, and detection: A way forward to smart agriculture. *Agronomy* **2023**, *13*, 1524. [CrossRef]

49. Shahrooz, M.; Talaeizadeh, A.; Alasty, A. Agricultural spraying drones: Advantages and disadvantages. In Proceedings of the 2020 Virtual Symposium in Plant Omics Sciences (OMICAS), Bogotá, Colombia, 23–27 November 2020; pp. 1–5.
50. Chen, P.; Douzals, J.P.; Lan, Y.; Cotteux, E.; Delpuech, X.; Pouxviel, G.; Zhan, Y. Characteristics of unmanned aerial spraying systems and related spray drift: A review. *Front. Plant Sci.* **2022**, *13*, 870956. [CrossRef]
51. Ugwu, J.A. Evaluation of food lures for oriental fruit fly *Bactrocera dorsalis* (Diptera: Tephritidae) trapping on *Chrysophyllum albidum* in Ibadan, Nigeria. *J. For. Res. Manag.* **2020**, *17*, 72–81.
52. Toukem, N.K.; Yusuf, A.A.; Dubois, T.; Abdel-Rahman, E.M.; Adan, M.S.; Mohamed, S.A. Landscape vegetation productivity influences population dynamics of key pests in small avocado farms in Kenya. *Insects* **2020**, *11*, 424. [CrossRef]
53. Showler, A.T.; Harlien, J.L.; de Léon, A.A.P. Effects of laboratory grade. limonene and a commercial limonene-based insecticide on *Haematobia irritans* (Muscidae: Diptera): Deterrence, mortality, and reproduction. *J. Med. Entomol.* **2019**, *56*, 1064–1070. [CrossRef]
54. Hassanali, A.; Herren, H.; Khan, Z.R.; Pickett, J.A.; Woodcock, C.M. Integrated pest management: The push-pull approach for controlling insect pests and weeds of cereals, and its potential for other agricultural systems including animal husbandry. *Philos. Trans. R. Soc. B* **2008**, *363*, 611–621. [CrossRef]

agronomy

MDPI

*Review*

# From Surveillance to Sustainable Control: A Global Review of Strategies for Locust Management

**Christina Panopoulou and Antonios Tsagkarakis ***

Laboratory of Sericulture and Apiculture, Agricultural University of Athens, 11855 Athens, Greece; stud118114@aua.gr
* Correspondence: atsagarakis@aua.gr; Tel.: +30-2105294578

**Abstract:** Locusts represent a persistent global agricultural pest, responsible for significant crop losses and socio-economic repercussions. The initiation of chemical control measures dates back to the late 19th century, with the use of poisoned baits, before advancing in the mid-20th century with the introduction of organochlorines, such as dieldrin. Despite their efficacy, the associated environmental, ecological, and human health risks led to the prohibition of dieldrin by the United States and the FAO by 1988. The demand for insecticides with reduced persistence and toxicity prompted the establishment of international organizations to coordinate locust research and management. In recent decades, chemical control has transitioned towards compounds with diminished persistence and selective agents. Concurrently, research has progressed in the development of bioinsecticides, notably *Metarhizium acridum*, and has reinforced preventive strategies. Emerging technologies, including remote sensing and machine learning, have facilitated early monitoring and predictive modeling, thereby enhancing outbreak forecasting. These tools support proactive, targeted interventions and are consistent with Integrated Pest Management principles, promoting more sustainable and ecologically responsible locust control strategies.

**Keywords:** locust control; preventive management; biological control; remote sensing; early warning systems

## 1. Introduction

Locusts represent a specific subset of grasshoppers within the Acrididae family [1–3], distinguished by their ability to undergo density-dependent phase polyphenism, a phenomenon initially described by Uvarov (1921) [4–6]. Under favorable environmental conditions [1,2], locusts transition from a solitary phase to a gregarious one, forming highly mobile hopper bands and swarms capable of consuming their body weight in vegetation daily and traveling hundreds of kilometers [1,2,7,8]. Unlike most orthopterans, these species exhibit polyphagous feeding behavior during outbreaks, targeting a wide array of crops and wild plants [2,3,8,9], thereby ranking among the most destructive insect pests globally [1,10,11]. Of the approximately twelve recognized locust species, the migratory locust (*Locusta migratoria*), desert locust (*Schistocerca gregaria*), Moroccan locust (*Dociostaurus maroccanus*), and Italian locust (*Calliptamus italicus*) are the most economically significant [1,8,10,11]. These species thrive in arid and semi-arid environments where alternating droughts and rainfall promote breeding and gregarization [1,2,8,11]. Historical records of locust plagues span Europe, Africa, and Asia, leading to recurrent food insecurity and socioeconomic disruption [1–3,12]. In the past two decades, significant events such as the 2003–2005 West African desert locust plague, which impacted over 8 million people [1,8,13] and incurred costs exceeding $400 million in control operations [14], and the 2019–2021 East African upsurge, one of the most severe in recent history, have underscored the persistent global threat posed by these insects [3,12].

Over the past century, strategies for managing locust populations have evolved significantly, transitioning from initial mechanical methods to the extensive use of chemical spraying in the mid-20th century [1]. Although chemical insecticides continue to be the primary method of control [1,2,11], they are predominantly reactive and present substantial risks to human health, biodiversity, and ecosystems [1–3,15]. These limitations have prompted a gradual shift towards integrated pest management, incorporating the operational use of biocontrol agents such as *Metarhizium acridum*, as well as the implementation of new technologies such as remote sensing technology, unmanned aerial vehicles (UAVs), and machine learning to improve forecasting and early warning systems [2,11,16].

Despite these advancements, most control campaigns remain reactive [1,2], initiated only when swarms pose a direct threat to crops [17–19]. Preventive control, as originally conceptualized by Boris Uvarov, is often misconstrued as synonymous with "outbreak prevention" [20–23]. In practice, preventive control involves early interventions in outbreak areas to safeguard major agricultural zones, whereas complete outbreak prevention is seldom feasible [5,11,22,24,25]. This misunderstanding, along with financial, logistical, and political constraints, has hindered the consistent adoption of preventive strategies [5,6,19,26,27].

This review critically examines the evolution of locust management, from historical mechanical and chemical approaches to contemporary biological and technological innovations. By integrating historical and recent evidence, we highlight both key advances and persistent challenges, particularly the difficulty of transitioning from reactive crisis responses to sustainable preventive control. We argue that effective locust management in the 21st century must integrate biocontrol, climate- and AI-informed forecasting, and strong institutional cooperation into a coherent preventive framework that safeguards food security in vulnerable regions.

## 2. Literature Search and Selection Strategy

This article is organized as a narrative review that integrates research on locust management strategies, with a specific emphasis on methodologies endorsed or assessed by the Locust Pesticide Referee Group (LPRG) of the Food and Agriculture Organization (FAO). Although it does not adhere to a strict systematic review protocol, efforts were undertaken to ensure transparency and comprehensiveness in identifying pertinent studies.

Search Databases and Keywords

Between 10 February and 12 September 2025, a comprehensive literature search was systematically conducted across prominent scientific databases, including Google Scholar, ResearchGate, ScienceDirect, Scopus, and Web of Science. The search strategy employed a combination of keywords such as:

*"locust control*";*

*"integrated locust management";*

*"biological control of locusts";*

*"chemical control of locusts";*

*"locust monitoring";*

*"outbreak prevention";*

*"remote sensing".*

We also used species-specific terms, such as *Locusta migratoria*, *Schistocerca gregaria*, and *Dociostaurus maroccanus*.

Inclusion and Exclusion Criteria

Eligible sources comprised peer-reviewed articles, review papers, technical reports, and guidelines from international organizations such as the FAO, EFSA, and EPA. Excluded from consideration were studies not available in English, conference abstracts lacking full texts, and publications that did not directly pertain to locust management. An initial pool of 112 records underwent title and abstract screening, followed by a full-text evaluation, resulting in a final selection of sources that informed this review.

Data Extraction and Synthesis

Data were synthesized narratively rather than quantitatively, emphasizing

- The historical evolution of control strategies;
- The efficacy and ecological impacts of chemical, biological, and preventive methods;
- The role of emerging technologies such as biocontrol, remote sensing technology, UAVs, and AI-driven forecasting.

## 3. The Institutionalization of Locust Control: Historical Milestones and Global Cooperation

Locusts have been acknowledged as significant agricultural pests since ancient times [1,10,11]; however, international initiatives to formalize their control commenced only in the early 20th century. In 1905, Jules Künckel d'Herculais orchestrated one of the initial international campaigns in Algeria [4,6,26], culminating in the 1916 global locust report by the International Institute of Agriculture, which underscored the pressing necessity for cross-border collaboration [4]. The inaugural international conference on locust control, convened in Rome in 1920, formally signified the commencement of institutionalized management [4,26,28].

During this period, pivotal scientific advancements were achieved. In 1921, Boris Uvarov identified phase polyphenism [4–6], a groundbreaking discovery that revolutionized locust research and continues to underpin management strategies [29]. By the 1930s, cooperation expanded through a series of international conferences (Rome 1931, Paris 1932, London 1934, Cairo 1936, Brussels 1938), which highlighted the necessity for global action [4,6,17,24].

Institutional frameworks soon emerged. France established the Comité d'Etudes de la Biologie des Acridiens in Algeria (1932) [4,6], which evolved into the ONAA (1943) [6,30], while Britain founded the Anti-Locust Research Centre (ALRC) in 1945 [6,24,31], later reorganized into several development institutes before becoming today's Natural Resources Institute [6]. The establishment of the FAO in 1945 was a pivotal moment, providing a permanent platform for global coordination [6,26]. Major initiatives included the Desert Locust Control Committee (DLCC, 1954), the Desert Locust Information Service (DLIS, 1978) [5,6], and the Emergency Prevention System (EMPRES, 1994), which emphasized early warning and preventive action across Africa and the Near East [6,28,32,33].

More recently, regional commissions such as CLCPRO (2000) have reinforced collaboration among frontline states [26,28]. Today, FAO-led systems integrate satellite imaging, AI-driven models, drones, and GIS platforms for real-time monitoring and forecasting. These advancements represent the culmination of a century-long institutional trajectory [5,6,13,26,28].

Despite these accomplishments, international coordination remains limited [6]. Control programs in numerous affected countries are hindered by inadequate funding, fragile governance, and insufficient preparedness, which delay responses [6,11,17,34]. Donor funding remains cyclical—peaking during plagues but diminishing during economic downturns—thereby undermining the long-term preventive ethos originally envisioned by Uvarov [6,10,17].

Furthermore, the uneven adoption of modern technologies such as satellite-based early warning systems, UAVs, and predictive modeling highlights capacity disparities between well-resourced and resource-poor countries. Political instability and bureaucratic inertia further impede effective cross-border coordination [6,17].

Critically, while the institutional framework for locust control has become increasingly sophisticated, its effectiveness still relies more on sustained political will and financing than on technical capacity. Without addressing these structural bottlenecks, global locust control remains susceptible to the same reactive cycles documented for over a century.

## 4. Locust Control Approaches

The management of locust populations has traditionally been categorized into two primary strategies: the reactive approach and the preventive approach [18,23]. Although recent literature has introduced the term "proactive" as an intermediate concept, this has frequently led to confusion [21,23,35]. Consistent with Uvarov's original framework and the prevailing consensus among locust specialists [5,22,24,25], we identify only two principal approaches—reactive and preventive—while elucidating the limitations of "outbreak prevention" as a distinct concept [21–23].

### *4.1. Reactive Approach*

The reactive approach pertains to interventions initiated only after swarming has occurred and substantial populations are already in motion. This approach characterized much of the 20th century and continues to be the standard response during emergency plague situations [17–19,23]. Typically, reactive operations involve large-scale aerial or ground spraying of broad-spectrum insecticides over extensive infested areas [1,2,11], necessitating considerable financial resources and logistical coordination [5,10,27,36,37]. The efficacy of such operations in mitigating immediate crop damage has been consistently demonstrated; however, they are accompanied by several significant drawbacks: high operational costs [1,2,14], reliance on international donor funding [6,10,17,34,36], risks to human health and non-target biodiversity, and the challenge of containing highly mobile swarms that may traverse political borders within days [1–3,15]. Consequently, reactive campaigns are frequently criticized as being untimely, costly, and environmentally unsustainable, although they remain indispensable when early control efforts fail or are not implemented in a timely manner [2,14,16].

### *4.2. Preventive Approach*

Preventive control, as initially conceptualized by Uvarov in the early 20th century, represents a strategic shift from reactive crisis management to proactive early intervention [5,22,23,37]. The core principle of this approach is the timely identification and management of nascent outbreaks in established breeding areas to prevent their escalation into widespread plagues [5,22,24,25]. It is crucial to distinguish preventive control from the absolute prevention of outbreaks, as outbreaks are an inherent aspect of locust ecology and cannot be entirely averted in practice [5,20,21,23]. Instead, as emphasized by Uvarov, the objective is to manage outbreaks and safeguard key agricultural regions through ongoing monitoring, forecasting, and swift localized interventions [5,22,23,37].

The practical application of preventive control is founded on three main components: (1) early warning systems that integrate field surveys with meteorological and remote-sensing data to detect breeding conditions [6,38–41]; (2) institutionalized monitoring and coordination, often facilitated by the FAO or regional organizations [4–6,17,27,31]; and (3) targeted interventions, typically employing insect growth regulators or biological agents, applied in barrier treatments rather than extensive blanket spraying [1,5,6,11,39,40,42,43]. This model has been progressively institutionalized since the mid-20th century and is regarded by many experts as the "best practice" in locust management [11].

The efficacy of preventive control has varied considerably across different locust species and regions. For species with relatively confined outbreak areas, such as the red locust (*Nomadacris septemfasciata*) in southern Africa, the African migratory locust (*Locusta migratoria migratorioides*) in the Sahel, and the South American locust (*Schistocerca cancellata*), preventive measures have often been successful in preventing plagues for extended durations, sometimes spanning decades. In these instances, the ability to focus surveillance and interventions within limited outbreak zones significantly enhances the likelihood of success [10,36,37].

In contrast, for more widespread and highly mobile species such as the desert locust (*Schistocerca gregaria*) and the Australian plague locust (*Chortoicetes terminifera*), preventive control measures can only partially mitigate outcomes. Although preventive programs have succeeded in reducing the size, duration, and frequency of plagues in these species, complete prevention remains elusive. Environmental variability, extensive breeding ranges, and the rapid mobility of swarms often exceed the capabilities of even well-funded surveillance systems [10,36,37].

Thus, distinguishing between preventive control and outbreak prevention is crucial. While outbreak prevention has occasionally been proposed as the ultimate objective, it has proven unattainable in practice, particularly for species with continental distributions. Misunderstandings of these terms have led to debates regarding program effectiveness and resource allocation [5,20–23]. As Hunter, 2024 emphasizes, the true measure of success lies not in preventing outbreaks entirely, but in mitigating their impact and protecting major agricultural areas through accurate forecasting and timely, targeted interventions [11].

Over recent decades, preventive control has largely met this benchmark in numerous regions. Uvarov's original vision—safeguarding key agricultural zones from devastating plagues—has been realized through the establishment of early warning systems, regionally coordinated monitoring, and increasingly sustainable intervention strategies [3,5,13,23,40]. Nonetheless, challenges persist, particularly the reliance on organophosphate insecticides, the underutilization of biological agents [23,25,26,41,44], and the susceptibility of preventive programs to funding shortages and political instability [6,17,37].

The dual framework of reactive versus preventive control remains the most effective conceptual model for locust management [37]. While reactive measures will continue to be necessary during crisis situations [1,2,17–19,23,37], preventive control represents the most sustainable approach, contingent upon adequate resourcing, scientific support, and institutional integration [6,17–19,23,36,37]. Although outbreak prevention may be an aspirational goal, in practice, the true measure of success lies in impact mitigation and the protection of agriculture, rather than the ecological suppression of the outbreaks themselves [5,20–23].

## 5. Preventive Methods

The foundation of modern preventive locust control strategies can be traced back to 1921, when Uvarov introduced the phase polyphenism theory to elucidate the recession and upsurge of locust populations [22,23,45]. This theory significantly enhanced the understanding of locust biology and laid the groundwork for the development of effective preventive measures [5,22,23]. Over a decade later, Uvarov proposed the original plague prevention strategy to avert crop damage in the principal agricultural regions of Africa, the Near East, Iran, and Indo-Pakistan [5,22,24,25]. According to this original preventive strategy, crop damage could be mitigated by managing hopper bands and swarms in outbreak areas [5,22,24,25]. Thus, preventive control and outbreak prevention should not be conflated [22]. On one hand, preventive control methods aim to reduce crop damage, as Uvarov proposed [20,22,24]. On the other hand, outbreak prevention endeavors to maintain locust population densities at a minimal level indefinitely [21–23].

Historically, preventive campaigns relied on long-residual insecticides such as dieldrin, which remained active in the soil for months and effectively suppressed hopper development. However, increasing awareness of their environmental persistence and ecological impacts led to their prohibition [5,46]. This shift created a critical gap: while the principle of preventive control remains valid, its implementation has become increasingly challenging without comparable alternatives [5].

Contemporary preventive strategies prioritize early detection, targeted biocontrol, and environmentally sustainable interventions. Technological advancements in satellite imaging, remote sensing, unmanned aerial vehicles (UAVs), and machine-learning models have significantly enhanced the precision in identifying potential outbreak zones [3,5,13,23,47]. Biopesticides, such as *Metarhizium acridum*, offer operationally viable and safer alternatives [48–51]; however, their limited persistence, elevated costs, and logistical challenges, including cold-chain storage and slower kill rates, constrain their scalability in preventive programs [26,41,51,52].

Despite its recognition as the most sustainable approach to locust management, preventive control remains the least consistently implemented strategy. The primary obstacles are not rooted in scientific knowledge but rather in structural and operational limitations [5,19,26]. Financial instability, characterized by donor-driven funding cycles that dissipate between plagues, undermines long-term preparedness [6,17,34]. Logistical constraints, such as the necessity for year-round surveillance, rapid deployment capacity, and cold-chain infrastructure for biopesticides, render implementation particularly challenging in resource-limited regions [6, 11,17]. Technical bottlenecks also persist: although biopesticides and precision-monitoring tools are available, their adoption in the field is impeded by formulation, storage, and cost barriers [6, 41,44]. Furthermore, political and institutional fragility, especially in frontline countries, restricts cross-border cooperation and delays coordinated action [6,17,37]. Consequently, preventive control remains more of an aspirational framework than an operational reality, with most countries continuing to rely on proactive or reactive strategies. Bridging this gap necessitates sustained financing, international commitment, and the translation of technological advances into field-ready solutions [5,27,37].

## 6. Mechanical Control Methods

Mechanical control constitutes the earliest method of locust management, preceding the advent of chemical and biological strategies [25]. Historically, communities employed direct physical interventions to mitigate damage from hopper bands and swarms [25]. Common techniques included excavating trenches to trap and bury hoppers, incinerating infested vegetation [25,33,51,53,54], and utilizing manual tools to crush or disorient locusts [24,25,33]. In certain regions, specialized collection machinery was introduced [25,54], while other practices involved the use of fire, smoke, or loud noise to repel swarms [24,35,53–55].

Paraffin-based sprays were occasionally applied to impair locust mobility and reproduction [54], while the destruction of egg pods through plowing or digging was employed as a preventive measure [24], albeit often with limited success [24,25]. Other localized measures included protective nets treated with natural repellents, such as garlic or neem oil, to safeguard nurseries and small agricultural plots [33,53,56].

Although mechanical methods were indispensable prior to the pesticide era, they are labor-intensive, locally constrained, and unsuitable for large-scale infestations [9,57]. Their historical significance lies in providing immediate protection to farmers and laying the groundwork for subsequent, more advanced control practices. Presently, they retain value primarily for smallholder farmers and localized outbreaks, particularly in contexts where chemical or biological options are unavailable or unaffordable. However, their scalability and sustainability are severely limited, precluding them from substituting modern integrated approaches. Instead, their relevance is best understood as complementary measures within broader Integrated Pest Management (IPM) strategies, where community-based action can reinforce more systematic interventions [1,2,25,33,37].

## 7. Chemical Control

Since the mid-20th century, chemical insecticides have constituted the primary strategy for locust management, facilitating rapid and extensive suppression of outbreaks [25,26,44]. Initial approaches utilized poisoned baits containing arsenic or organochlorines; however, these methods were found to be either inefficient or environmentally persistent [24,51,54,58]. The introduction of dieldrin in the 1950s marked a significant advancement in locust control due to its prolonged residual activity, which enabled effective "barrier spraying" against hopper bands [25,31,54]. Nevertheless, its persistence, bioaccumulation, and toxicity to wildlife [54,59] resulted in international prohibitions by the 1980s [40].

In response, the Food and Agriculture Organization (FAO) established the Locust Pesticide Referee Group (LPRG) in 1989 [41,47] to assess insecticides for efficacy, safety, and environmental impact [41,47]. This group formulated standardized guidelines, emphasizing barrier treatments over blanket applications and advocating for the adoption of less toxic alternatives, such as insect growth regulators (IGRs) and biological pesticides [32,43,46,48–50,60–63]. Despite these initiatives, the majority of operational campaigns continue to rely on neurotoxic insecticides, including organophosphates and pyrethroids, primarily due to their availability and cost-effectiveness [43].

*Main Pesticide Groups*

Organochlorines, including compounds such as DDT, dieldrin, aldrin, and BHC, were the initial synthetic insecticides extensively employed in locust control from the late 1940s onwards [25,54]. Their prolonged persistence and stomach-poison action facilitated barrier treatments and large-scale aerial spraying, thereby revolutionizing control campaigns [25,31,51,53,54,64]. However, due to their bioaccumulation, ecological toxicity, and associations with human health risks [41,65], widespread bans were initiated in the 1970s [25,54], with the FAO officially prohibiting the use of dieldrin in 1988 [40].

The Locust Pesticide Referee Group (PRG), established by the FAO in 1989 [41,47] and renamed LPRG in 2021 [41], systematically reviewed insecticides for efficacy and environmental safety [41,47]. Over time, it approved organophosphates, pyrethroids, carbamates, insect growth regulators, fipronil, and the biopesticide *Metarhizium acridum*, gradually guiding locust management towards safer compounds. Despite this progress, campaigns continue to rely heavily on organophosphates and pyrethroids due to their cost-effectiveness, and few new products have been introduced in recent decades [32,43,46,48–50,60–63].

Organophosphates (OPs), such as fenitrothion, malathion, and chlorpyrifos, emerged as the primary substitutes for organochlorines [25,61] and remain the cornerstone of modern locust control [43]. These compounds function as fast-acting cholinesterase inhibitors, providing high efficacy in both hopper and adult stages [41]. They are widely available and relatively inexpensive, making them particularly valuable for large-scale emergency campaigns [43]. However, they also pose acute toxicity risks to applicators, pollinators, and aquatic organisms, and chronic exposure has been linked to human health effects [41,48,66–68]. Increasing regulatory restrictions and public concern underscore the urgent need for safer alternatives in future control programs [41,69,70].

Pyrethroids, including synthetic variants such as deltamethrin and lambda-cyhalothrin, are preferred for emergency spraying due to their rapid knockdown effect on locust populations [62, 71]. These compounds are frequently applied in ultra-low-volume (ULV) formulations, which allow for effective coverage with minimal quantities of active ingredient [25,26]. Despite their operational appeal, the high toxicity of pyrethroids to non-target insects, particularly bees and other pollinators, as well as their detrimental effects on aquatic organisms, poses significant challenges to their long-term sustainability [49,50,71]. Additionally, sublethal effects on insect physiology and the development of resistance in certain pest species further exacerbate these concerns [72]. Nonetheless, pyrethroids remain among the most commonly utilized compounds in reactive locust control efforts globally [41,71].

Insect Growth Regulators (IGRs), specifically benzoylurea-based IGRs, have been endorsed over the years for locust control strategies by the LPRG [62]. The insecticidal action of IGRs is predicated on the inhibition of chitin synthesis [41,64], rendering them effective solely for hopper control, as adult locusts are not significantly impacted [51,62]. Due to their prolonged foliar persistence, IGRs have been deemed suitable for barrier spray treatments since 1994 [23,62,73]. However, their slow mode of action renders them inadequate for emergency responses during large swarms [63], and regulatory prohibitions in the European Union have further restricted their availability [41,74]. Despite these limitations, IGRs continue to play a crucial role in preventive control, particularly when employed in targeted barrier treatments during the early stages of outbreaks [41].

Phenylpyrazoles: Fipronil, a phenylpyrazole insecticide, has demonstrated remarkable efficacy against locusts at minimal dosages, rendering it a promising candidate for barrier treatments [32,47,63,75]. However, research has underscored its significant toxicity to aquatic arthropods, soil invertebrates, and other non-target organisms, thereby raising substantial ecological concerns [47–50]. Consequently, its operational application has been confined to limited use in non-crop areas [43,47,49,50]. Although fipronil remains on the FAO's list of recommended insecticides for locust control, its role has been considerably diminished as international agencies increasingly prioritize environmental safety and sustainability. Its future utilization is likely to decline further unless new, safer formulations become available [41,74].

Neonicotinoids: Compounds such as imidacloprid and thiamethoxam were evaluated for locust management during the late 1990s and early 2000s [48], initially showing promise due to their systemic action and relatively novel mode of activity [49,76]. However, evidence from both field and ecotoxicological studies has revealed risks to pollinator health, particularly affecting honeybees and wild bees, as well as concerns regarding soil and water contamination [41,76]. As a result, neonicotinoids were never widely adopted for locust control and remain excluded from the FAO's recommended list of operational insecticides [49]. Their use has largely been abandoned in favor of safer compounds [76].

Spinosad: Spinosad, derived from the soil actinomycete Saccharopolyspora spinosa [77], was first evaluated for locust control in 2004 as a bio-derived alternative with a favorable ecotoxicological profile [49]. It acts primarily on the nicotinic acetylcholine receptor, offering rapid knockdown with low mammalian toxicity [49,77]. Field trials have shown encouraging results in terms of efficacy and safety, but its integration into large-scale operational campaigns has so far been limited [50]. Despite spinosad not being approved as a control agent against locusts, the LPRG recommended further efficacy trials in 2023 [43].

The progression of chemical control strategies in locust management illustrates an ongoing equilibrium between effectiveness, cost-efficiency, and sustainability [23,40]. Organochlorines, such as dieldrin, achieved near-complete control but were discontinued due to their ecological toxicity [40,54]. Although organophosphates and pyrethroids are cost-effective and efficient, they continue to pose health and environmental concerns and are subject to increasing regulatory constraints [41,43,48–50,66–68]. Insect growth regulators (IGRs), fipronil, and spinosad represent the potential of more selective compounds; however, their adoption has been hindered by slow action, ecological side effects, or high costs [41,43,47,49,50,62,74,75,78]. The Food and Agriculture Organization's Locust and Other Migratory Pests Group (LPRG) has played a crucial role in advocating for safer and more sustainable options [28,32,41,43,48–50,60–63]. Nonetheless, the absence of new compounds under evaluation underscores a concerning reliance on outdated chemistries [43]. Ultimately, while chemical control remains essential in crisis response, excessive dependence on broad-spectrum insecticides perpetuates ecological risks [41,43]. Future initiatives must prioritize investment in safer alternatives, large-scale validation of bio-insecticides, and the integration of chemical tools into preventive, ecologically sound frameworks.

## 8. Biological Control

Biological control has gained recognition as a sustainable alternative to chemical pesticides, presenting reduced risks to ecosystems, non-target organisms, and human health [16,27,44,52,79]. Since the 1990s, the Food and Agriculture Organization (FAO) has advocated for research into biological agents [62], leading to the inclusion of *Metarhizium acridum* in the Locust and Other Migratory Pests Group (LPRG)'s list of approved control agents [43,78].

*Main Biological Approaches*

Natural enemies: Locust eggs, nymphs, and adults are subject to predation and parasitism by a variety of organisms, including birds, mammals, mites, and hymenopteran wasps such as *Scelio* spp., with parasitism rates occasionally surpassing 25%. Despite this, their effectiveness in curbing large-scale locust plagues is generally inadequate [68].

Botanical insecticides: Extracts from neem (*Azadirachta indica*), garlic, and cumin have demonstrated repellent and toxic properties [26,55,62,80], yet inconsistencies in their composition and delayed action reduce their dependability in field applications [26,62].

Semiochemicals: Pheromone-based methods hold promise for monitoring or disrupting locust aggregation, with compounds such as phenylacetonitrile being explored [26].

Entomopathogenic microorganisms: Since 1994, *Metarhizium flavoviride* has been identified as a promising biological control agent against locust hoppers and adults in small-scale settings, particularly in ecologically sensitive regions [32,47,62,68,78,81]. Since 1999, it has been referred to as *Metarhizium anisopliae* var. *acridum*, and IT is currently known as *Metarhizium acridum* [48]. This fungus has emerged as a leading biopesticide, achieving 80–90% mortality within 2–3 weeks under optimal conditions and exhibiting strong environmental safety [26,41]. Commercial products like Green Muscle® are now being utilized in certain areas [6,41,43,81,82]. Moreover, microsporidia (*Nosema locustae*), bacteria (*Serratia marcescens*, *Bacillus thuringiensis*), and viruses (Entomopoxvirus) have shown experimental success [68], but issues with formulation, stability, and large-scale production limit their practical application [56,62,63,83]. Furthermore, nematode species such as *Steinernema* sp. and *Heterorhabditis bacteriophora* have been shown to infect locusts [68,84,85], with laboratory studies indicating high mortality rates [85]. However, their susceptibility to UV radiation and desiccation restricts their use in arid environments. Biological control remains the most promising strategy for sustainable locust management, yet its widespread adoption is limited [84,85].

Operational efforts continue to rely heavily on chemical pesticides due to the slower action, logistical challenges, and higher costs associated with biological alternatives. The case of *M. acridum* illustrates both the potential and the limitations: while it is safe and effective in trials, its adoption has been slow outside of donor-funded initiatives [6,41]. Increased investment in formulation technologies, long-term field research, and integration with monitoring systems is essential to transition biological control from experimental to mainstream use [6,41,44].

## 9. Locust Harvesting for Consumption

Harvesting locusts for food and feed has been increasingly advocated as a sustainable complement to conventional management strategies, offering the dual benefits of reducing pest populations and providing a valuable protein source [35,51,86]. Locusts are highly nutritious, comprising up to 60% protein along with essential amino acids, fatty acids, vitamins, and minerals [86–90], rendering them suitable for animal feed in aquaculture, poultry, and pig production [87,89,91]. In humans, entomophagy has a long-standing tradition in Africa, Asia, and Latin America [86–88,90], where over 120 locust and grasshopper species are consumed [90]. Swarming behavior facilitates large-scale harvesting [35,86], particularly at night when the insects are less active [35]. Recognizing their nutritional and sustainability potential, the European Commission authorized migratory locust (*Locusta migratoria*) products as novel foods in 2022 [74].

While harvesting presents a promising dual benefit of pest control and food security, its large-scale adoption encounters several challenges. The outbreak-driven availability renders it an unpredictable protein source, cultural resistance in many regions impedes broader acceptance, and pesticide contamination in wild-caught locusts raises safety concerns [86–88,90]. Consequently, locust harvesting should be considered a complementary tool within Integrated Pest Management (IPM), rather than a substitute for preventive or chemical control strategies [90].

## 10. Future Challenges

Despite advancements in institutional frameworks and technological innovations, the control of locust populations continues to encounter substantial challenges. These challenges pertain not only to the biological characteristics of the pest but also to systemic issues related to governance, funding, and the adoption of technology [6,10,19,23,36,37,92].

### *10.1. Preventive Control and Early Warning Systems*

Preventive control is widely acknowledged as the most sustainable and cost-effective approach to locust management [6,23,40]. It is crucial to note, however, that preventive control does not equate to absolute outbreak prevention [5,20,22]. Outbreaks are an inherent aspect of locust ecology and cannot always be averted [1,4–8,22,23,45,93], particularly for species with extensive distributions such as the desert locust (*Schistocerca gregaria*) or the Australian plague locust (*Chortoicetes terminifera*) [11]. Instead, as originally conceptualized by Uvarov, preventive control emphasizes early detection and timely intervention in outbreak-prone areas to curtail population growth and, most importantly, to safeguard major agricultural regions from catastrophic losses [5,22–25].

The empirical evidence from preventive programs demonstrates both their potential and their limitations. For species with relatively confined outbreak areas, such as the red locust (*Nomadacris septemfasciata*), the African migratory locust (*Locusta migratoria migratorioides*), and the South American locust (*Schistocerca cancellata*), preventive strategies have successfully averted plagues for extended durations, in some instances for decades [10,11,36]. Conversely, for more widely distributed species, preventive management has mitigated the magnitude, frequency, and duration of plagues, though it has not eradicated them. This distinction highlights why equating preventive control with outbreak prevention fosters unrealistic expectations and risks undermining confidence in otherwise effective programs [3,5,11,23,40].

Despite its conceptual clarity, the implementation of preventive control continues to encounter systemic obstacles. Historically, Uvarov's strategy of continuous surveillance and early intervention has been compromised by a "boom-and-bust" funding cycle: donor resources increase during crisis periods but diminish during recessions, leaving national programs inadequately resourced for sustained surveillance. This financial instability, combined with governance challenges and political instability in several frontline states, undermines the capacity to maintain the constant vigilance necessary for effective prevention [6,10,17,36,37].

Concurrently, technological advancements have significantly enhanced the tools available for early warning systems. Innovations such as satellite imagery, UAVs, GIS mapping, and machine learning now provide unparalleled capabilities to monitor breeding conditions, detect hopper bands, and forecast gregarization risk [2,3,5,13,15,23,40,94]. However, the operational adoption of these technologies remains inconsistent. Numerous affected countries lack the necessary infrastructure, connectivity, and technical expertise to fully implement these tools, and in regions affected by conflict, even basic field surveys may be disrupted [6,10,17,36].

Looking ahead, the primary challenge is not merely to prevent outbreaks but to consolidate and enhance preventive control as a comprehensive system of early warning and early action. Three priorities are essential:

- Institutional reforms are necessary to ensure stable, inter-plague funding for surveillance and control, rather than relying solely on emergency aid.
- It is imperative to bridge the technology–capacity gap through training, infrastructure development, and equitable access to advanced monitoring systems.
- The integration of predictive analytics into decision-making processes, utilizing models that incorporate ecological, climatic, and socio-economic variables, is essential for guiding timely interventions.

As [11] emphasizes, the true measure of success lies not in the eradication of outbreaks, but in the protection of agriculture and the mitigation of socio-economic impacts through accurate forecasting and early intervention. Without these systemic changes, preventive control will remain underutilized, and the global community will continue to oscillate between neglect during recessions and crisis-driven reactive campaigns during plagues [11].

### 10.2. Future Research in Chemical Control

Chemical insecticides are anticipated to remain integral, particularly in emergency situations [6]. Nonetheless, the dependence on organophosphates and pyrethroids is problematic due to concerns about ecotoxicity, the development of resistance, and regulatory limitations [41,49,50,69,70]. Although spinosad and anthranilic diamides are under investigation as more selective alternatives [43], progress in this area remains gradual. Future research should focus on the development of insecticides characterized by reduced environmental persistence, targeted specificity, and compatibility with integrated pest management strategies.

### 10.3. Future Challenges in Biological Control

Biopesticides, such as *M. acridum*, have demonstrated safety and efficacy in trials; however, their slow action, limited shelf life, and high costs constrain their large-scale application [41,44]. Other agents, including *N. locustae*, entomopathogenic bacteria, and nematodes, remain predominantly experimental [84,95]. Future research should prioritize enhancing formulations, scaling up production, and investigating synergies among various biocontrol agents [6,41,44]. Integration with modern surveillance tools could facilitate more targeted and cost-effective utilization [6,40].

### *10.4. Critical Synthesis*

Locust management is currently at a pivotal juncture. While preventive and biological strategies are increasingly acknowledged as crucial for sustainable practices, operational realities continue to necessitate a reliance on reactive chemical control measures. To bridge this gap, three significant shifts are required: (i) the institutionalization of stable funding mechanisms that prioritize prevention over emergency aid, (ii) the assurance of equitable access to advanced monitoring technologies, and (iii) the acceleration of research on selective chemical and biological agents suitable for application at the field scale. Without these systemic changes, locust control is at risk of remaining reactive, perpetuating the costly cycles of outbreak and emergency response.

## 11. Conclusions

Locust management continues to pose a significant global challenge, exacerbated by the climate crisis, which is anticipated to expand outbreak ranges and intensify swarm dynamics. Following the prohibition of organochlorine insecticides such as dieldrin, significant advancements have been achieved in the development of chemical alternatives, biological agents, and preventive frameworks [5,23,25,26]. The implementation of Ultra-Low Volume (ULV) spraying, UAV-based applications, and novel compounds with enhanced ecotoxicological profiles has mitigated some risks associated with older chemical formulations [23,40,81]. Concurrently, the successful development of *M. acridum* and advancements in monitoring and forecasting models have contributed essential components for more sustainable control [2,6,23,40,41,44].

The institutionalization of locust control, through the establishment of FAO-led frameworks and regional commissions, has facilitated international collaboration and the standardization of protocols [6,43]. Nevertheless, significant weaknesses persist. Preventive strategies, although conceptually robust, remain underfunded and inconsistently applied, resulting in many countries being caught in a reactive cycle of "recession neglect" followed by costly crisis response [6,17]. Biological agents, despite their potential, encounter challenges related to cost, slow action, and production logistics, hindering their large-scale adoption [6,37,41,44].

In considering future directions, the primary challenge lies not merely in the discovery of novel tools but in the structural reorganization of locust management systems. Three priorities are evident: the establishment of stable financing mechanisms to support preventive surveillance during economic downturns; bridging the technology–capacity gap by ensuring that innovations such as remote sensing and machine learning are complemented by local infrastructure and training; and the integration of selective chemical and biological tools within preventive frameworks to enable environmentally safer and more targeted interventions. In summary, while locust management has made significant advancements over the past century, it remains reactive and fragmented. Without sustained institutional commitment and investment in sustainable technologies, future responses will likely perpetuate the costly cycle of crisis and emergency control. Transforming management into a proactive, integrated, and ecologically grounded system represents both the greatest challenge and the most urgent priority.

**Author Contributions:** Conceptualization, C.P. and A.T.; methodology, C.P.; validation, C.P. and A.T.; investigation, C.P.; writing—original draft preparation, C.P.; writing—review and editing, C.P. and A.T.; supervision, A.T.; project administration, A.T. All authors have read and agreed to the published version of the manuscript.

**Funding:** This research received no external funding.

**Data Availability Statement:** No new data were created or analyzed in this study.

**Conflicts of Interest:** The authors declare no conflicts of interest.

## References

1. Latchininsky, A.V. Locusts and Remote Sensing: A Review. *J. Appl. Remote Sens.* **2013**, *7*, 075099. [CrossRef]
2. Zhang, L.; Lecoq, M.; Latchininsky, A.; Hunter, D. Locust and Grasshopper Management. *Annu. Rev. Entomol.* **2019**, *64*, 15–34. [CrossRef] [PubMed]
3. Peng, W.; Ma, N.L.; Zhang, D.; Zhou, Q.; Yue, X.; Khoo, S.C.; Yang, H.; Guan, R.; Chen, H.; Zhang, X.; et al. A Review of Historical and Recent Locust Outbreaks: Links to Global Warming, Food Security and Mitigation Strategies. *Environ. Res.* **2020**, *191*, 110046. [CrossRef] [PubMed]
4. Buj, A. International experimentation and control of the locust plague. In *Les Sciences Hors d'Occident au XXe Siècle*; ORSTOM: Paris, France, 1995; pp. 93–105.
5. Sword, G.A.; Lecoq, M.; Simpson, S.J. Phase Polyphenism and Preventative Locust Management. *J. Insect Physiol.* **2010**, *56*, 949–957. [CrossRef]
6. Word Ries, M.; Adriaansen, C.; Aldobai, S.; Berry, K.; Bal, A.B.; Catenaccio, M.C.; Cigliano, M.M.; Cullen, D.A.; Deveson, T.; Diongue, A.; et al. Global Perspectives and Transdisciplinary Opportunities for Locust and Grasshopper Pest Management and Research. *J. Orthoptera Res.* **2024**, *33*, 169–216. [CrossRef]
7. Khairov, K.S.; Lazutkaite, E.; Latchininsky, A.V. Distribution, Population Dynamics, and Management of Moroccan Locust Dociostaurus Maroccanus (Thunberg, 1815) (Orthoptera, Acrididae) in Tajikistan. *Insects* **2024**, *15*, 684. [CrossRef] [PubMed]
8. Le Gall, M.; Overson, R.; Cease, A. A Global Review on Locusts (Orthoptera: Acrididae) and Their Interactions With Livestock Grazing Practices. *Front. Ecol. Evol.* **2019**, *7*, 263. [CrossRef]
9. Latchininsky, A.; Piou, C.; Franc, A.; Soti, V. Applications of Remote Sensing to Locust Management. In *Land Surface Remote Sensing*; Elsevier: Amsterdam, The Netherlands, 2016; pp. 263–293, ISBN 978-1-78548-105-5.
10. Gay, P.; Lecoq, M.; Piou, C. The Limitations of Locust Preventive Management Faced with Spatial Uncertainty: Exploration with a Multi-agent Model. *Pest Manag. Sci.* **2020**, *76*, 1094–1102. [CrossRef]
11. Hunter, D. World's Best Practice Locust and Grasshopper Management: Accurate Forecasting and Early Intervention Treatments Using Reduced Chemical Pesticide. *Agronomy* **2024**, *14*, 2369. [CrossRef]
12. Klein, I.; Cocco, A.; Uereyen, S.; Mannu, R.; Floris, I.; Oppelt, N.; Kuenzer, C. Outbreak of Moroccan Locust in Sardinia (Italy): A Remote Sensing Perspective. *Remote Sens.* **2022**, *14*, 6050. [CrossRef]
13. Waldner, F.; Babah Ebbe, M.; Cressman, K.; Defourny, P. Operational Monitoring of the Desert Locust Habitat with Earth Observation: An Assessment. *ISPRS Int. J. Geo-Inf.* **2015**, *4*, 2379–2400. [CrossRef]
14. Piou, C.; Marescot, L. Spatiotemporal Risk Forecasting to Improve Locust Management. *Curr. Opin. Insect Sci.* **2023**, *56*, 101024. [CrossRef]
15. Klein, I.; Oppelt, N.; Kuenzer, C. Application of Remote Sensing Data for Locust Research and Management—A Review. *Insects* **2021**, *12*, 233. [CrossRef]
16. Valverde-Garcia, P.; Santiago-Álvarez, C.; Tomas, M.B.; Maranhao, E.A.A.; Garrido-Jurado, I.; Quesada-Moraga, E. Sublethal Effects of Mixed Fungal Infections on the Moroccan Locust, Dociostaurus Maroccanus. *J. Invertebr. Pathol.* **2019**, *161*, 61–69. [CrossRef]
17. Lecoq, M. La lutte préventive contre le criquet pèlerin. Progrès récents et leçons des dernières invasions. In *Compte-Rendu de la Réunion du 27 Janvier 2007 Portant sur la Lutte Préventive Contre le Criquet Pèlerin Dans Les Zones Sèches de l'afrique de l'Ouest*; Eremos: Dourdan, France, 2007; pp. 11–20.
18. Showler, A.T. Desert Locust Control: The Effectiveness of Proactive Interventions and the Goal of Outbreak Prevention. *Am. Entomol.* **2019**, *65*, 180–191. [CrossRef]
19. Mseba, A. Politics, Techno-Science, and the Environment: The Late Twentieth-Century Challenges of Locust Control in Post-Colonial Southern Africa. *J. Hist. Med. Allied Sci.* **2024**, *10*, 1–16. [CrossRef]
20. Babah, M.A.O. Strategy for Controlling the Desert Locust in Mauritania. In *New Strategies in Locust Control*; Birkhäuser: Basel, Switzerland, 1997.

21. Showler, A. Proaction: Strategic Framework for Today's Reality. In *New Strategies in Locust Control*; Birkhäuser: Basel, Switzerland, 1997; ISBN 978-3-0348-9943-7.
22. Magor, J.I.; Lecoq, M.; Hunter, D.M. Preventive Control and Desert Locust Plagues. *Crop Prot.* **2008**, *27*, 1527–1533. [CrossRef]
23. Showler, A.T. Early Intervention against Desert Locusts: Current Proactive Approach and the Prospect of Sustainable Outbreak Prevention. *Agronomy* **2021**, *11*, 321. [CrossRef]
24. Uvarov, B.P. The Locust Plague. *J. R. Soc. Arts* **1943**, *91*, 109–118. Available online: https://www.jstor.org/stable/41361608 (accessed on 13 June 2025). [CrossRef]
25. Sharma, A. Locust Control Management: Moving from Traditional to New Technologies–An Empirical Analysis. *Entomol. Ornithol. Herpetol. Curr. Res.* **2014**, *4*, 1–7. [CrossRef]
26. Lecoq, M. Recent Progress in Desert and Migratory Locust Management in Africa. Are Preventative Actions Possible? *J. Orthoptera Res.* **2001**, *10*, 277–291. [CrossRef]
27. Peveling, R. Environmental Conservation and Locust Control—Possible Conflicts and Solutions. *J. Orthoptera Res.* **2001**, *10*, 171–187. [CrossRef]
28. Lecoq, M. Desert Locust Threat to Agricultural Development and Food Security and FAO/International Role in Its Control. In Proceedings of the Eight Arab Congress of Plant Protection, Al-Bayda, Libya, 12–16 October 2003.
29. Lecoq, M.; Cease, A. What Have We Learned after Millennia of Locust Invasions? *Agronomy* **2022**, *12*, 472. [CrossRef]
30. Peloquin, C. Locust Swarms and the Spatial Techno-Politics of the French Resistance in World War II. *Geoforum* **2013**, *49*, 103–113. [CrossRef]
31. Lomer, C.J.; Bateman, R.P.; Johnson, D.L.; Langewald, J.; Thomas, M. Biological Control of Locusts and Grasshoppers. *Annu. Rev. Entomol.* **2001**, *46*, 667–702. [CrossRef]
32. FAO. Evaluation of Field Trial Data on the Efficacy and Selectivity of Insecticides on Locusts and Grasshoppers. In Proceedings of the 6th Meeting of the Pesticide Referee Group, Rome, Italy, 10–12 December 1996; p. 22.
33. Showler, A.T.; Shah, S.; Sulaiman; Khan, S.; Ullah, S.; Degola, F. Desert Locust Episode in Pakistan, 2018–2021, and the Current Status of Integrated Desert Locust Management. *J. Integr. Pest Manag.* **2022**, *13*, 1. [CrossRef]
34. Uvarov, B.P. Locusts and a Rational Anti-Locust Policy. *EAST Afr. Agric. J.* **1935**, *1*, 198–202. [CrossRef]
35. Samejo, A.A.; Sultana, R.; Kumar, S.; Soomro, S. Could Entomophagy Be an Effective Mitigation Measure in Desert Locust Management? *Agronomy* **2021**, *11*, 455. [CrossRef]
36. Gay, P.; Lecoq, M.; Piou, C. Improving Preventive Locust Management: Insights from a Multi-agent Model. *Pest Manag. Sci.* **2018**, *74*, 46–58. [CrossRef] [PubMed]
37. Showler, A.T.; Lecoq, M. Desert Locust Management Is Plagued by Human-Based Impediments. *Agronomy* **2025**, *15*, 2102. [CrossRef]
38. Magor, J.I.; Pender, J. Desert Locust Forecasters' GIS: A Researchers' View. In *New Strategies in Locust Control*; Birkhäuser: Basel, Switzerland, 1997; pp. 21–25.
39. Ceccato, P.; Cressman, K.; Giannini, A.; Trzaska, S. The Desert Locust Upsurge in West Africa (2003–2005): Information on the Desert Locust Early Warning System and the Prospects for Seasonal Climate Forecasting. *Int. J. Pest Manag.* **2007**, *53*, 7–13. [CrossRef]
40. Matthews, G.A. New Technology for Desert Locust Control. *Agronomy* **2021**, *11*, 1052. [CrossRef]
41. Mullié, W.C.; Prakash, A.; Müller, A.; Lazutkaite, E. Insecticide Use against Desert Locust in the Horn of Africa 2019–2021 Reveals a Pressing Need for Change. *Agronomy* **2023**, *13*, 819. [CrossRef]
42. Löw, F.; Waldner, F.; Latchininsky, A.; Biradar, C.; Bolkart, M.; Colditz, R.R. Timely Monitoring of Asian Migratory Locust Habitats in the Amudarya Delta, Uzbekistan Using Time Series of Satellite Remote Sensing Vegetation Index. *J. Environ. Manag.* **2016**, *183*, 562–575. [CrossRef]
43. FAO. Locust Pesticide Referee Group (LPRG). In Proceedings of the 11th Meeting of the Locust Pesticide Referee Group (LPRG), Nairobi, Kenya, 13–17 March 2023; p. 6.
44. Wakil, W.; Ghazanfar, M.U.; Usman, M.; Hunter, D.; Shi, W. Fungal-Based Biopesticide Formulations to Control Nymphs and Adults of the Desert Locust, Schistocerca Gregaria Forskål (Orthoptera: Acrididae): A Laboratory and Field Cage Study. *Agronomy* **2022**, *12*, 1160. [CrossRef]
45. Uvarov, B.P.; Zolotarevsky, B.N. Phases of Locusts and Their Interrelations. *Bull. Entomol. Res.* **1929**, *20*, 261–265. [CrossRef]
46. FAO. Evaluation of Field Trial Data on the Effectiveness of Insecticides to Locusts and Grasshoppers. In Proceedings of the 3rd Meeting of the Pesticide Referee Group, Rome, Italy, 30 March–1 April 1992; p. 21.
47. Matthews, G.A. The Pesticide Referee Group of FAO and Its Contribution to Locust Control. *J. Orthoptera Res.* **2005**, *14*, 203–206. [CrossRef]

48. FAO. Evaluation of Field Trial Data on the Efficacy and Selectivity of Insecticides to Locusts and Grasshoppers. In Proceedings of the 8th Meeting of the Pesticide Referee Group, Rome, Italy, 11–14 October 1999; p. 36.
49. FAO. Evaluation of Field Trial Data on the Efficacy and Selectivity of Insecticides to Locusts and Grasshoppers. In Proceedings of the 9th Meeting of the Pesticide Referee Group, Rome, Italy, 18–21 October 2004; p. 39.
50. FAO. Evaluation of Field Trial Data on the Efficacy and Selectivity of Insecticides to Locusts and Grasshoppers. In Proceedings of the 10th Meeting of the Pesticide Referee Group, Gammarth, Tunisia, 10–12 December 2014; p. 66.
51. Price, R. Alternative Strategies for Controlling the Brown Locust, Locustana Pardalina (Walker). *Agronomy* **2021**, *11*, 2212. [CrossRef]
52. Prior, C.; Streett, D.A. Strategies for the use of entomopathogens in the control of the desert locust and other acridoid pests. *Mem. Entomol. Soc. Can.* **1997**, *129*, 5–25. [CrossRef]
53. Shrestha, S.; Thakur, G.; Gautam, J.; Acharya, N.; Pandey, M.; Shrestha, J. Desert Locust and Its Management in Nepal: A Review. *J. Agric. Nat. Resour.* **2021**, *4*, 1–28. [CrossRef]
54. Gunn, D.L. Systems and Management: Strategies, Systems, Value Judgements and Dieldrin in Control of Locust Hoppers. *Philos. Trans. R. Soc. Lond. B Biol. Sci.* **1979**, *287*, 429–445. [CrossRef]
55. Katel, S.; Mandal, H.R.; Neupane, P.; Timsina, S.; Pokhrel, P.; Katuwal, A.; Subedi, S.; Shrestha, J.; Shah, K.K. Desert Locust (Schistocerca Gregaria Forskal) and Its Management: A Review. *J. Agric. Appl. Biol.* **2021**, *2*, 61–69. [CrossRef]
56. Zhang, L.; Hunter, D. Management of Locusts and Grasshoppers in China. *J. Orthoptera Res.* **2017**, *26*, 155–159. [CrossRef]
57. Latchininsky, A.V.; Sivanpillai, R. Locust Habitat Monitoring and Risk Assessment Using Remote Sensing and GIS Technologies. In *Integrated Management of Arthropod Pests and Insect Borne Diseases*; Ciancio, A., Mukerji, K.G., Eds.; Springer: Dordrecht, The Netherlands, 2010; pp. 163–188. ISBN 978-90-481-2463-3.
58. Gunn, D.L. Locust Plagues and the Dieldrin Controversy. *PANS Pest Artic. News Summ.* **1975**, *21*, 148–154. [CrossRef]
59. Maccuaig, R.D. Pesticides for Locust Control. *Philos. Trans. R. Soc. Lond. B Biol. Sci.* **1979**, *287*, 447–455.
60. FAO. Report of the FAO Expert Meeting on the Evaluation of Pesticides for Locust Control. In Proceedings of the 1st Meeting of the Pesticide Referee Group, Rome, Italy, 2 February 1989; p. 25.
61. FAO. Evaluation of Field Trial Data on the Effectiveness of Insecticides to Locusts and Grasshoppers. In Proceedings of the 2nd Meeting of the Pesticide Referee Group, Rome, Italy, 26–29 November 1990; p. 19.
62. FAO. Evaluation of Field Trial Data on the Effectiveness of Insecticides to Locusts and Grasshoppers. In Proceedings of the 4rd Meeting of the Pesticide Referee Group, Rome, Italy, 19–21 September 1994; p. 19.
63. FAO. Evaluation of Field Trial Data on the Effectiveness of Insecticides to Locusts and Grasshoppers. In Proceedings of the 5th Meeting of the Pesticide Referee Group, Rome, Italy, 20–22 September 1995; p. 27.
64. van Huis, A. New Development in Desert Locust Management and Control. *Proc. Neth. Entomol. Soc. Meet.* **1992**, *3*, 2–18.
65. Maccuaig, R.D. The Toxicity of Insecticides to Adult Locusts. *J. Sci. Food Agric.* **1958**, *9*, 330–342. [CrossRef]
66. Walker, P.W.; Story, P.G.; Hose, G.C. Comparative Effects of Pesticides, Fenitrothion and Fipronil, Applied as Ultra-Low Volume Formulations for Locust Control, on Non-Target Invertebrate Assemblages in Mitchell Grass Plains of South-West Queensland, Australia. *Crop Prot.* **2016**, *89*, 38–46. [CrossRef]
67. Everts, J.W.; Ba, L. Environmental Effects of Locust Control: State of the Art and Perspectives. In *New Strategies in Locust Control*; Birkhäuser: Basel, Switzerland, 1997; pp. 331–336.
68. Yasin, M.; Khan, A.; Qayyum, M.A.; Yousuf, H.M.B.; Mehfooz, A.; Hunter, D. Biological Control of Locusts and Grasshoppers: A Review. *J. Orthoptera Res.* **2024**, *33*, 289–304. [CrossRef]
69. European Food Safety Authority (EFSA). Updated Statement on the Available Outcomes of the Human Health Assessment in the Context of the Pesticides Peer Review of the Active Substance Chlorpyrifos-methyl. *EFSA J.* **2019**, *17*, 5908. [CrossRef]
70. EPA. Ingredients Used in Pesticide Products: Chlorpyrifos. 2025. Available online: https://www.epa.gov/ingredients-used-pesticide-products/chlorpyrifos (accessed on 26 February 2025).
71. Abuagla, M.I.B.; Iqbal, J.; Raweh, H.S.A.; Alqarni, A.S. Olfactory Learning Behavior and Mortality of the Honey Bee Apis Mellifera Jemenitica in Response to Pyrethroid Insecticide (Deltamethrin). *Toxics* **2023**, *12*, 25. [CrossRef] [PubMed]
72. Fang, Y.; Xu, W.; Zhang, W.; Guang, C.; Mu, W. Microbial Elimination of Pyrethroids: Specific Strains and Involved Enzymes. *Appl. Microbiol. Biotechnol.* **2022**, *106*, 6915–6932. [CrossRef] [PubMed]
73. Coppen, G.D.A.; Jepson, P.C. Comparative Laboratory Evaluation of the Acute and Chronic Toxicology of Diflubenzuron, Hexaflumuron and Teflubenzuron against II Instar Desert Locust (*Schistocerca Gregaria*) (Orthoptera: Acrididae). *Pestic. Sci.* **1996**, *46*, 183–190. [CrossRef]
74. European Commission. Commission Implementing Regulation (EU) 2022/801. *Off. J. Eur. Union* **2022**, *143*, 7–10.

75. Maute, K.; French, K.; Story, P.; Bull, C.M.; Hose, G.C. Short and Long-Term Impacts of Ultra-Low-Volume Pesticide and Biopesticide Applications for Locust Control on Non-Target Arid Zone Arthropods. *Agric. Ecosyst. Environ.* **2017**, *240*, 233–243. [CrossRef]
76. El-Samad, L.M.; El-Gerbed, M.S.; Hussein, H.S.; Flaven-Pouchon, J.; El Wakil, A.; Moussian, B. Imidacloprid-Induced Pathophysiological Damage in the Midgut of Locusta Migratoria (Orthoptera: Acrididae) in the Field. *Environ. Sci. Pollut. Res.* **2022**, *29*, 57644–57655. [CrossRef]
77. Chio, E.H.; Li, Q.X. Pesticide Research and Development: General Discussion and Spinosad Case. *J. Agric. Food Chem.* **2022**, *70*, 8913–8919. [CrossRef]
78. FAO. Evaluation of Field Trial Data on the Efficacy and Selectivity of Insecticides to Locusts and Grasshoppers. In Proceedings of the 7th Meeting of the Pesticide Referee Group, Rome, Italy, 2–6 March 1998; p. 26.
79. Zhang, L.; Lecoq, M. Nosema Locustae (Protozoa, Microsporidia), a Biological Agent for Locust and Grasshopper Control. *Agronomy* **2021**, *11*, 711. [CrossRef]
80. Abdelatti, Z.A.S.; Hartbauer, M. Plant Oil Mixtures as a Novel Botanical Pesticide to Control Gregarious Locusts. *J. Pest Sci.* **2020**, *93*, 341–353. [CrossRef]
81. Price, R.E.; Bateman, R.P.; Brown, H.D.; Butler, E.T.; Müller, E.J. Aerial Spray Trials against Brown Locust (Locustana Pardalina, Walker) Nymphs in South Africa Using Oil-Based Formulations of Metarhizium Flavoviride. *Crop Prot.* **1997**, *16*, 345–351. [CrossRef]
82. Retkute, R.; Hinton, R.G.K.; Cressman, K.; Gilligan, C.A. Regional Differences in Control Operations during the 2019–2021 Desert Locust Upsurge. *Agronomy* **2021**, *11*, 2529. [CrossRef]
83. Lockwood, J.A.; Bomar, C.R.; Ewen, A.B. The History of Biological Control with Nosema Locustae: Lessons for Locust Management. *Int. J. Trop. Insect Sci.* **1999**, *19*, 333–350. [CrossRef]
84. Şahin, Y.S.; Bouchari, A.; Ulu, T.C.; Sadıç, B.; Sursuluk, İ.A. New Application Method for Entomopathogenic Nematode Heterorhabditis Bacteriophora (Poinar, 1976) (Rhabditida: Heterorhabditidae) HBH Strain against Locusta Migratoria (Linnaeus, 1758) (Orthoptera: Acrididae). *Turk. J. Entomol.* **2018**, *42*, 305–312. [CrossRef]
85. Fathy, Z.; El-Rahman, R.M.A. Effect of Entomopathogenic Nematodes Steinernema Species (Steinernematidae: Rhabditida) and Heterorhabditis Bacteriophora (Heterorhabditidae: Rhabditida) on the Digestive Enzymes and Midgut Histology of the African Migratory Locust Locusta Migratoria Migratorioides (Acrididae: Orthoptera). *Int. J. Trop. Insect Sci.* **2023**, *43*, 727–736. [CrossRef]
86. Kietzka, G.J.; Lecoq, M.; Samways, M.J. Ecological and Human Diet Value of Locusts in a Changing World. *Agronomy* **2021**, *11*, 1856. [CrossRef]
87. Egonyu, J.P.; Subramanian, S.; Tanga, C.M.; Dubois, T.; Ekesi, S.; Kelemu, S. Global Overview of Locusts as Food, Feed and Other Uses. *Glob. Food Secur.* **2021**, *31*, 100574. [CrossRef]
88. Mmbone, S.; Gohole, L.; Fredrick Wanjala, F.M. Current Knowledge and Perception of Edible Grasshopper/Locusts' Consumption in Western Kenya. *Psyche J. Entomol.* **2023**, *2023*, 1–16. [CrossRef]
89. Ochiai, M.; Inada, M.; Horiguchi, S. Nutritional and Safety Evaluation of Locust (*Caelifera*) Powder as a Novel Food Material. *J. Food Sci.* **2020**, *85*, 279–288. [CrossRef]
90. Siddiqui, S.A.; Ghisletta, M.; Yunusa, B.M.; Jiddum, F.A.; Saraswati, Y.R.; Fernando, I.; Nagdalian, A.A.; Gvozdenko, A.A.; Shah, M.A.; Lorenzo, J.M.; et al. Grasshoppers and Locusts as Human Foods–a Comprehensive Review. *J. Insects Food Feed* **2023**, *9*, 1247–1264. [CrossRef]
91. Makkar, H.P.S.; Tran, G.; Heuzé, V.; Ankers, P. State-of-the-Art on Use of Insects as Animal Feed. *Anim. Feed Sci. Technol.* **2014**, *197*, 1–33. [CrossRef]
92. Mullié, W.C.; Cheke, R.A.; Young, S.; Ibrahim, A.B.; Murk, A.J. Increased and Sex-Selective Avian Predation of Desert Locusts Schistocerca Gregaria Treated with Metarhizium Acridum. *PLoS ONE* **2021**, *16*, e0244733. [CrossRef]
93. Tucker, C.J.; Hielkema, J.U.; Roffey, J. The Potential of Satellite Remote Sensing of Ecological Conditions for Survey and Forecasting Desert-Locust Activity. *Int. J. Remote Sens.* **1985**, *6*, 127–138. [CrossRef]
94. Piou, C.; Jaavar Bacar, M.E.H.; Babah Ebbe, M.A.O.; Chihrane, J.; Ghaout, S.; Cisse, S.; Lecoq, M.; Ben Halima, T. Mapping the Spatiotemporal Distributions of the Desert Locust in Mauritania and Morocco to Improve Preventive Management. *Basic Appl. Ecol.* **2017**, *25*, 37–47. [CrossRef]
95. Muhammad, J.; Fathy, Z.; Moussa, S. Biological Control of Schistocerca Gregaria and Locusta Migratoria Migratorioides Using Entomopathogenic Bacteria. *Sci. Rep.* **2025**, *15*, 4601. [CrossRef] [PubMed]

MDPI AG
Grosspeteranlage 5
4052 Basel
Switzerland
Tel.: +41 61 683 77 34

*Agronomy* Editorial Office
E-mail: agronomy@mdpi.com
www.mdpi.com/journal/agronomy

www.ingramcontent.com/pod-product-compliance
Lightning Source LLC
Chambersburg PA
CBHW040837250926
51850CB00007BA/1339